SHALE OIL AND GAS HANDBOOK

SHALE OIL AND GAS HANDBOOK

Theory, Technologies, and Challenges

SOHRAB ZENDEHBOUDI, PhD
*Department of Process Engineering (Oil & Gas Program)
Faculty of Engineering and Applied Science, Memorial
University St. John's, NL, Canada*

ALIREZA BAHADORI, PhD, CEng
*School of Environment, Science & Engineering, Southern
Cross University, Lismore, NSW, Australia
Managing Director of Australian Oil and Gas Services,
Pty Ltd, Lismore, NSW, Australia*

Amsterdam • Boston • Heidelberg • London
New York • Oxford • Paris • San Diego
San Francisco • Singapore • Sydney • Tokyo

Gulf Professional Publishing is an imprint of Elsevier

Gulf Professional Publishing is an imprint of Elsevier
50 Hampshire Street, 5th Floor, Cambridge, MA 02139, United States
The Boulevard, Langford Lane, Kidlington, Oxford, OX5 1GB, United Kingdom

Copyright © 2017 Elsevier Inc. All rights reserved.

No part of this publication may be reproduced or transmitted in any form or by any means, electronic or mechanical, including photocopying, recording, or any information storage and retrieval system, without permission in writing from the publisher. Details on how to seek permission, further information about the Publisher's permissions policies and our arrangements with organizations such as the Copyright Clearance Center and the Copyright Licensing Agency, can be found at our website: www.elsevier.com/permissions.

This book and the individual contributions contained in it are protected under copyright by the Publisher (other than as may be noted herein).

Notices
Knowledge and best practice in this field are constantly changing. As new research and experience broaden our understanding, changes in research methods, professional practices, or medical treatment may become necessary.

Practitioners and researchers must always rely on their own experience and knowledge in evaluating and using any information, methods, compounds, or experiments described herein. In using such information or methods they should be mindful of their own safety and the safety of others, including parties for whom they have a professional responsibility.

To the fullest extent of the law, neither the Publisher nor the authors, contributors, or editors, assume any liability for any injury and/or damage to persons or property as a matter of products liability, negligence or otherwise, or from any use or operation of any methods, products, instructions, or ideas contained in the material herein.

Library of Congress Cataloging-in-Publication Data
A catalog record for this book is available from the Library of Congress

British Library Cataloguing in Publication Data
A catalogue record for this book is available from the British Library

ISBN: 978-0-12-802100-2

For information on all Gulf Professional Publishing publications
visit our website at https://www.elsevier.com/

  Working together
to grow libraries in
developing countries

www.elsevier.com • www.bookaid.org

Publisher: Joe Hayton
Acquisition Editor: Katie Hammon
Editorial Project Manager: Kattie Washington
Production Project Manager: Mohana Natarajan
Designer: Mark Rogers

Typeset by TNQ Books and Journals

CONTENTS

About the Authors   xi

1. Shale Gas: Introduction, Basics, and Definitions   1
1. Introduction   1
2. Natural Gas and Gas Reservoir Basics   3
3. Types of Natural Gas   5
4. What Are Shale and Shale Gas?   9
5. Types and Origin of Shale Gas   11
6. Occurrence and History of Shale Gas   12
7. Important Parameters in the Shale Gas Context   14
8. Shale Gas Reserves   17
9. Shale Gas Production Trend   23
10. Key Challenges in Shale Gas Production and Exploration   24
11. Shale Gas Importance   24
12. Applications of Shale Gas   25
References   25

2. Shale Gas Characteristics   27
1. Introduction   27
2. Characterization of Gas Shale: Background   29
3. Characterization of Gas Shale: Methods   31
4. Petrophysical Characteristics of Gas Shales   36
5. Composition of Shale Gas   43
6. Ranges of Temperature, Pressure, and Depth   44
7. Shale Gas Viscosity and Density   47
8. Thermal Properties of Gas Shales   47
9. PVT Behavior of Shale Gas Mixture   54
10. Petrology and Geochemistry of Gas Shale   55
11. Estimation of Gas in Place Volume   57
12. Geological Description of Gas Shale Formations   58
13. Porosity and Permeability: Theory and Experimental   60
14. Porosity and Permeability Measurements: Practical Methodologies   64
15. Pore Size Distribution in Gas Shales   68
16. Challenges in Shale Gas Characterization   70
17. Hot Topic Research   73
References   77

3. Exploration and Drilling in Shale Gas and Oil Reserves — 81

1. Introduction — 81
2. Exploration Techniques — 81
3. Advantages and Disadvantages of Exploration Techniques — 82
4. Challenges and Risks in Shale Gas and Oil Formations — 85
5. Typical Exploration Costs — 85
6. Surface Mining — 87
7. Underground Mining — 87
8. Drilling Technologies — 88
9. Horizontal, Vertical, and Directional Drilling — 90
10. Hydraulic Fracturing — 97
11. Exploration Wells — 111
References — 119

4. Shale Gas Production Technologies — 123

1. Introduction — 123
2. Production of Shale Gas — 124
3. Rock Properties — 129
4. Production Methods — 131
5. Drilling Methodology — 135
6. Optimization — 135
7. Limitations — 136
8. Gas Liquid Separation — 137
9. Corrosion Issues — 139
10. Transportation and Storage — 140
11. Mathematical Formulas for Transport Flow — 142
12. Shale Gas Production — 144
13. Research and Development — 145
14. Future Prospects — 146
15. Economic Considerations — 148
16. Conclusions — 150
References — 151

5. Shale Gas Processing — 153

1. Introduction — 153
2. Shale Gas Processing: Background — 154
3. Description of Gas Processing Stages — 156
4. Hydrate Formation and Inhibition — 158
5. Gas Dehydration Process and Technologies — 160
6. Gas Sweetening — 167

		7. Process Design of Gas Processing Plants	181
		8. Modeling and Optimization of Equipment or/and Units in Gas Processing Plants	183
		9. Transport Phenomena in Gas Processing	185
		10. Corrosion in Gas Processing Plants	185
		11. Typical Costs for Shale Gas Processing	188
		References	190

6. Shale Oil: Fundamentals, Definitions, and Applications   193

1. Introduction   193
2. Types of Crude Oil and Oil Reservoirs   195
3. Shale Oil   196
4. Shale Oil Composition   198
5. Kerogen and Its Composition   201
6. Types and Source of Shale Oil   204
7. Occurrence and History of Shale Oil   206
8. Definitions of Main Factors and Parameters in Shale Oil Framework   213
9. Oil Shale Reservoirs   214
10. Production History of Shale Oil Reservoirs   218
11. Estimates of Recoverable Shale Oil Resources   220
12. Importance of Oil Shale   221
13. Short Descriptions of Main Companies Involved in Oil Shale Development   221
14. Energy Implication of Shale Oil   225
References   228

7. Properties of Shale Oil   231

1. Introduction   231
2. Shale Oil Utilization   232
3. Oil Shale Formations   233
4. Kerogen: Types, Structure, and History   235
5. Characterization Methods for Shale Oil   238
6. Extraction Processes for Shale Oil   249
7. Characteristics of Shale Oil   254
8. Characteristics of Oil Shales   270
9. Pressure on Configuration of Oil and Gas Markets   280
References   281
Further Reading   282

8. Production Methods in Shale Oil Reservoirs — 285

1. Introduction — 285
2. History of Shale Oil Production Development — 287
3. Production Methods for Oil Reservoirs — 289
4. Production Techniques for Shale Oil Reservoirs — 291
5. Effect of Rock Properties on Oil Production From Oil Shales — 297
6. Shale Oil Wellhead and Gathering — 298
7. Limitations for Production From Oil Shale: Operational Problems — 300
8. Examples of Production Techniques Implication in Real Oil Shale Cases — 301
9. Governing Equations to Model Shale Oil Production Methods — 301
10. Modeling and Optimization of Production Techniques in Shale Oil Reservoirs — 303
11. Screen Criteria for Oil Production in Oil Shale — 306
12. Advantages and Limitations of Oil Production Technology in Shale Oil Reservoirs — 307
13. Technical and Economic Aspects of Oil Production in Shale Oil Reservoirs — 308
14. Typical Costs to Conduct Oil Production Processes in Oil Shale — 310
15. Environmental/Public Support Issues With Shale Oil — 311
16. Opposing Views Regarding Shale Oil — 314
17. Future Prospects of Oil Shale Production — 315
18. Research and Technology Development in Shale Oil Production — 316
19. Conclusions and Recommendations — 317
References — 317

9. Shale Oil Processing and Extraction Technologies — 321

1. Introduction — 321
2. Description of Oil Shale Processing — 322
3. What Is Oil Shale Retorting? — 322
4. Chemistry of Oil Shale Retorting — 322
5. Chemistry of Kerogen Decomposition — 324
6. Chemistry of Carbonate Decomposition — 324
7. Pyrolysis or Retorting of Oil Shale: Experiments, Apparatus, Methodology — 325
8. Optimal Retorting Conditions — 326
9. Kinetics of Pyrolysis or Retorting of Oil Shale — 328
10. Isothermal and Nonisothermal Kinetics Measurement and Expressions for Shale Oil — 330
11. Ex Situ Retorting Techniques — 332
12. Advantages and Disadvantages of Ex Situ Processes — 337
13. In Situ Retorting Techniques — 338

14.	Wall Conduction	339
15.	Externally Generated Hot Gas	339
16.	ExxonMobil Electrofrac	340
17.	Volumetric Heating	340
18.	Advantages and Disadvantages of In Situ Processes	341
19.	Shale Oil Refining and Upgrading Processes	342
20.	Advantages and Disadvantages of Refining and Upgrading Techniques	343
21.	Supercritical Extraction of Oil From Shale	344
22.	Supercritical CO_2 Extraction of Oil Shale: Experiments, Apparatus, and Procedure	344
23.	Supercritical Methanol/Water Extraction of Oil Shale	345
24.	Continuous Supercritical Extraction	346
25.	Advantages and Disadvantages of Supercritical Extraction Methods	346
26.	Mathematical Modeling of Oil Shale Pyrolysis	347
27.	Parametric Study of Oil Shale Pyrolysis	349
28.	Economic Considerations in Oil Shale Processing	351
29.	Theoretical, Practical, and Economic Challenges in Oil Shale Processing	352
30.	Hot Topic Research Studies in Oil Shale Processing and Extraction Technologies	353
	References	354
	Further Reading	355

10. Shale Oil and Gas: Current Status, Future, and Challenges   357

1.	Introduction	357
2.	Political Implications	360
3.	Federal and Provincial (or State) Regulations	363
4.	Environmental Issues/Aspects	367
5.	Geomechanics Challenges	371
6.	Comparison of Conventional and Shale Oil and Gas Reserves	374
7.	Management Rules in Development of Oil and Gas Shale	377
8.	Technical and Economic Constraints	380
9.	Economic Challenges	387
10.	Research Needs in Oil and Gas Shale	390
11.	Past and Current Status of Oil and Gas Shale	392
12.	Future Prospects of Oil and Gas Shale	395
13.	Current Projects for Oil and Gas Shale	397
	References	401

Index   405

ABOUT THE AUTHORS

Sohrab Zendehboudi, PhD, PEng, is currently an Assistant Professor and Statoil Chair in Reservoir Analysis Department of Process Engineering (Oil & Gas Program), Faculty of Engineering and Applied Science, Memorial University St. John's, NL, Canada. He specializes in shale oil and gas, process systems, enhanced oil recovery, and transport phenomena in porous media. Previously, he was a Post-Doctoral Fellow at Massachusetts Institute of Technology (MIT) and the University of Waterloo, where he received his PhD from the Department of Chemical Engineering. He earned his MSc in Chemical Engineering from Shiraz University in Iran, ranking first in his class. He received a BSc in Chemical Engineering from the Petroleum University of Technology in Iran. Dr. Zendehboudi has taught multiple petroleum engineering courses and has been employed by multiple oil and gas institutions, such as the National Petrochemical Company in Tehran, Shiraz Petrochemical Company, and the Petroleum University of Technology. Dr. Shorab has written for multiple prestigious journals including Elsevier's Journal of Natural Gas Science and Engineering.

Alireza Bahadori, PhD, CEng, MIChemE, CPEng, MIEAust, RPEQ, NER, is a research staff member in the School of Environment, Science and Engineering at Southern Cross University, Lismore, NSW, Australia, and managing director and CEO of Australian Oil and Gas Services, Pty. Ltd. He received his PhD from Curtin University, Perth, Western Australia. During the past 20 years, Dr. Bahadori has held various process and petroleum engineering positions and was involved in many large-scale oil and gas projects. His multiple books have been published by multiple major publishers, including Elsevier. He is a Chartered Engineer (CEng) and Chartered Member of the Institution of Chemical Engineers, London, UK (MIChemE), Chartered Professional Engineer (CPEng), Chartered Member of the Institution of Engineers Australia, Registered Professional Engineer of Queensland (RPEQ), Registered Chartered Engineer of the Engineering Council of United Kingdom, and Engineers Australia's National Engineering Register (NER).

CHAPTER ONE

Shale Gas: Introduction, Basics, and Definitions

1. INTRODUCTION

Hydrocarbons in the forms of oil and gas phases are the primary energy sources which humans around the globe depend on to provide fuel for the advanced technologies that we rely on to make our lives easier. Thus, the demand for energy from fossil fuels is constantly on the rise to meet our increasingly energy-intensive lifestyles [1–3].

Natural gas is a fossil fuel which is derived from living organisms that are buried under the earth's crust. Over time, heat and pressure convert the organisms into oil and gas. It is one of the cleanest and most efficient sources of energy. Due to its high calorific value and no ash content, it is widely employed as a major fuel in several sectors such as automobile, refining, house heating, and so on [1–3].

Natural gas has been used as an energy source in Canada since the 1800s, but it did not become a common energy source until the late 1950s. Following the construction of the Trans Canada Pipeline, it started gaining popularity. After the price hike of crude oil in the late 1970s, its demand grew very quickly. The oil crisis resulted in long line-ups outside gas stations, which caused decision-makers to consider natural gas. The environment safety concern has also added to its popularity because burning of natural gas is cleaner, compared to other fossil fuels [1–3].

Natural gas comes from both conventional and unconventional formations. The key difference between conventional and unconventional natural gases is the method, ease, and cost associated with technology of extraction/production [1–3].

Shale gas is natural gas, one of several forms of unconventional gas. Shale gas is trapped within shale formations with low permeability, which is fine-grained sedimentary rock. The rock acts as its source as well as a reservoir. The shale rock appears to be the storage material and also the creator of the gas through the decomposition of organic matters. Therefore, the techniques used at one well may not result in success at another shale gas location [2,4,5].

Shale Oil and Gas Handbook
ISBN: 978-0-12-802100-2
http://dx.doi.org/10.1016/B978-0-12-802100-2.00001-0

© 2017 Elsevier Inc.
All rights reserved.

Shale reserves discovered across the world consist of several billion of tons of trapped oil and gas, making them fossil fuel resources of the century [1,2]. It is estimated that approximately 456×10^{12} m^3 of shale gas are available globally [6]. Development of economic, eco-friendly and safer drilling technologies to access trapped gas, made shale resources the next big reliable source of energy in the world, particularly in North America [7]. The US Department of Energy projects that shale gas will occupy 50% of total energy produced in the country by 2035, that is, around 340 billion cubic meters/year [8]. In addition to the production of natural gas, other fuels like NGLs (natural gas liquids; propane and butane) are simultaneously produced from the shale reservoirs [3,7].

The gas in many US shale formations such as Antrim shale formation (Michigan) and New Albany Shale formation (Illinois) has been created in the last 10,000—20,000 years [9]. In 1825, the first extraction of shale gas was performed in Fredonia (NY) in shallow and low-pressure fractures. In naturally fractured Devonian shales, the development of the Big Sandy gas field commenced in Floyd County, Kentucky, 1915 [10]. Until 1976, the field extended over 1000 square miles of southern West Virginia and into eastern Kentucky, with production from the Cleveland Shale and the Ohio Shale, together called the "Brown Shale," where there are 5000 wells in Kentucky alone. By the 1940s, to stimulate the shale wells, explosive down the hole operations had been utilized. In 1965, other efficient techniques, such as hydraulic fracturing (including 42,000 gallons of water and 50,000 pounds of sand), were developed for production wells, particularly those with low recovery rates [10]. The average production per-well was small since the flow rate was mainly dependent on the existence of natural fractures; however, the field had a final gas recovery of 2×1012 ft3. In the 1920s, there were other widespread commercial gas production basins such as Michigan, Appalachian, and Illinois basins in the Devonian-age shale, though the production was typically insignificant [10].

The discovery and exploitation of shale oil and gas present a major innovation with economic and political implications for developing countries. In recent years, the rapid expansion of shale gas development and production has had a profound impact on the current and future of the global energy market. The advancement of natural gas production from shell formations is revolutionizing the energy industry in general and the oil and gas and petrochemical sectors in particular. North America, especially the United States of America (USA) is leading the development of this new type of hydrocarbon resources. Innovations in extraction (or/and production) technologies have made access to these vast amounts of natural gas resources technically and economically feasible.

Depending on the downstream use of the natural gas, shale gas may have a net negative or positive impact on greenhouse gas (GHG) emissions [10–12]. In the USA, the abundance of cheap natural gas from fracking will likely replace coal as the preferred fuel for energy generation. This will likely decrease the American energy sector's GHG emissions. In other regions such as the United Kingdom, however, natural gas might supersede fledgling renewable energy operations and have a net negative impact on climate change [10–12].

As the global significance of shale gas increases, there is a need for better understanding of the shale characteristics, shale gas production and processing, and potential, environmental, social and economic impacts within its value chain.

2. NATURAL GAS AND GAS RESERVOIR BASICS

Natural gas is a type of fossil fuel that forms when several layers of buried animals, gases, and plants (trees) during thousands of years are exposed to high pressure and heat. The initial energy of the plants originated from the sun is stored in natural gases in the form of chemical bonds/links [1–3]. In general, natural gas is considered as a nonrenewable energy form as it does not return over a fairly acceptable timeframe. Natural gas mainly includes a high concentration of methane and low percentages of other alkanes, hydrogen sulfide, nitrogen, and carbon dioxide. The main utilization of natural gas is electricity generation, cooking, and heating [1–3]. Other uses of natural gases can be raw materials for various chemicals/materials (e.g., petrochemical products, plastics, and polymers) and car fuel. Natural gas can only be used as a fuel if it is processed to remove impurities, including water, to meet the specifications of marketable natural gas. The byproducts of this processing operation include ethane, propane, butanes, pentanes, and higher-molecular-weight hydrocarbons, water vapor, carbon dioxide, hydrogen sulfide, and sometimes helium and nitrogen [1–3].

A natural gas reservoir is defined as a naturally occurring storage space formed by rock layers such as anticline structures deep inside the earth's crust. These are often referred to as reservoir rocks. Reservoir rocks are permeable and porous, to store the gases within the pores and allow them to move through the permeable membranes. To trap natural gas, reservoir rocks require to be capped by an impervious rock in order to seal the storage area and prevent gas from escaping. Reservoir rocks are sedimentary rocks like sandstone, arkoses, and limestone, which have high porosity and permeability. Impervious rocks are less permeable rocks (e.g., shales) [1−3,13−15].

Naturally, gas is formed in two ways; either directly from organic matters or by thermal breakdown of oil at very elevated temperatures. In addition, formation of natural gas can occur through bacterial processes from organic sedimentary rocks at shallow depth [1−3]. The bacteria which act on organic substances are anaerobic in nature and the gas produced is named biogenic gas. The volume of biogenic gas produced per unit volume of sediment is lower, compared to other types of natural gases. Biogenic gas can be found at depths lower than 2200 ft [1−3,13−15].

A gas reservoir is formed by the natural occurrence of four geological sequences at a time, namely, (1) source rock, (2) reservoir rock, (3) seal, and (4) trap [1−3].

Oil and gas phases produced in source (or sedimentary) rocks migrate to nearby reservoir rocks and are accumulated due to the trap formed by seal rocks like shale. Migration takes place through permeable membranes and the pressure difference between pores. In the context of transport phenomena in porous media, capillary pressure plays an important role in the movement of oil and gas.

Most reservoir rocks contain saturated water. Due to the density difference, the gas moves up and occupies the space above oil so that water being denser remains below the oil layer to create an aquifer. The arrangement of oil, gas, and water phases is depicted in Fig. 1.1 [1−3,16,17].

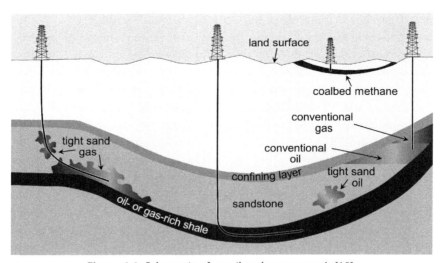

Figure 1.1 Schematic of an oil and gas reservoir [18].

There are two types of traps, (1) stratigraphic and (2) structural. Stratigraphic traps form the reservoir rocks which are covered by seal rocks from above and below like a coastal barrier island. The deposition of rock layers is in the manner where reservoir rocks create a discontinuous layer. Structural traps are formed as a result of deformation of rocks due to folding of rock layers and faults that occurred a long time back. The structure favors storage of oil and gas with a seal formed by shale rock on the top of the trap [1–3,16,17].

A gas or oil resource can be defined as the totality of the gas or oil originally existing on or within the earth's crust in naturally occurring accumulations, and includes discovered and undiscovered, recoverable and unrecoverable. This is the total estimate, which is irrespective of whether the gas or oil is commercially recoverable [1–3,16,17].

"Recoverable" oil or gas refers to the portion of the total resource that can be commercially extracted by utilizing a specific technically feasible recovery project, a drilling plan, fracking program, and other related project requirements. In order to create a clearer picture of the value of these resources, the industry breaks them into three categories [1–3]:
- Reserves, which are discovered and commercially recoverable;
- Contingent resources, which are discovered and potentially recoverable but subcommercial or noneconomic in today's cost–benefit regime;
- Prospective resources, which are undiscovered and only potentially recoverable.

Using a similar approach to the industry, the Potential Gas Committee (PGC), which is responsible for developing the standards for USA gas resource assessments, also divides resources into three categories of technically recoverable gas resources, including shale gas [1–3]:
- Probable;
- Possible;
- Speculative.

3. TYPES OF NATURAL GAS

Natural gas is generally categorized in two main groups in terms of method of production and type of rock; namely conventional and unconventional [19,20].

The definition of conventional oil and gas according to the US Department of Energy (EIA) is "oil and gas produced by a well drilled into a geologic formation in which the reservoir and fluid characteristics permit the oil and natural gas to readily flow to the wellbore." Conventional gas can be extracted using traditional methods from reservoirs with permeability greater than 1 millidarcies (mD) [19,21]. Currently as a result of the availability of resources and low cost of extraction, conventional gas represents the largest share of global gas production.

Unconventional gas on the other hand is found in reservoirs with below permeability and as such cannot be extracted using conventional techniques. The production process is more complex as the geological unconventional formations have low permeability and porosity. They also contain fluids that might have density and viscosity very different from water. (They generally have high viscosity and density.) Thus, conventional techniques cannot be employed to produce, refine, and transport them. As a result, it costs much more to extract unconventional petroleum, compared to conventional oil and gas [19−21].

Conventional gas is typically free gas trapped in multiple, relatively small, porous zones in naturally occurring rock formations such as carbonates, sandstones, and siltstones. The exploration and extraction of conventional gas is easy compared to unconventional gas. Unconventional gas has composition or/and components similar to conventional natural gas. It is the unusual characteristics of the reservoir that contain unconventional gas. It is also usually more difficult to produce, compared to conventional gas. Unconventional gas reservoirs include tight gas, coal bed methane, shale gas, and methane hydrates [1−3,20].

Unconventional reservoirs contain much larger volumes of hydrocarbons than conventional reservoirs [22]. Fig. 1.2 illustrates different types of unconventional gas reservoirs in comparison with convectional reservoirs in terms of permeability, volume, improve, technology development, and cost.

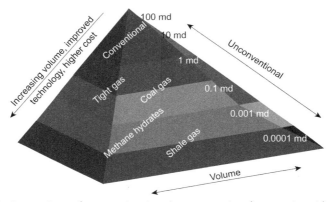

Figure 1.2 Comparison of conventional and unconventional reservoirs with respect to properties and reserves volume [22].

As expected, shale gas is found in the largest volumes but has the lowest permeability of about 0.0001 mD compared to conventional which has a permeability of 100 mD [22]. According to Fig. 1.2, it is also concluded that the cost of producing from unconventional reservoirs is much more expensive since the formation needs to be treated in order to increase its porosity and permeability [22].

Most of the growth in supply from current recoverable gas resources is found in unconventional configurations. The technological improvements in the field of drilling, especially horizontal drilling and fracturing, have made shale gas and other unconventional gas supplies commercially feasible and have brought a revolution in the field of natural gas in Canada and the USA [19,20].

In addition, a different categorization for natural gas has been introduced so that there are various types of natural gas reserves based on the gas formation mechanisms and rock properties as follows [2,23–25]:

3.1 Biogas

The fermentation of organic matter in the absence of oxygen results in the production of a form of gas which is called biogas. This phenomenon is known as an anaerobic decomposition process. It occurs in landfills or where materials such as animal waste, industrial byproducts, and sewage, are decayed. This type of gas is biological, originating from living and nonliving plants and animals. Combustion of materials such as forest residues results in generation of a renewable energy source. In comparison with natural gas, biogas comprises less methane, but it can be processed and then employed as a viable energy source [2,23–25].

3.2 Deep Natural Gas

Another type of unconventional gas is deep natural gas. Deep natural gas can be found in deposits at least 15,000 feet beneath the earth surface, while a majority of conventional gas resources are just a few thousand feet deep. From an economical point of view, drilling of deep natural gas formations is not practical in most cases, though various methods to produce deep natural gas have been developed and further improvement is currently sought through implementation of several research and engineering activities [2,23–25].

3.3 Shale Gas

Another category of unconventional gas is shale gas. Shale as a sedimentary rock is fine-grained and does not disjoint in water. Some scholars argue that shales are very impermeable so that we can suppose marble as sponge compared to them. Natural gas is usually placed between thick shale layers. The most recognized production methods for shale gas are hydraulic fracturing and horizontal drilling. Hydraulic fracturing is a process that open up fractures in the rocks through injecting water with a high pressure, and then keeping them open using small particles/grains including silica, sand, and glass. Horizontal drilling describes a drilling method such that it first starts with drilling straight down into the ground, then sideward, and eventually parallel with the surface direction [20,23−25].

3.4 Tight Gas

There is another kind of unconventional natural gas, called tight gas, which is trapped in a very low-permeable underground formation that makes its extraction very difficult. More specifically, it is defined by having less than 10% porosity and less than 0.1 mD. Gas extraction from tight rocks generally needs difficult and costly processes including acidizing and fracking. Acidizing involves injecting an acid (commonly hydrochloric acid) into the gas well to dissolve the tight rock, leading to unblocking the gas and facilitating its flow. Fracking is similar to acidizing operations [2,23−25].

3.5 Coal Bed Methane

Coal bed methane is considered as another form of unconventional natural gas which can be a popular energy source. In common, coal bed methane is found in various underground coals. Natural gas can be released over coal mining, collected, and then used for various purposes such as heating, cooking, and electricity generation [23−25].

3.6 Gas in Geopressurized Zones

Unconventional natural gas can also be found in geopressurized areas. These zones are generally in the range of 10,000−25,000 feet (3000−7600 meters) underground. The geopressurized zones are created when clay layers hastily accumulate and compact above the material which is more porous (e.g., silt and sand). As the gas phase is pushed out from the compressed clay, it is placed within silt, sand, and other absorbents under elevated pressures. Mining operations for geopressurized zones are very complicated and expensive; however, these areas might hold a significant amount of natural gas. A majority of geopressurized zones in the USA are located in the Gulf Coast region [23−25].

3.7 Methane Hydrates

Another category of unconventional natural gas is methane hydrate. Methane hydrates were found lately only in permafrost areas of the Arctic and ocean sediments. Methane hydrates are normally produced at high pressures and low temperatures, about 32°F or 0°C. Methane hydrates are released into the atmosphere when a considerable change in environmental conditions takes place. According to the United States Geological Survey (USGS), it is forecasted that methane hydrates can comprise twice the extent of carbon in all of the oil, coal, and conventional natural gas together across the globe [2,23−25].

In sea sediments, methane hydrates are created on the interior slant as different microorganisms that sank to the sea floor and disintegrate in the sediment. Methane, caught inside the sediments, can concrete the loose sediments and retain the interior holder stable. Nonetheless, the methane hydrates break down when the water temperature goes up. This leads to natural gas release and underwater landslide. Methane hydrates form in permafrost ecosystems so that bodies of water freeze and water molecules make singular enclosures (cages) around every molecule of methane. The gas that is caught in the water lattice has a greater density than its gas form. The methane leaks if the cages of ice defrost [2,19,22−24].

The quality and composition of natural gas is further defined using the following terms [2,20]:
- Lean gas—gas in which methane is the major component;
- Wet gas—this gas composition has considerable amounts of hydrocarbons with higher molecular weight;
- Sour gas—the mixture contains hydrogen sulfide;
- Sweet gas—none or a very small amount of hydrogen sulfide is present;
- Residue gas—natural gas from which the hydrocarbons with higher molecular weight have been extracted;
- Casing head gas—this gas is derived from petroleum and separated at the well head.

4. WHAT ARE SHALE AND SHALE GAS?

This section briefly introduces shale rock and shale gas.

4.1 Shale

Shale is a sedimentary rock, that was once deposited as mud (clay and silt) and is generally a combination of clay, silica (quartz), carbonate (calcite or

Figure 1.3 Schematic of light and dark shale rocks [24].

dolomite), and organic material. Mainly shale is a composite of a large amount of kerogen, which is a mixture of organic compounds. As a primary composition, it has kerogen, quartz, clay, carbonate, and pyrate. Uranium, iron, vanadium, nickel, and molybdenum are present as secondary components. From this rock, the shale hydrocarbons (liquid oil and gas) are extracted [19,20].

Various shales exist; namely, black shale (dark) and light shale as shown in Fig. 1.3. Black shale has a rich content of organic matters, while light shale has much less, relatively. Black shale formations were buried under little or no presence of oxygen and this preserved the organic matters from decay. This organic matter may produce oil and gas through a heating process. Many shale formations in the USA are black shale formations which give natural gas via heating [26,27].

Shale oil is the substitute for the synthetic crude oil, but extraction from the oil shale is costly in comparison with conventional crude oils. The composition of the crude oil is not the same throughout the world; it depends on geographical structure and other factors (e.g., depth, temperature). Its feasibility is strongly affected by the cost of the conventional crude oil. If its price is greater than that of conventional crude oil, it is uneconomical [20,24].

4.2 Shale Gas

Shale gas refers to natural gas that is trapped within shale formations. Shales are fine-grained sedimentary rocks that can be rich sources of petroleum and natural gas (see Fig. 1.4). Shale gas is trapped within the pores of this

Figure 1.4 A gas shale outcrop with the layered structure clearly visible [26].

sedimentary rock. Gas is normally stored through three ways in gas shales [24,26]:
1. Free gas: The gas is within the rock pores and natural fractures;
2. Adsorbed gas: The gas is adsorbed on organic materials and clay;
3. Dissolved gas: The gas is dissolved in the organic materials.

Over the past decade, the combination of horizontal drilling and hydraulic fracturing has allowed access to large volumes of shale gas that were previously uneconomical to produce. The production of natural gas from shale formations has rejuvenated the natural gas industry [24,25].

The supply chain for shale gas includes: wells, a collection network, pretreatment, NGL extraction and fractionation facilities, gas and liquid transporting pipelines, and storage facilities.

5. TYPES AND ORIGIN OF SHALE GAS

There are two types of shale gas generated and stored in the reservoir. Biogenic gas is formed at low depths and temperatures through anaerobic bacterial decay of alluvial organic matters by microbes [28]. Thermogenic gas is formed at elevated depths and temperatures by thermal cracking of oil into gas [29].

Like coal, conventional gas and oil, shale gas is found in the earth's crust. It is derived from source rock which has been formed many years ago through deposition of organic matters at the bottom of lakes and oceans. Over time, the sediments gradually became compacted and more deeply buried. Heat and pressure cause formation of hydrocarbons from the organic matters.

After the gas is formed in the source rock, most of the hydrocarbons migrate to the reservoir rocks and are trapped by seal rocks within nonporous and nonpermeable formations. The hydrocarbons, which remain trapped in the source rock, make shale oil and gas [29].

In general, there are three different classifications in terms of oil and gas shale origin as follows [20,30]:

- Terrestrial shale: Organic precursors (also referred to as cannel coal) of terrestrial hydrocarbon shale are found in stagnant oxygen-depleted water deposits or peat-forming swamps and bogs [20,30]. These deposits exist in small sizes; however the grade quality is very high [20,30]. In this category, the oil- and gas-generating rich organic matter (cannel coal) is derived from plant resins, pollen, plant waxes, spores, as well as corky tissues of vascular plants [20,30]. Cannel coal exists in brown to black color [20,30].
- Lacustrine shale: The lipid-rich oil- and gas-generating organic matter is obtained from algae that existed in freshwater, brackish, or saline lakes [20,30].
- Marine shale: The lipid-rich organic matter in marine shale deposits are derived from marine algae, unicellular organisms, and marine dinoflagellates [20,30].

6. OCCURRENCE AND HISTORY OF SHALE GAS

The newer sources of oil and gas being developed are from more difficult to produce reservoirs. These sources are referred to as unconventional resources, as they usually require different or unique technologies for recovery of the oil or gas.

Natural gas from shale reservoirs has gained popularity in recent years, but there are instances of shale gas production in the past. In the early cases of shale gas production, there was sufficient natural fracturing of the shale to allow economic recovery. This gas was typically produced through shallow vertical wells producing at low rates over long periods of time [20]. Natural gas has been produced from shale formations in the Appalachian Mountains of the USA since the late 1800s [20]. In 1920, oil from fractured shale deposits was discovered at Norman Wells in Canada's Northwest Territories [20]. In southeast Alberta and southwest Saskatchewan, gas has been produced from the Second White Speckled Shale for decades [20].

Shale gas has also been produced from the Antrim Shale in the Michigan Basin since the late 1940s [20].

Technologies have been developed for the production of hydrocarbons from unconventional reservoirs. These technologies were created to help in improving flow characteristics of productive hydrocarbon reservoirs. The most dramatic technological advance occurred after World War II with the development of hydraulic fracturing techniques [15,20].

The petrochemical industry has been developing more enhanced and cost-effective methods for stimulating fractures in reservoirs for more than 60 years [15,19]. Improvements to hydraulic fracturing include advances in the areas of fracturing fluids, surface and down-hole equipment, computer applications and modeling of fracture treatments, and the science of fracture creation in relation to tectonic stresses [15,19,20].

Commercial applications of hydraulic fracturing began in the late 1940s. The first commercial hydraulic fracturing job was at Velma, Oklahoma, in 1949 [15,19,20]. The first application of hydraulic fracturing in Canada was in the Cardium oil field in the Pembina region of Alberta in the 1950s [4]. Since then, more than a million wellbores have been drilled and stimulated using hydraulic fracturing [20,31].

Barnett shale in Texas, USA, was developed in 1981. This was the first shale gas commercially developed. Production of shale gas in the USA considerably rose from almost negligible in 2000 to 10 bcfd in 2010. There are many shale regions in the world which can serve as potential shale gas sources [15,19,20].

Shale formations are characterized as porous systems with very low permeability and small pore sizes which make them tough for fluids to move through the rock. While the occurrence of shale formations was acknowledged even in the early 1980s; however, it was only a little after (less than two decades) that drilling corporations came up with new techniques for extracting the trapped oil and natural gas from gas shales [20]. Shale gas is found at approximately 1500—3000 m below the earth in source rocks. Source rock is usually a sedimentary rock argillaceous in nature and rich in organic matters. The rock undergoes changes due to heat and pressure and is transformed into laminated and fine-grained rock called shale rock.

The gas is trapped in tiny pores between the grains and is tightly attached to the matrix of the rock. Therefore, the extraction of shale gas is a very difficult task [29].

As shown in Fig. 1.5, the majority of the shale reserves are in the USA, Canada, China, Australia, and India [29].

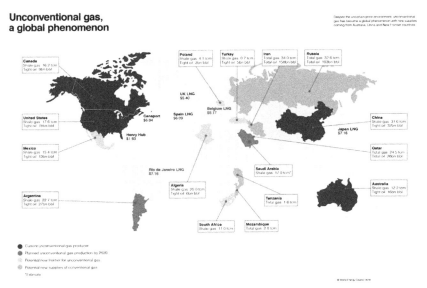

Figure 1.5 Occurrence of shale gas around the world [29].

7. IMPORTANT PARAMETERS IN THE SHALE GAS CONTEXT

7.1 Type of Shale

There are different types of shale in terms of material content and origin. The shales are marine or nonmarine. Marine shales have low clay content. They are high in brittle minerals such as quartz, feldspar, and carbonates. Hence, they respond better to hydraulic stimulation [20,30].

7.2 Depth

In general, the depth is directly related to the amount of natural and generated hydrocarbons deposited in the formation. For instance, gas as a biogenic gas is formed through anaerobic microorganisms throughout the early stage of the burial process or/and thermogenic breakdown of kerogen at higher temperatures and depths. The typical depth for shale gases ranges from 1000–5000 m. The shale formations shallower than 1000 m normally experience lower gas concentrations and pressures, while regions with a depth greater than 5000 m commonly have reduced permeability which is translated into higher costs for drilling activity and field development [20,28,30].

7.3 Adsorbed Gas

Adsorbed gas is the gas accumulated on the surface of a solid material, such as a grain of a reservoir rock, or more particularly the organic particles in a shale reservoir. Measurement of adsorbed and interstitial gas (existing in pore spaces) allows calculation of gas in place in a reservoir [20,28,30].

7.4 Organic Maturity

This is expressed in terms of vitrinite reflectance (% Ro). The range of 1.0–1.1% indicates that organic matter is adequately mature to generate gas. Generally, higher gas in-place resources should be produced by more mature organic matter [20,28,30].

7.5 Permeability

The permeability of any type of porous media is defined as the ability of fluid (i.e., gas, oil, water) flow due to a pressure difference through the porous system. Therefore, it implies the fluid transmissivity and storage features of a shale formation. Permeability of shales is generally very low ($<10^{-3}$ mD). Hence, artificial stimulations (particularly hydraulic fracturing) are needed to ease the hydrocarbon flow toward the well. If natural fractures exist in the shale formation, it is crucial to map the orientation and intensity of the open fractures. If the fractures are poorly cemented or open, stimulation will open these formerly created regions of weakness. In some circumstances, compactly cemented fractures can create barriers of fractures which are measured in millidarcies (mD) [20,28,30].

7.6 Porosity

Porosity is the percentage of void space versus solid rock, which is the space where gas is potentially trapped. The porosity of shale reserves is normally lower than 10% [20,24].

7.7 Reservoir Thickness

This defines the vertical extent thickness of the productive portion of a reservoir. The formation thickness varies from one shale reserve to another shale reservoir. The typical range of the thickness is 2–5 m [20,28,30].

7.8 Total Organic Content (TOC)

This is expressed as the total amount of organic material present in the rock (a percentage by weight). The higher the TOC, the higher is the potential for hydrocarbons (HCs) production. Typical values are equal to or greater than 1%. The TOC and thermal maturity of source rocks are assessed by means of lab analysis [20,28,30].

7.9 Thermal Maturity

This is the measure of the extent to which organic matters contained in the rock have been heated over time and potentially converted to liquid and/or gaseous HCs. The indicator for this measure is called vitrinite reflectance and has typical values ranging from 1% to 3% [20,28,30].

7.10 Viscosity

Viscosity is a measure of how easily oil will flow. Inside the reservoir, viscosity is measured in poises (P); it is normally measured in centistokes (cS) outside the reservoir [20,28,30].

7.11 Mineralogy

The mineral structure in shale formations is complicated. One ought to endeavor to obtain a few cores for an underlying assessment in a new region. Electron catch spectroscopy (ECS) logs give a good estimation of mineralogy, but not as the mineral (e.g., granular against cryptocrystalline), which plays a significant role in brittleness behavior. One effective strategy is to build a ternary outline of total aggregate carbonate, total aggregate mud, and quartz and guide it to elastic parameters (e.g., E [Young's modulus] and Poisson's ratio), leading to a brittleness template. Such layouts can then be adjusted to production logs, microseismic event area, and production itself to evaluate the ductility or brittleness of the rock and how well the induced fractures have actuated it [20,28,30].

7.12 Fluid in Place

In general, fluid in place is determined using TOC, porosity, temperature, and pressure data for economical evaluation of the shale [20,28,30].

7.13 Free Gas Quantification

Adsorption phenomenon is considered as a more effective mechanism at gas storage under low pressures, whereas free gas signifies the main amount of gas at elevated pressures. The free gas percentage in shale gases varies in the range of 15—80%, depending on gas saturation, porosity, and reservoir pressure. Hence, determination of free gas is essential to describe/characterize gas shales. Thus, the quantification of free gas is also necessary to characterize gas shale. This parameter is expressed by the following relationship:

$$Gcfm = \frac{1}{B_g}(\phi_{eff}(1 - S_w))\frac{\psi}{\rho_b} \qquad (1.1)$$

in which, $Gcfm$ represents the free gas volume (standard cubic feet per ton or scf/ton), B_g is the gas formation volume factor (reservoir cf/scf), ϕ_{eff} is the symbol for effective porosity (vol/vol), S_w implies the water saturation (vol/vol), ρ_b shows the bulk density (g/cm^3), and ψ refers to the conversion constant (32.1052) [20,28].

7.14 Productibility

Productibility is the product of permeability multiplied by formation thickness. The important parameter in tight gas shale formations to effectively design stimulation processes and accurately estimate production rate and recovery is permeability. In general, two permeabilities, including matrix and entire system, are needed for such design/operation purposes. The permeability of matrix in shales typically varies from 10^{-8} to 10^{-4} mD. There are various measurement techniques, such as core analysis and log evaluation (if the developed local calibration is available), to determine the matrix permeability with acceptable precision. The system permeability corresponds to the matrix permeability plus the involvement of open fractures in flow conductivity. System permeability cannot be measured/determined through conventional logs as they are not sensitive to fractures. The common method to identify and map fractures that cross the borehole in shales is the full bore formation microimager. The size of fracture aperture can also be estimated [20,28,30].

8. SHALE GAS RESERVES

Table 1.1 lists the main countries that contain large amounts of recoverable shale gas. As is clear from Table 1.1, China holds the first rank, followed by Argentina, Algeria, the USA, and Canada.

Giving more detail, proved and unproved shale gas and other resources in the world are listed in Table 1.2. This table also presents the increase in total gas resources due to the existence of shale gas.

In recent years, the development of unconventional shale gas resources in North America has had a major impact on the overall energy landscape of the region. This rapid expansion is now altering the global energy supply. The USA in particular is on the forefront of what some call the shale gas revolution. This has all become possible due to significant advances in hydraulic fracturing and horizontal drilling technology, which allows access and recovery of unconventional resources that were considered economically and technically nonrecoverable just a few years ago.

Table 1.1 Top 10 Countries With Recoverable Shale Gas Resources [32]

Rank	Country	Amount of Shale Gas (Trillion Cubic Feet)
1	China	1115
2	Argentina	802
3	Algeria	707
4	USA	665 (1161)
5	Canada	573
6	Mexico	545
7	Australia	437
8	South Africa	390
9	Russia	285
10	Brazil	245
	Total	**7299 (7795)**

Table 1.2 Proved and Unproved Gas Resources in the World [32]

USA	Amount of Wet Natural Gas (Trillion Cubic Feet)
Shale gas proved reserves	97
Shale gas unproved reserves	567
Other gas proved resources	220
Other gas unproved reserves	1546
Total	2431
Increase in total gas resources due to shale gas	38%
Share of shale gas in total	27%
Outside USA	
Shale gas unproved reserves	6634
Other gas proved resources	6521
Other gas unproved reserves	7269
Total	20,451
Increase in total gas resources due to shale gas	48%
Share of shale gas in total	32%
Total World	
Shale gas proved reserves	97
Shale gas unproved reserves	7201
Other gas proved resources	6741
Other gas unproved reserves	8842
Total	22,882
Increase in total gas resources due to shale gas	47%
Share of shale gas in total	32%

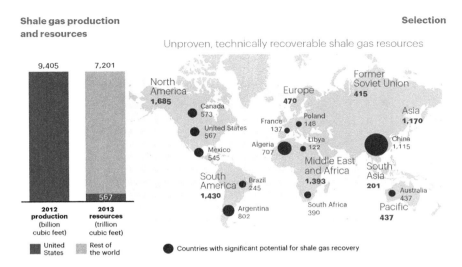

[1] Annual estimate based on daily U.S. production

Figure 1.6 Contribution of various regions/countries in shale gas resources and production [32]. *US Energy Information Administration study of 42 countries; A.T. Kearney analysis.*

8.1 World/Global

The shale gas revolution in the USA and also the availability of new technologies for access and recovery of these resources have caught the attention of many countries around the world. A number of regions that previously had total dependency on foreign supply of energy have now the potential to become net exporters. In addition to that, the availability and production of shale gas resources have a significant impact on the energy security of countries such as the USA and China. As a result, investment in shale gas production has grown exponentially all over the globe. As demonstrated in Fig. 1.6, some of the countries with the highest estimates of shale gas reserves include Algeria, Argentina, Australia, Canada, China, Mexico, and the USA.

8.2 USA

As mentioned earlier, the USA is leading the way in the production of shale gas. According to the several years' experience of shale gas extraction in the USA, the analysis shows that shale gas production has grown at an unprecedented rate in the past few years. Shale gas now accounts for almost 40% of US natural gas production. Currently about 80% of all shale gas production

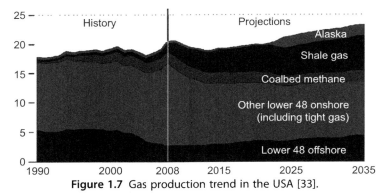

Figure 1.7 Gas production trend in the USA [33].

comes from five plays; however some of them are in decline which poses an important challenge [32,33]. Providing further information, history and projected gas production rate are illustrated in Fig. 1.7.

Since shale gas wells lower at very rapidly declining rates, it requires continuous inputs of capital in order to maintain production. The capital requirement is estimated at around $42 billion per year. From a technical perspective, drilling of an additional 7000 wells is also needed.

The area of Devonian Ohio shale off the Appalachian basin, with more than 30,000 gas wells and annual production of 120 bcf, contains a majority of shale gas wells in the USA [32,33]. Some other major shale gas regions are

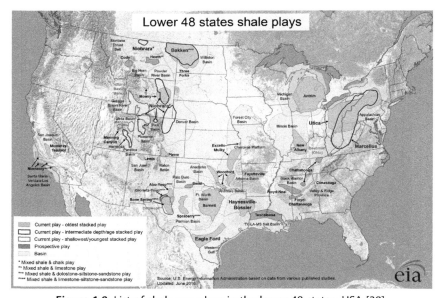

Figure 1.8 List of shale gas plays in the lower 48 states, USA [39].

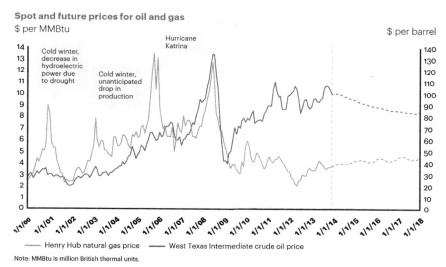

Figure 1.9 Spot and future prices for West Texas oil and Henry Hub gas [32].

namely, Barnett, Haynesville, Fayetteville, and Marcellus. All shale gas reserves in the lower 48 states of the USA are depicted in Fig. 1.8.

In the United States, the development of shale gas has drastically reduced the price of natural gas (NG) and decoupled it from oil prices. For instance, the history and future tend of Henry Hub natural gas price and west Texas intermediate oil price is presented in Fig. 1.9 in the time period of 2000–2018 [32]. As seen in Fig. 1.9, the development of shale gas considerably affects the gas price.

8.3 Canada

While Canada is already a major producer of conventional gas, recently there has been an increased focus on developing natural gas from unconventional resources such as shale gas. This is in line with the Canadian economic policy objective as it will help to offset the decline in conventional natural gas production, since new conventional resources become harder to find.

Similar to what took place in the USA in the last decade, the Canadian gas industry is currently undergoing a transformation, significantly focusing on shale gas production. There are considerable shale reserves all over the country, with the most momentous shale basins being located in northeastern British Columbia. Other potential regions are Alberta, Ontario,

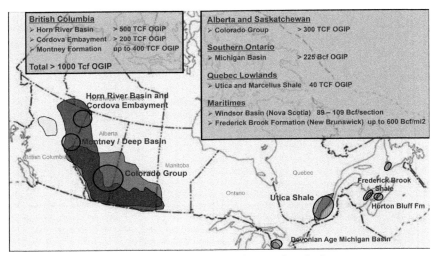

Figure 1.10 Canadian shale gas plays [15].

Quebec, and the Maritimes [15]. The complete list of Canadian shale gas reserves is given in Fig. 1.10. While large-scale commercial production of shale gas has not yet begun in Canada, this is likely to alter in the coming years as the industry transforms itself. For example, already $2 billion has been invested in order to establish land positions in the Horn River Basin and the Montney Trend area in northeast British Columbia [15].

Providing potential Canadian natural gas, Canadian Society for Unconventional gas recently reported that the total Canadian gas in place is predicted to be almost 4000 trillion cubic feet (tcf) as listed in Table 1.3.

As of January 1, 2010, recoverable shale gas resources of the world were estimated to be 7060 tcf which is about 25% of total recoverable global natural gas, as illustrated in Table 1.4. It is noted that estimates are uncertain and will be likely higher when accurate/more information becomes available [5,34].

Table 1.3 Canada's Gas in Place (GIP) Resources [15]

Type of Gas	Amount of Gas in Place (tcf)
Conventional (remaining GIP)	692
Natural gas from coal/coal bed methane	801
Tight gas	1311
Shale gas	1111
Total	**3915**

Petrel Robertson/CUSG Study, 2010.

Table 1.4 Recoverable Natural Gas Resources by Region [5,34]

Region	Total Gas (tcf)[a]	Shale Gas (%)
Eastern Europe and Eurasia	8119	0
Middle East	4907	10
Asia Pacific	4095	44
OECD North America	4836	40
Latin America	2577	48
Africa	2330	44
OECD Europe	1341	42
Total world	**28,205**	**25**

[a]Trillion cubic feet.

9. SHALE GAS PRODUCTION TREND

The projected trend of gas production in the time period of 2000–2035 is illustrated in Fig. 1.11. It is important to note that change in the gas production rate is dependent on a variety of factors such as political matters, war, social and economic aspects, development of new drilling, production and processing technologies, and exploration of new gas resources. According to Fig. 1.11, the gas is categorized into four groups including conventional, tight, coal bed methane, and solution gas. The trends clearly show that the production of all gas categories except conventional (nontight) will increase from 2020 [35].

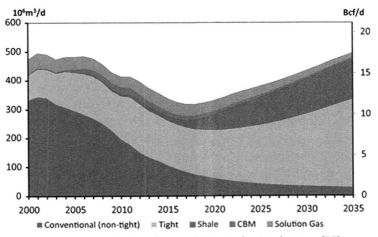

Figure 1.11 Variation of natural gas production by type [35].

10. KEY CHALLENGES IN SHALE GAS PRODUCTION AND EXPLORATION

The rise of unconventional shale gas production will increase a sense of energy security in certain markets; however, it also brings a complex set of challenges at global and local levels, while changing the power generation to natural gas from other sources has a severe impact on the climate, as well as fugitive methane emissions during production. The main issues concerning shale gas production and exploration are listed below, in brief [33–35]:

- Utilization of a large amount of water and toxic chemicals used in the hydraulic fracking process may not only become the cause of contamination but also a source of threat to drinking water.
- The massive use of chemicals, associated emissions, and truck traffic has a considerable impact on the environment, biodiversity, and ecosystems.
- A number of social, cultural and economic consequences for the local communities arise from the different factors like landscape impacts, high volume of truck traffic, and the consequences of an influx of new work forces into an area.
- A lot of challenges are involved to operating companies pertaining to the scale and multiple operators, and contractors working in a single area, raising issues for coordination and the anticipation and management of risks, including accidents and occupational health hazards.
- In countries having weak governance and a history of corruption and poor contractor management, the major concern is to provide effective oversight for considering all complex potential impacts.

11. SHALE GAS IMPORTANCE

Historically, the protection of oil supplies has been referred to as energy security. Since 2000, rapid development has been seen in the rise of shale gas in North America. Cumulative production of the USA and Canada is 25% of the global natural gas production. The contribution of shale gas will keep on an increasing trend in the future [36].

Generally, the importance of shale gas is listed below [36,37]:

- It is able to provide approximately a quarter of overall US energy;
- It can be used for the generation of electricity as it initiated a 35% increment in natural gas-fired power plant generations between 2005 and 2012;
- It can offer heat for over 56 million businesses and residences;
- Combined with the continued displacement/retirement of coal power plants, greater shale gas use has helped the US achieve approximately 70% of the CO_2 reductions targeted under the Kyoto Protocol as of 2012;

- It is able to provide 35% of energy and feedstock to US industries;
- It can create employment opportunities for a population of over two million;
- It can generate more than $250 billion/year government revenue.

Aside from these benefits, shale gas production means increased tax and royalty receipts for state and federal governments, and royalty and bonus payments to landowners.

12. APPLICATIONS OF SHALE GAS

Shale gas is the same as conventional natural gas in terms of utilization. Hence, shale gas can be used wherever natural gas is employed (e.g., electricity production and direct use). Direct use involves industrial use (e.g., furnaces and NH_3 production) and utilization in homes for heating and cooking [38].

REFERENCES

[1] http://naturalgas.org/overview/background/.
[2] Mokhatab S, Poe WA, Speight JG. Handbook of natural gas transmission and processing. (Amsterdam, The Netherlands): Elsevier; 2006.
[3] Speight JG. The chemistry and technology of petroleum. 5th ed. (Boca Raton, FL): CRC Press, Taylor & Francis Group; 2014.
[4] Scott Institute & Carnegie Mellon University. Shale gas and the Environment. (Pittsburgh, PA): Wilson E. Scott Institute for Energy Innovation; March 2013.
[5] Linley D. Fracking under pressure: the environmental and social impacts and risks of shale gas development. (Toronto): Sustainalytics; August 2011.
[6] Jing W, Huiqing L, Rongna G, Aihong K, Mi Z. A new technology for the exploration of shale gas reservoirs. Petroleum Science and Technology 2011;29(23):2450–9. http://dx.doi.org/10.1080/10916466.2010.527885.
[7] Clark CE, Burnham AJ, Harto CB, Horner RM. The technology and policy of hydraulic fracturing and potential environmental impacts of shale gas development. 2012.
[8] Vengosh A, Warner N, Jackson R, Darrah T. The effects of shale gas exploration and hydraulic fracturing on the quality of water resources in the United States. Procedia Earth and Planetary Science 2013;7:863–6.
[9] Scott AR, Kaiser WR, Ayers WB. Thermogenic and secondary biogenic gases, San Juan Basin, Colorado and New Mexico—Implications for Coalbed gas producibility. American Association of Petroleum Geologists 1994;78(8):1186–209.
[10] US Energy Information Administration. World shale gas resources: an initial assessment of 14 regions outside the United States. April 2011. Washington, DC.
[11] Friends of the Earth. Shale gas: energy solution or fracking hell?. (London): Friends of the Earth; 2012.
[12] Bolle L. Shale gas overview: challenging petrophysics and geology in a broader development adn production context. (Houston): Baker Hughes; 2009.
[13] National Energy Board. A primer for understanding Canadian shale gas. November 2009 [Online]. Available: http://www.neb.gc.ca/clf-nsi/rnrgynfmtn/nrgyrprt/ntrlgs/prmrndrstndngshlgs2009/prmrndrstndngshlgs2009-eng.pdf.
[14] Government of Alberta. Shale gas. 2013 [Online]. Available: http://www.energy.alberta.ca/NaturalGas/944.asp.

[15] Canadian Society for Unconventional Gas. Understanding Hydraulic Fracturing [Online]. Available: http://www.csur.com/images/CSUG_publications/CSUG_HydraulicFrac_Brochure.pdf.
[16] Gas Reservoir, http://www.britannica.com/EBchecked/topic/226468/gas-reservoir.
[17] Wikipedia, http://en.wikipedia.org/wiki/Natural_gas.
[18] https://upload.wikimedia.org/wikipedia/commons/5/5d/Schematic_cross-section_of_general_types_of_oil_and_gas_resources_and_the_orientations_of_production_wells_used_in_hydraulic_fracturing.jpg.
[19] Canadian Association of Petroleum Engineers, http://www.capp.ca/environmentCommunity/airClimateChange/Pages/SourGas.aspx.
[20] Speight JG. Shale gas production processes. Gulf Professional Publishing; June 11, 2013. Science.
[21] EIA. US Energy Information Administration — EIA — Independent statistics and analysis. EIA; 2013. N.p, http://www.eia.gov/forecasts/aeo/er/early_production.cfm.
[22] Rahim Z, Al-Anazi H. Improved gas recovery—1: maximizing Postfrac gas flow rates from conventional, tight reservoirs. Login to Access the Oil & Gas Journal Subscriber Premium Features. Saudi Aramco. 2012.
[23] Alberta Energy, http://www.energy.alberta.ca/NaturalGas/944.asp.
[24] Geology.com, http://geology.com/rocks/shale.shtm.
[25] Cascading Shale Rock, http://www.pbase.com/camera0bug/image/16090687.
[26] Maiullari G. Gas shale reservoir: characterization and modelling play shale scenario on wells data base. (San Donato Milanese, Italy): ENI Corporate University; 2011.
[27] Boak J, Birdwell J. An introduction to issues in environmental geology of oil shale and tar sands. 2010.
[28] Martini AM, Walter LM, Budai JM, Ku TCW, Kaiser CJ, Schoell M. Genetic and temporal relations between formation waters and biogenic methane: upper Devonian Antrim Shale, Michigan Basin, USA. Geochimica et Cosmochimica Acta May 1998; 62(10):1699—720.
[29] http://gastoday.com.au/news/us_shale_gas_story_and_australian_lng_revolutionising_the_global_gas_market/101282.
[30] Speight JG. Shale oil production processes. Gulf Professional Publishing; 2012. Technology & Engineering.
[31] Petroleum Technology Alliance Canada. The modern practices of hydraulic fracturing: a focus on Canadian resources. November 2012 [Online]. Available: http://www.capp.ca/canadaIndustry/naturalGas/ShaleGas/Pages/default.aspx.
[32] Analysis & Projections, http://www.eia.gov/analysis/studies/worldshalegas/.
[33] EIA. Annual energy outlook. 2010.
[34] World Energy Outlook 2011. Are we entering a golden age of gas ?. (Paris): International Energy Agency; 2011.
[35] National Energy Board, https://www.neb-one.gc.ca/nrg/ntgrtd/ftr/2013/index-eng.html.
[36] Bonakdarpour M, Flanagan B, Holling C, Larson JW. The economic and employment contributions of shale gas in the United States. IHS Global Insight. America's Natural Gas Alliance; 2011.
[37] Natural Gas From Shale, http://energy.gov/sites/prod/files/2013/04/f0/why_is_shale_gas_important.pdf.
[38] Louwen A. Comparison of the life cycle greenhouse gas emissions of shale gas, conventional fuels and renewable alternatives from a Dutch perspective. (The Netherlands): Utrecht University; 2011.
[39] https://www.eia.gov/maps/maps.htm.

CHAPTER TWO

Shale Gas Characteristics

1. INTRODUCTION

Gas shale is the name given to a shale gas reservoir (play). Shale gas reservoirs are spread over large areas up to 500 m. They are characterized by low production rates. Shale gas reservoirs are fine-grained and rich in organic carbon content that signifies large gas reserves [1]. There are disparities in lithology in gas shales which point toward the fact that natural gas is stored in the reservoir in a broad array of lithology and textures such as non-fissile shale, siltstone, and fine-grained sandstone (not only shale). Often, shale laminations or beds are interbedded in siltstone- or sandstone-dominant basins [1,2].

Shale gas usually occurs between 1 and 5 km below the earth's surface. Regions which are less than a depth of 1 km tend to have lower pressures and gas concentrations and regions deeper than 5 km often have high density and lower permeability, which results in higher expenses for development stages such as drilling operations [1–3].

Natural shale gas is mainly composed of methane, although it might also contain compounds that energy companies have to separate from methane to make the gas usable commercially. These impurities may be different in each well and reservoir [1–3]. The other compounds found in shale gas include natural gas liquids, which are hydrocarbons of a heavier nature that will be separated in processing plants as liquids [1–3]. These liquids include heptane, hexane, pentane, butane, and propane. Shale gas also includes condensates and water. The gaseous components of raw shale gas include sulfur dioxide, hydrogen sulfide, helium, nitrogen, and carbon dioxide. Mercury may also be found in smaller concentrations in most reservoirs where natural gas is obtained. The mercury found will be lowered in concentration until it goes below the detectable threshold of one part per trillion (ppt) [1–3].

The petrophysical data analysis for shale formations is the same as that for unconventional reservoirs (e.g., gamma rays, resistivity, porosity, and acoustic, along with addition of neutron capture spectroscopy data). The petrophysical analysis of shale oil starts with gamma ray log. This indicates the presence of organic-rich shale. The organic matter contains higher levels of naturally occurring radioactive materials than the ordinary mineral reservoirs. Petrophysicists use the gamma ray count to identify organic-rich shale formations.

Permeability and porosity of shale reservoirs depend on natural fractures which are usually very low for shale rocks. When natural fractures are absent or do not allow gas to be produced, stimulation techniques like hydraulic fracturing are applied to produce shale gas [2–4].

The presence of a variety of rock types with dissimilar geochemical and geological features requires unique techniques to extract the natural gas [5]. Gas shales exhibit a broad variety of mechanical properties, extent, and anisotropic distribution of clay and solid organic materials.

Some of the clay minerals that occur in shale formations absorb or adsorb large amounts of natural gas, water, ions, or other substances [2,4]. This can serve to selectively and steadfastly hold or freely release fluids or ions. The relative contributions and combinations of free gas from matrix porosity and from desorption of adsorbed gas is a vital parameter affecting the well production profile. After the rapid decline of the initial gas production due to depletion of gas from the fracture network, the depletion of gas stored in the matrix becomes the primary process for production [2,4]. Supplementary to this depletion process is desorption, i.e., release of the gas due to a decrease in the reservoir pressure. Reservoir pressure is a vital factor to the rate of gas production by desorption. Since alterations in pressure transmit slowly through the rock because of low permeability, tight well spacing is sometimes needed to cause release of a significant amount of adsorbed gas. The total recovery with all the processes combined is around 28–40% [2,4].

Shale gas reservoirs have low recovery factors, compared to conventional reservoirs. Some of the important parameters which characterize shale gas reservoirs are thickness, maturation, organic richness, or total organic carbon content (TOC), brittleness, mineralogy, gas–in place, pressure, permeability, porosity, and pore pressure [2,4]. For a shale gas reservoir to become a successful shale gas play, these properties need to be considered. Besides all these, the depth of the shale gas formation is also important as it has a bearing on the economics of the gas recovery. An optimum combination of these factors leads to favorable productivity. As different shale gas reservoirs have different properties, it is imperative to study them before any exploitation plan is put in place. Another important point to take into account is that such properties can be determined at the location of the wells where the well log and core data are available. However, different geophysical workflows need to be used on 3D surface seismic data to characterize the shale gas formations [2,4].

Many of the petrologic and geochemical tools needed for shale-gas evaluation/characterization are unfamiliar to scientists and engineers trained in conventional petroleum geology. The purpose of this chapter is to introduce these tools and explain how they are employed to evaluate shale-gas reservoirs and resources, with specific examples from and emphasis on the Marcellus Shale, other Devonian black shales, and the Utica Shale in the Appalachian basin [6]. This chapter will also summarize the main characterization workflows.

2. CHARACTERIZATION OF GAS SHALE: BACKGROUND

The mechanisms of storage of gas and its flow in gas shales are not trivial and difficult to classify by conventional core studies. These days, analysis methods from petrology, petrophysics, and material science are collectively used to conduct characterization of shale gas specimens [7].

Representative analysis methods for shale-gas reservoir rocks include vitrinite reflectance, TOC, fluid saturation, X-ray diffraction, porosity, permeability, adsorbed/canister gas, detailed core and thin-section descriptions, and optical and electron microscopy [7,8]. The results from these studies can be used in conjunction with results from well-log studies such as gamma ray log, sonic log, density log, neutron log, resistivity logs, NMR data, and borehole images for a full characterization of gas shales [8].

Overall, a comprehensive range of geological, geochemical, geophysical, and mechanical data is vital for the examination of shale gas/oil potential and evaluation of its resources as demonstrated in Fig. 2.1 [9].

There are many different methodologies (workflows) adopted by the oil companies for characterization of different gas shales. One typical workflow is depicted in detail in Fig. 2.2 [10]. According to this figure, Slatt et al. developed a systematic integrated characterization methodology to characterize sedimentary rock for horizontal well placement and artificial fracture treatment [10].

The characterization is normally carried out in the following manner.

Mineralogic analyses are conducted by standard X-ray powder diffraction (XRD) and Fourier transform infrared spectroscopy (FTIR) techniques, complemented by chemical studies. These analyses, coupled with TOC by combustion, provide the foundation for identifying lithofacies based on compositional and fabric features [10].

Porosity and permeability measurements are conducted using standard techniques. Pores and their connectivity can be directly observed and

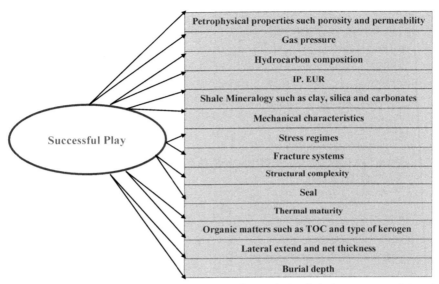

Figure 2.1 Typical set of geological required for analysis of shale gas potential and resource well assessment. *Modified after Elgmati MM, Zhang H, Bai B, Flori RE, Qu, Q. Submicron-pore characterization of shale gas plays. In: North American unconventional gas conference and exhibition. Society of Petroleum Engineers; 2011.*

somehow quantified by using scanning electron microscopy (SEM) and field emission-scanning electron microscopy (FE-SEM) as shown in Figs. 2.3 and 2.4 [10–14]. It is also possible to determine the grain morphologies and elemental compositions (for mineral identification of grains) from energy-dispersive X-ray (EDX) analysis [10–14]. Tight rock analysis pyrolysis technique is also used to quantify ultra-low permeability [14].

ROCK-EVAL is a proper tool for characterization of source rock quality which gives information on quantity of recoverable organic materials in the source rock and the residual kerogen. A uniaxial rock mechanic test can be employed to measure Young's modulus (E) and Poisson's ratio (v). These two parameters affect the wellbore stability and hydraulic fracturing [11].

Researchers have also concluded that high pressure (up to 60,000 psi) mercury porosimetry analysis (MICP) can find out the pore size distribution. A robust, detailed tomography procedure using SEM-focused ion beam (FIB) can efficaciously describe the pore structures below a micrometer. SEM can explore many sorts of porosities [2,7].

Giving more information, Table 2.1 summarizes the coring and logging methods/data to determine the main target properties such as porosity, permeability, and TOC [15].

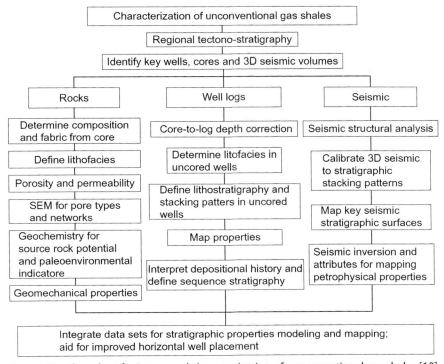

Figure 2.2 Flow chart for integrated characterization of unconventional gas shales [10].

3. CHARACTERIZATION OF GAS SHALE: METHODS

There are a variety of characterization methods for shale gas reservoirs. The main inorganic petrographic analysis techniques and the information they yield are given in Table 2.2 [2,12].

Figure 2.3 Scanning electron micrographs of Woodford shale. Microchannels (*red* (gray in print versions) *arrows*), HC migration pathways [10].

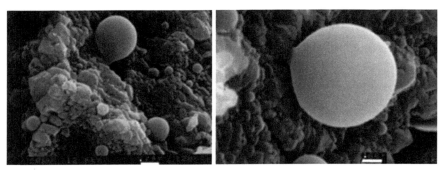

Figure 2.4 Scanning electron micrographs image of Woodford shale. Oil droplets exuding from rock into open microchannels [10].

Key geochemical analysis techniques and the information they give are shown in Table 2.3.

Many particle testing agencies, like Micromeritics Analytical Services, provide services for physical characterization. A summary of the techniques and information provided is tabulated in Table 2.4 [2,13].

Until now, the focus of this section has been on laboratory characterization techniques. Well logging techniques to attain petrophysical data will be investigated later in this chapter. Slatt and his team utilized a core-to-log-depth correction factor to relate geological observation and laboratory-derived petrophysical properties to well log-derived characteristics [10].

Table 2.1 Various Types of Properties Extracted From Core Data and Log Data [15,16]

Property of Interest	Core Data	Log Data
Porosity	Crushed dry rock He porosimetry	Density (mostly)
TOC	LECO or RockEval	GR, density, resistivity
Water saturation	As-received retort or Dean—Stark	Resistivity, kerogen corrected porosity
Mineralogy	XRD, FTIR, XRF	Density, neutron, Pe, ECS-type logs
Permeability	Pulse decay on crushed rock	(This is tough.)
Geomechanics	Static module	DTC, DTS, RHOB, and synthetic substitutes
Geochemistry	Ro, S1-S2-S3, etc.	Resistivity (sort of)

Table 2.2 Major Inorganic Petrographic Analysis Techniques and Corresponding Information Attained [2,12]

Technique	Information
Core gamma log	Depth shift corrections
Whole core CT scanning	Density profile, bedding heterogeneity, identify natural or induced fractures
Thin sections	Mineralogy, texture, cementation, pore geometry, classification
FIB/SEM imaging	Mineral morphology, pore geometry, microfabric texture, energy-dispersive spectra (EDS) mineral mapping, 3D volume modeling
XRD	Bulk mineral composition, identification of fluid-sensitive clays, grain density correlations
Fluid sensitivity analysis	Frack fluid selection, fines generation and migration, filtrate—rock interactions

3.1 Well Logging Characterization Methods

3.1.1 Well Log Data

As resistivity increases appreciably in mature rocks because of the generation of nonconducting hydrocarbons, it can be identified easily on the resistivity log curves [3,17]. The $\Delta \log R$ technique, which is valid over a wide range of maturities, is employed for measuring TOC in shale gas formations (see Fig. 2.5). The resistivity curve and the transit time curves are scaled so

Table 2.3 Common Geochemical Analyses Techniques and the Information They Provide [2,12]

Technique	Information
Programmed pyrolysis	Present-day thermal maturity, existing volatile hydrocarbons, and remaining hydrocarbon generative potential
Isotope geochemistry	Genetic origin, thermal maturity, interformational mixing, correlation of hydrocarbons to their source, and reservoir compartmentalization
Vitrinite reflectance and kerogen assessment	Thermal maturity, kerogen type, and original hydrogen index (HI)
Oil fingerprinting	Types of oil components and indirect evidence of HC transport potential through shale matrix
Biomarkers	Thermal maturity (useful only in the oil window), paleoecology, lithology, age
Yield calculations	Gas in place (GIP) and oil in place (OIP) estimates

Table 2.4 A Number of Physical Characterization Techniques [2,13]

Technique	Information
Mercury intrusion porosimetry	Pore size information, pore volume, porosity, and density measurements
Brunauer–Emmett–Teller (BET) surface area using the gas adsorption technique	Predict the quantity of free gas stored within pores, the amount of adsorbed gas or dissolved gas on the surface or in pores, and kinetics for rate of gas production
Gas displacement pycnometers	Skeletal volume of shale. When combined with other density measurements, skeletal volume can be used to determine porosity of both crushed and intact shale samples
High-pressure gas adsorption isotherms	Model kinetic data and determine volume adsorbed at simulated shale depth conditions

that their relative scaling is equal to 100 ms/ft per two logarithmic resistivity cycles [4,17]. In this respect, the two curves cover each other over a broad depth interval which in turn is a function of maturity and is linearly related to TOC. However, there is an exception when we deal with the organic-rich formations in which they would show a detached trend by $\Delta \log R$. Using the sample analysis, the level of maturity (LOM) can be generally determined in which the vitrinite reflectance (Ro%) approach is employed to attain the goal. Depending on the type of organic matter, LOM values calculated from R% normally vary from 6 or 7 to 12. Number 7 indicates

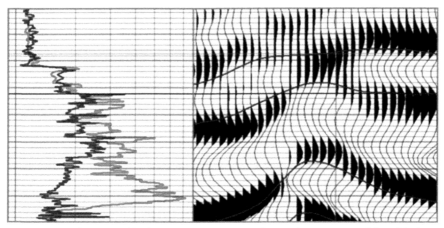

Figure 2.5 Indication of high TOC by log data in form of transit time (blue (black in print versions)) and resistivity (red (gray in print versions)) [4,17].

the onset of maturity for oil-prone kerogen and number 12 implies the onset of overmaturity for oil-prone kerogen [4,17]. The left panel of Fig. 2.5 includes the sonic and resistivity curves indicating the $\Delta\log R$ separation. Attributes generated from the available log curves (sonic, density, resistivity, and porosity) need to be cross-plotted to gain more insight into the shale reservoir and determine which factors would be suitable to distinguish the reservoir from the nonreservoir zones [3,4,17,18].

3.1.2 Seismic Data

TOC variations in shale formations influence VP, VS, anisotropy, and density and consequently should be identified on the basis of response of the seismic [18]. For a given kerogen content and layer thickness, by increasing the angle, the PP reflection coefficient decreases. This suggests that if the far and near stack are assessed for a specified seismic data volume, the base of the reservoir region will exhibit a separate positive reflection that will also lower with offset, leading to an increase to a class-I AVO response. In the same way, at the reservoir rock top, the negative amplitudes on the near stack are seen as lowered on the far stack, revealing a class-IV AVO response [3,18]. This is feasible since the acoustic impedance for shale reservoir rocks with TOC >4% is lower than the same rocks without TOC. When the impedance curves from wells, which penetrate the shale layers plotted on the same graph, show the changes of the reservoir quality. Zones linked with lower acoustic impedance are associated with the higher organic content and could be chosen from such a display [3,4,18]. Brittleness of a rock formation can be approximated using Young's modulus and Poisson's ratio versus the corresponding depth. This proposes a procedure for approximating brittleness using 3D seismic data, through concurrent prestack inversion that results in ZS, ZP, Poisson's ratio, and VP/VS. Zones with low Poisson's ratio and high Young's modulus are those that have better reservoir quality as well as being brittle [3,4]. This approach is suitable for good-quality data.

3.1.3 Hybrid Workflow

This requires the relative acoustic impedance from thin-bed reflectivity along with generation of curvature volumes. The reflection coefficient volume gained from the input seismic data through spectral inversion is defined as thin-bed reflectivity [18–20]. The relative acoustic impedance obtained from this volume has a higher level of detail. This is mainly because in this volume there is not a seismic wavelet. In a graphical illustration, it exhibits a direct demonstration of a horizon cut via a k1 most-positive principal curvature volume with a vertical cut through the conforming seismic amplitude volume, which specifies an overlay of the hypothesized fracture network over the relative impedance display and lineaments correlated to fractures at that level [18–21]. It should be mentioned that the different curvature lineaments appear to fall into the proper high-impedance compartments splitting them from low-impedance compartments, and may imply cemented fractures [18–20].

Natural fractures can provide permeability pathways in shale formations and so require to be described if they exist. Whereas there are various shale reservoirs that may not include such natural fractures, others such as the Woodford Shale have this characteristic. The Muskwa-Otter-ParkeEvie Shale package and the Eagle Ford Shale in the Horn River Basin also exhibit natural fractures.

Such fractures can be categorized by using prestack processes (for instance VVAz/AVAz/AVO) or by processing the discontinuity attributes on poststack data. Using the analysis of azimuthal variations of the impedance field/velocity, the natural fractures in the shale formations can be identified [3,19–21].

3.1.4 Seismic Waveform

This classification is another fast and easy-to-use approach that holds potential as an application for shale-gas reservoir description. It basically categorizes seismic waveforms based on their shapes in the interval of interest. Typically, the traces are grouped into different categories and exhibited a variety of colors so that each group has its own color. The subsequent map shows seismic facies change and facilitates separation of desirable regions from the others. Depending on how the waveforms are segmented, the waveform grouping process can be constrained or unconstrained. Fig. 2.6 depicts the outcome of unconstrained waveform grouping and compares it with a corresponding demonstration from the relative acoustic impedance obtained from the thin-bed reflectivity [18–20].

4. PETROPHYSICAL CHARACTERISTICS OF GAS SHALES

In gas shales, the four major areas to attain information on petrophysics are the type of rock, amount of volume in the reservoir rock, type of fluid, and how well the fluid can flow. Different measurements and techniques are

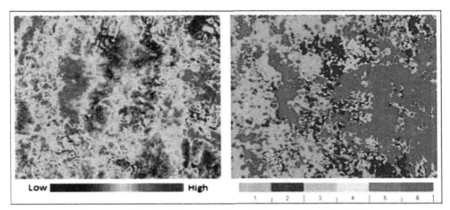

Figure 2.6 Reflectivity inversion of 3D seismic data [4,17].

utilized to characterize the rock [15,16]. This section is described through a real case study.

4.1 Rock Type by Coring

A section of a reservoir rock is presented in Fig. 2.7 and zones are labeled with number for simplicity to illustrate the characterization techniques later on.

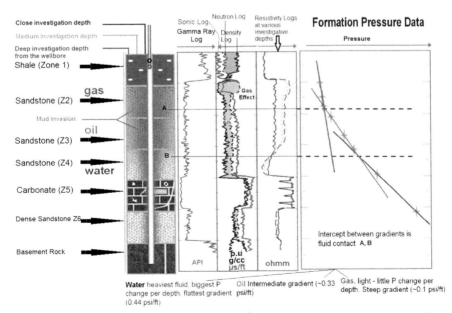

Figure 2.7 Gas shale cross-section with various well logs and formation pressure data [15,16].

Mudrock (shale) is on the top (Zone 1), sandstone rocks below (Z2, Z3, Z4), then carbonate (Z5), some more sandstone (Z6), and then a basement rock or granite is demonstrated [15,16].

It is common to drill into the reservoir and then stop, change the drill bit and then go in with the device called a corer and take some core, i.e., a cylinder of rock and remove it out of the reservoir. Then, it is taken to the lab and analyzed under a microscope or under the naked eye. Therefore, in coring, a rock sample is recovered to the surface, and a description and further analysis can be made [15,16].

The same piece of core is shown in Fig. 2.8. The images on the left-hand side are in the normal white light. It is observed that this rock is sandstone stained dark with black oil and the other rocks are mudrocks. It is also clear that oil stays between the mudrocks in the reservoir. On the right-hand side, the same interval of rock is laid out on the table with a photograph under ultraviolet (UV) light. Under UV light, oil and HCs tend to fluoresce [15,16].

The mudrocks, which do not contain the oil, appear quite dark and petrophysical properties can be directly measured. The small holes are called core plugs that are taken out of the reservoir, taken out of the core, and property measurements (e.g., porosity) are made. In some cases, it is possible to even measure directly the fluid that is in the core. Thus, a lot of information about reservoir characteristics can be obtained from the core data, such

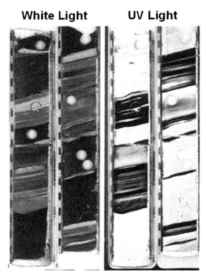

Figure 2.8 Core samples visualized under white light and UV light [15,16].

as the sedimentology (the geology). For instance, geologists can specify whether the rock was deposited in a swamp or deep under the sea or a river. When such data are attainable, the best location of the shale rock in the reservoir can be predicted [15,16].

Measurements can be also made of the hardness and mechanical properties of the rocks. This work phase helps to understand the best strategies to drill the shale reserves in the future. Various chemicals that might be put in during the course of the oil field development can be also tested on the rocks to see how they react, i.e., will they cause any swelling and damage to the wellbores or are they effective at removing scaling? Coring is a direct measurement. It is not practical to core all of the rock so that only small sections can be studied. It is expensive to drill into the reservoir, stop, cut a core, come out, go back in, cut a core, and repeat the process [15,16]. In reality, only a few cores are cut and direct measurement data are taken. In general, other types of measurements should be conducted [15,16].

4.2 Rock Type by Gamma Ray Logging

The first element that is of interest is the rock type. It is possible to log 4000—5000 feet of reservoir interval by placing in tools where drilling has been done. Gamma rays use a passive device so that they listen to the signals given off by the reservoirs [15,16]. Rocks contain naturally occurring radioactive elements, Th, K, and U, which all emit gamma rays. Clays and mudrocks have a much higher Th and K content and they offer a high gamma ray response. The carbonate rocks are usually less radioactive. Thus, they have a cleaner response. Igneous, volcanic, and metamorphic rocks have a varied mineralogy and they can have a high rate of radioactivity. Therefore, the tool is put down the hole at the end of the wire line. As the tool is pulled out, a measurement is made. Fig. 2.7 shows an example of a gamma ray log. Increasing radioactivity is from left to right. This is called a wire line log. In fact, it is a trace of a measurement made in a wellbore from which many assumptions can be made and calculations can be done for the type of rock [4,15,16].

4.3 Volume Estimation by Porosity Logging

Measurements need to be made to quantify how much pore space is available to hold fluid. A different type of wire line tool is put in the wellbore to conduct porosity logging. There are a number of different measurements through active device means. This means that they stimulate the formation in some way and then measure the response to the stimulus [15,16].

4.3.1 Sonic Log

In this technique, the sound waves are sent to the rock. Then, it is monitored how long the sound waves take to go from the transmitter to the receiver. The amount of time that takes for sound to travel through the rock provides some information about the pore properties of the rock. A longer time means a lot of holes are in the way since sound travels quicker through solids, compared to liquids or air. According to Fig. 2.7, sonic log is faster in the carbonate and sandstone. In the shale region, the log is slower. It might be interpreted as very high porosity. However, shales do not have a very high effective porosity [16]. They may have a high total porosity and may contain quite a bit of water, i.e. less rock. Since porosity for shale formations is not connected most of the time, it is not called effective porosity for oil field purposes [15,16].

4.3.2 Density Log

The active device beams gamma rays at the formation. If the radioactive gamma rays encounter a lot of solid rock in the way, they will be bounced back. Hence, the lower the porosity, the more the rock, and less counts are registered with the detector. The density log measured is scaled in a similar way. From the log in Fig. 2.7, it can be concluded that the upper sandstone has a lower density than the lower sandstone interval. A low-density rock has higher porosity. For shales, more rock compared to pore volume is seen [15,16].

4.3.3 Neutron Log

In this method, neutrons are emitted into the formation and neutrons react with hydrogen present in the water and oil. Thus, this log is a fluid-affected log. Depending on how much fluid is in the way, this controls what signal reaches the detector [15,16]. Therefore it is able to determine how much fluid is in the way or/and how many pores exist. The neutron log in Fig. 2.7 implies that the porosity is lower in the carbonate. It is also found that the upper sandstone has a reasonable porosity and quite a bit of water is in the shales, so that there is an increase in the apparent neutron porosity. Hence, it can be called fluid porosity log. Sonic log takes the combination of fluid and rock into consideration and density log reflects more the rock [15,16].

4.4 Fluid Type by Resistivity Logging

Knowing the amount of pore space available, it seems important to figure out what type of fluid is in the pore space. Sometimes when wells are drilled, it is found that it is water wells (instead of oil ones) that are being logged. This is performed by resistivity logging. The tool is placed into the wellbore and it has a transmitter and receiver. In general, there are two types of resistivity measurements [15,16]. One is called direct measurements where an electric current is beamed directly into the formation. There is conductivity between the tool and formation. The resistance of the rock to the electric current is measured. The other is called an indirect measurement where an induction tool stimulates a current in the reservoir. It has an alternating current (AC), which creates a fluctuating magnetic field around the tool downhole. This magnetic field induces a current to flow in the formation and movement of this current around the tool again induces a current back in the receiver coil within the tool. This current is proportional to the current flow in the formation. Based on this concept, the conductivity of the formation is obtained and its inverse gives the resistivity [16].

It is well known that water conducts electricity if it is saline. Commonly, all natural waters have some salinity. The more porosity there is in the system, the bigger the volume of water, the easier it is for electric current to flow. Hence, water-bearing rocks with more porosity are more conductive. Oil is nonconductive and if oil is there in the rock, it tends to restrict the current flow. In general, there is always some water in the rock that makes some sort of conductive pathway over the grain of the rock. It is very rare that the rock and oil would be completely nonconductive [16].

While interpreting logs, it should be noted that higher resistivity may not be the result of oil saturation because sometimes it just happens that the rock is tight. Hence, it is important to have the porosity log done before interpretation of the resistivity log to distinguish low-porosity/high-resistivity rocks from high-porosity oil-bearing rocks [15,16].

The resistivity log is commonly scaled to increase toward the right. While interpreting Fig. 2.7 (for the deep investigative depth from the wellbore) from the bottom up, one can notice that there is some resistivity in the basement zone. The resistivity drops down a little bit into the conductive sand (Z6) which has some intermediate resistivity. As limestone/carbonate cemented interval (Z5) is encountered, the resistivity increases. Before the presence of hydrocarbon is confirmed, the porosity log should be examined, which will clearly show that there is very low porosity, implying an increase in resistivity [15,16].

In the porous zone of Z4, resistivity drops due to big pore space and lots of water volume. As more oil is encountered upwards in zone Z3, resistivity increases. This is interesting because there is a high-porosity region which demonstrates the presence of hydrocarbons there, based on the porosity logs. According to the gas-filled zone Z2, resistivity may or may not increase. In the shale region (mudrock) which usually contains a lot of water, resistivity lowers. Hence, once the porosity is known and the resistivity measurement is made, it can be concluded somehow that there is a quantity of HC in the field [16].

Drilling mud is used while drilling the well. It helps to stop the pressurized fluids underground from falling into the well bore. It also keeps the drill bit cooled. It also seeps into the rock like a sponge and an invasion profile is delivered. As seen from Fig. 2.7, in the poorer-quality rock, there is invasion to some depth and in the better-quality rock the mud filtrate is pushing the oil away and seeping in deeper. However, deeper invasion characteristics might be also noticed in the poorer-quality rock because rather than building up a nice mud cake that stops further invasion, there is not good permeability to develop a mud cake so that the mud constantly seeps in [16].

In Fig. 2.7, the red log represents the resistivity measurements made away from the invaded region. If measurements are made at other depths of investigation: from wellbore neighborhood and intermediately away from the wellbore, different resistivity profiles are attained, depending on the amount of invasion. This gives an idea of the permeability, too [16].

In the zone Z6, resistivity measured at the medium investigative depth (from the wellbore) overlies the one at deepest investigative depth. This means that the invasion has not reached out till the medium depth. There might be an increase in the resistivity due to the mud cake build-up. In the tight nonpermeable carbonate cemented zone, all the curves overlay. The borehole wall fractures are indicated by the small peaks signaling drops in resistivity [16].

In the acceptable porosity/high-permeability water-bearing sand, there is low resistivity due to conductive sand. In the oil-bearing sandstone, mud filtrate has gone deeper and has increased the resistivity at the medium depth. The resistivity is even greater due to mud cake interaction at the medium and deep locations (from the well bore). At the shallow depths, the relatively conductive mud filtrate has displaced the resistive oil away. Hence the resistivity profile is lower. The separation can easily be seen for zone Z3. It is found that the oil is movable, indicating a good rough estimate of permeability [16].

Gas generally has a low hydrogen index and it is less dense and has low neutron porosity. On the gas log, this looks like the gas effect. Hence, it is concluded that the presence of a fluid type can have an effect on the log [16].

4.5 Fluid Type by Formation Pressure Data

A tool is sent downhole that is a pump attached to the side of the borehole wall, a small amount of the fluid is then suck into the tool and the pressure the fluid exerts on the tool is measured. This provides information about the permeability and mobility of the fluid, and the pressure at each point. The pressure is dependent on the type of fluid above the point where pressure is being measured. Yellow stars in Fig. 2.7 represent a series of stationary pressure measurements in the well bore [16]. At each point, a reading is taken and it is seen that pressure increases quickly in the water leg because water is the heaviest. Considering the oil gradient, i.e., the pressure points measured in the oil leg, it is observed that the pressure does not increase as quickly as the water leg. Similarly to the gas pressure, the increase is not much down the wellbore. The intercepts give effective oil/water contact and gas/oil contact. The density of the fluid is reflected in the pressure gradient [16].

4.6 Flow Type and Permeability by Nuclear Magnetic Resonance

Nuclear magnetic resonance (NMR) is carried down the whole wellbore and an NMR signal is obtained. As the protons return to the ground state, the signal decays off. The larger the pore size, the smaller is the decay, since the protons do not interact with the pore walls. In small pore sizes, the protons are quickly returned to their ground state by interaction with the pore walls and the signal decays more rapidly. The pore system characterization is employed to estimate permeability [15,16].

5. COMPOSITION OF SHALE GAS

Composition of shale gas is similar to that of natural gas in conventional reservoirs (see Table 2.5). Shale gas is typically a dry gas and it contains 60—95% v/v of methane and nitrogen, and sometimes trace quantities of ethane, propane, the noble gases, oxygen, and carbon oxide. However, no traces of harmful hydrogen sulfide have been found in shales [2,22]. Dry gas is classified as having less than 0.1 gal/1000 cf of gasoline vapor (higher-molecular-weight paraffins). When the presence of heavier HCs is greater, it is called wet gas. The detailed composition of shale gas is shown in Table 2.5 [2].

As a sample, Table 2.6 presents the composition of Marcellus shale gas.

Table 2.5 Composition of Shale Gas [2]

Name	Vol%
Methane	>85
Ethane	3−8
Propane	1−5
Butane	1−2
Pentane	1−5
Carbon dioxide	1−2
Hydrogen sulfide	1−2
Nitrogen	1−5
Helium	<0.5

Table 2.6 Marcellus Shale Gas Composition (Volume %) [2,22]

Well No.	CH_4	C_2H_6	C_3H_8	CO_2	N_2
1	79.4	16.1	4.0	0.1	0.4
2	82.1	14.0	3.5	0.1	0.3
3	83.8	12.0	3.0	0.9	0.3
4	95.5	3.0	1.0	0.3	0.2

A number of produced shale gases include an extensively greater magnitude of ethane, compared to the historical average of transmission-grade gas, while others consists of hexanes (or diluents) and heavier hydrocarbons over the historical average.

Not only are shale gases unlike with respect to the historical transmission-quality gases, they differ from one reservoir to another, and even within the same formation. Major differences in concentration of propane, hexane, ethane, and heavier components, and diluents (principally CO_2 and N_2) have been noticed among the different shale formations as tabulated in Table 2.7. These, in turn, lead to noteworthy dissimilarities in the Wobbe number, heating value, and other factors which direct costumer applications of natural gas [15].

6. RANGES OF TEMPERATURE, PRESSURE, AND DEPTH

6.1 Pressure

Shale gas studies place particular emphasis on identifying areas with overpressure, which enables a higher concentration of gas to be contained within a fixed reservoir volume. A conservative hydrostatic gradient of 0.433 psi per foot of depth is utilized when actual pressure data are

Table 2.7 Main Compositional Characteristics of Shale Gas at Different Sites of the USA [2,15]

Component (Vol.%)	US Mean Value	Site 1	Site 2	Site 3	Site 4	Site 5	Site 6	Site 7	Site 8	Site 9
CH_4	94.3	79.4	82.1	83.8	95.5	95.0	80.3	81.2	91.8	93.7
C_2H_6	2.7	16.1	14.0	12.0	3.0	0.1	8.1	11.8	4.4	2.6
C_3H_8	0.6	4.0	3.5	3.0	1.0	0.0	2.3	5.2	0.4	0.0
C_4H_{10}	0.2	Nil	Nil	Nil	Nil	Nil	Nil	Nil	Nil	Nil
C_5H_{12}	0.2	Nil	Nil	Nil	Nil	Nil	Nil	Nil	Nil	Nil
CO_2	0.5	0.1	0.1	0.9	0.3	4.8	1.4	0.3	2.3	2.7
N_2	1.5	0.4	0.3	0.3	0.2	0.1	7.9	1.5	1.1	1.0
Total inert ($CO_2 + N_2$)	2.0	0.5	0.4	1.2	0.5	4.9	9.3	1.8	3.4	3.7
Total	100	100	100	100	100	100	100	100	100	100
HHV (Btu/SCF)	1035	1188	1165	1134	1043	961	1012	1160	1015	992
Specific gravity	0.592	0.675	0.660	0.653	0.583	0.601	0.663	0.672	0.607	0.598
Wobbe Number (Btu/SCF)	1345	1445	1435	1404	1366	1239	1243	1415	1303	1284

unavailable. Pressure normally ranges from 2500 to 2800 psia for most shale gas reserves [2,22].

6.2 Temperature

A number of research studies assemble data on the temperature of shale formations, giving particular emphasis on identifying areas with higher than average temperature gradients and surface temperatures. A temperature gradient of 1.25°F per 100 feet of depth plus a surface temperature of 60°F is used when actual temperature data are unavailable. The temperature for a shale gas reservoir is usually between 75 and 160°F [2,22].

6.3 Depth

The depth for main shale plays is greater than 1000 meters but less than 5000 meters (3300 feet to 16,500 feet). Areas shallower than 1000 meters have lower reservoir pressure and thus lower driving forces for oil and gas recovery. In addition, shallow shale formations have risks of higher water content in their natural fracture systems. Areas deeper than 5000 meters have risks of reduced permeability and much higher drilling and development costs [2,22].

The USA is the highest shale gas-producing country. Temperature and pressure ranges for US gas shale reservoirs are depicted in Fig. 2.9 [23].

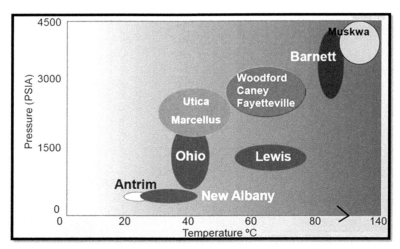

Figure 2.9 Pressure and temperatures ranges for some gas-producing US gas shales [23].

7. SHALE GAS VISCOSITY AND DENSITY

The Lee correlation is employed to determine gas viscosity (μ_g) at a specific temperature and pressure as [2,23]:

$$\mu_g = 10^{-4} K \exp(X\rho^Y) \tag{2.1}$$

in which, the unit of gas viscosity is cp and ρ is the gas density, g/cm³. K, X, and Y are defined as follows [2,23]:

$$K = \frac{(9.379 + 0.01607 M_w) T^{1.5}}{209.2 + 19.26 M_w + T} \tag{2.2}$$

$$X = 3.448 + \left[\frac{986.4}{T}\right] + 0.01009 M_w \tag{2.3}$$

$$Y = 2.447 - 0.2224 X \tag{2.4}$$

In Eqs. (2.2) and (2.3), M_w stands for the gas molecular weight, lb/lbmol. T also represents the temperature, °R.

Based on laboratory and field measurement data for brine-saturated rocks, Gardner et al. obtained an empirical equation that relates the rock density (ρ) to P-wave velocity (V) as follows [24]:

$$\rho = a V^m \tag{2.5}$$

in which, the units of density and P-wave velocity are g/cm³ and m/s, respectively. Also, a and m are the constants, depending on the type of rock, with the default magnitudes of 0.31 and 0.25, respectively. Eq. (2.5) provides reasonable approximation where the examined rocks are sandstones, shales, and carbonates.

In addition, the bulk density (ρ_b) is determined by the following relationship [2]:

$$\rho_b = \rho_m \nu_m + \rho_k \nu_k + \rho_{gk} \phi_k + \rho_{gm} \phi_m \tag{2.6}$$

In Eq. (2.6), ρ stands for the density, ν refers to the fractional volume relative to the rock, and ϕ is the symbol for porosity. Subscripts m, k, gk, and gm represent (inorganic) matrix, kerogen, intrakerogen gas, and intermatrix gas, respectively.

8. THERMAL PROPERTIES OF GAS SHALES

A material's response to addition or loss of heat is described using the thermal conductivity, l (W/mK); specific heat, cp (J/kgK); and thermal diffusivity, k (m^2/s) [25]. Thermal conductivity and specific heat are two vital properties, while the third can be determined from these and the density of the medium [26].

The technological and geological aspects have received a great deal of attention; there is a need to develop a sound fundamental base for an understanding of the basic thermophysical properties of shale gas. Thermophysical measurements of shale gas are particularly important in view of the inherent nature of the methods that are commonly employed for extraction of the shale organic matter; these are mostly based on the application of heat for the pyrolysis of the organic matter in the shale. The effect of temperature on the properties of shale is thus crucial for efficient process design [25,26].

The "thermophysical" term is utilized in this chapter to describe the factors which are directly or indirectly related to the absorption or release and transport of heat. Characteristics including specific heat, thermal diffusivity, and thermal conductivity fall logically into this classic categorization format. For substances which endure phase transformation or thermal decomposition (e.g., thermally active), it is also indispensable to characterize their thermal behavior by thermo-analytical methodologies such as differential thermal analysis (DTA) and thermogravimetry (TG). Electrical and mechanical properties have customarily become an integral part of thermophysical characterization in view of their extreme sensitivity to variations happening in the material on application of heat. Thus, thermophysical properties in general can be broadly divided into three categories including thermal, electrical, and mechanical, as described in Fig. 2.10. These properties in turn encompass a wide spectrum of measurement parameters which can be correlated to yield a self-consistent picture on the overall thermophysical behavior of the material of interest. The relevance of thermophysical measurements to some typical on-field applications in oil shale technology is also demonstrated in Fig. 2.10 [25,26].

The typical trend of thermal conductivity over the range of ambient temperature to 473K (as a function of temperature) for shale samples is demonstrated in Fig. 2.11 [27]. It is observed that the relationship between thermal conductivity and temperature (below 473K) exhibits three distinct regions. In the first zone, there is an increase in thermal conductivity with temperature up to 363K. The second region occurs between 363 and 423K. Over this temperature range, there is a decrease in conductivity which may be due to the loss of free, adsorbed, or absorbed water. If the shale has low water content, the decrease in thermal conductivity is not

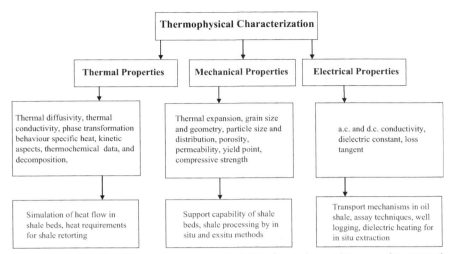

Figure 2.10 Classification of thermophysical properties and the relevance of measured parameters to on-field applications in shale technology. *After Waite WF, Santamarina JC, Cortes DD, Dugan B, Espinoza DN, Germaine J, Jang J, Jung JWT, Kneafsey T, Shin H, Soga K, Winters WJ, Yun T-S. Physical properties of hydrate-bearing sediments, Rev Geophys 2009;47; Moridis GJ. Challenges, uncertainties and issues facing gas production from gas hydrate deposits. Lawrence Berkeley National Laboratory; 2011.*

pronounced. The third region is over the temperature range of 423−473K, where conductivity increases with temperature. The relationship between conductivity and temperature appears to be more prominent in the first region, with conductivity generally increasing by 20% [27].

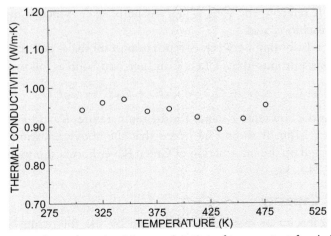

Figure 2.11 Thermal conductivity as a function of temperature for shale [27].

Average thermal conductivities for experiments conducted on many US shales are shown in Table 2.8. The number of data points and the standard deviation associated with the average thermal conductivity are also listed in Table 2.8 [27]. In addition, the minimum and maximum conductivities, as well as the temperatures at which they occurred, are presented. It is noticed that significant data scatter exists in the temperatures. The main reason is that some samples showed the maximum conductivity at a higher temperature rather than the minimum conductivity, while other samples exhibited the opposite trend. Indeed, the minimum and maximum conductivities are within two standard deviations from the average for all but three samples. Thus, the trends might be due to the effects of water movement and evaporation. Researchers concluded that the data presented were not sufficient to distinguish between the shale types. Collecting a broader temperature range may allow significant differences in thermal characteristics, particularly at temperatures above the regime where water evaporation apparently takes place [27].

8.1 Thermal Conductivity

The thermal conductivity of shales at ambient conditions varies in the range of 0.5 to 0.2 W/m·K. In addition, thermal conductivity varies from sample to sample and depends on composition [25].

Diment and Robertson correlation is used to define thermal conductivity (λ) in W/m·K as given below [25–27]:

$$\lambda = 2761 - 15R \tag{2.7}$$

in which, R refers to the weight percentage of shale which is insoluble in dilute hydrochloric acid.

Also, a relationship between composition of oil shales and thermal conductivity was introduced by Tihen, Carptner, and Sohn as follows [25–27]:

$$\lambda = C_1 + C_2 F + C_3 T + C_4 F^2 + C_5 T^2 + C_6 FT \tag{2.8}$$

where C_i is the constant; T stands for the temperature, K; and F is the shale Fisher assay, L/kg. It should be noted that the above equation has been obtained based on the information of Green River Formation with the assay of 0.04–0.24 L/kg.

8.2 Heat Capacity

The heat capacity of shale gas is expressed by the following correlation [25–27]:

Table 2.8 Thermal Conductivity Within the Temperature Range of the Ambient Temperature to 473K for Six US Shales [27]

Type of Shale	Depth (m)	No. Data Points	Average λ (W/m·K)	Standard Deviation	Minimum λ[a]	Maximum λ[a]
Ohio Cleveland[b]	137.2	8	0.73	0.06	0.67 (344)	0.81 (473)
Chagrin Three Lick Bed[b]	149.7	7	0.83	0.05	0.78 (369)	0.90 (456)
Upper Huron[b]	157.3	17	0.86	0.03	0.79 (429)	0.89 (404)
	157.6	6	0.79	0.04	0.75 (405)	0.86 (373)
	158.4	10	0.78	0.04	0.73 (426)	0.84 (373)
	158.5	9	0.91	0.05	0.84 (426)	1.02 (372)
	158.8	9	0.91	0.05	0.83 (430)	1.00 (374)
	160.9	9	0.85	0.03	0.81 (427)	0.93 (375)
	164	5	0.94	0.01	0.93 (413)	0.95 (389)
Middle Huron[b]	175.6	7	0.87	0.04	0.81 (442)	0.93 (341)
	176.2	6	1.09	0.04	1.02 (445)	1.14 (393)
	176.3	9	1.08	0.06	1.05 (461)	1.20 (374)
	176.3	10	1.05	0.06	0.93 (439)	1.17 (374)
	176.5	10	0.93	0.03	0.91 (431)	1.02 (374)
	176.8	8	0.94	0.03	0.90 (434)	0.98 (345)
Pierre	45.1	4	1.01	0.21	0.75 (441)	1.22 (348)
	45.3	9	0.68	0.05	0.63 (313)	0.76 (372)
	45.4	4	0.88	0.08	0.78 (417)	0.98 (370)
	45.7	9	0.84	0.10	0.76 (431)	0.95 (316)
	49.1	4	0.87	0.06	0.81 (406)	0.96 (375)
Oil[c]		8	1.07	0.09	0.98 (311)	1.23 (451)

[a]The value in parentheses is the temperature (K) at which the thermal conductivity was measured.
[b]From EGSP Kentucky No. 5 Well.
[c]From Green River Formation.

$$\ln(A) = \frac{4\pi\theta\lambda}{Q} + \gamma + \ln\frac{R^2}{4t} \tag{2.9}$$

In Eq. (2.9), A is the symbol for thermal diffusivity; λ is the thermal conductivity; θ stands for the temperature rise; Q is the heat input; γ is the Euler's constant; t represents the time; and the distance from line heat source is shown with R.

It is worth noting that the thermal diffusivity is defined as follows [25–27]:

$$A = \frac{\lambda}{\rho C_p} \tag{2.10}$$

No systematic study has been found which addresses variables affecting the thermal characteristics of several shale types. However, a review of all studies gives guidance to know the important parameters. These include composition, temperature, porosity, pressure, and anisotropy [25–27].

8.3 Effects of Composition

Thermal conductivity data on a variety of geologic media indicate a dependence on composition. Functionality of thermal conductivity on water content, clay content, Fisher assay of the shale (in gallons per ton), and inorganic mineral content have been examined by several research studies. It should be noted that the heterogeneity of natural material makes prediction of thermal conductivity difficult, because [25–27]:
- The conductivity of many constituent minerals is not accurately known;
- In geologic media, the constituents usually do not occur as discrete, unflawed crystals;
- The resistances are not always exclusively in series or in parallel but very often are in some combination of these or in a completely random pattern.

The impact of the heterogeneity is illustrated by Birch and Clark, who have reported differences in thermal conductivities as high as 50% between adjacent samples of coarse-ground rock cut 1 cm apart. Although, the authors presumed that the differences were due to variations in composition, correlations between thermal conductivity and composition have historically been empirical and need to be based on a large number of data. Diment and Robertson obtained a relationship between thermal conductivity and composition of shale as discussed before. Tihen, Carpenter, and Sohn also introduced an equation that relates thermal conductivity and composition of oil shales. Their correlation indicates that the effects of composition and temperature are interrelated for some shale samples [25—27].

8.4 Effects of Temperature

Thermal conductivity measurements on geologic media have exhibited a dependence on temperature. Conductivity of oil shales from the Green River Formation is reported to have a function of T^2. In some cases, this dependence resulted in a 50% decrease in thermal conductivity at 653K, compared with that at 353K. As discussed previously, thermal conductivity is dependent on temperature. The differences in the strength of the relationship between thermal conductivity and temperature for the two shale types may be due to compositional differences, as it is clear based on the empirical equation presented by Tihen, Carpenter, and Sohn. However, differences in the porosities of the samples might also contribute to the differences [25—27].

8.5 Effects of Porosity

For porous media, thermal conductivity is a strong function of porosity and the fluid contained in the voids. Numerous empirical models have been developed which relate thermal conductivity and porosity. Porosity seems a significant variable affecting thermal conductivity, considering some shale formations have porosities as high as 50%. According to the literature, thermal conductivities of water-saturated shale samples (at ambient temperature) are 3—50% higher than vacuum-dried samples. Shale samples from the Conasauga Group which were studied by Dell'hico, Captain, and Chanskyl had an average porosity of 1.45% (as determined by water uptake of vacuum-dried samples). Presumably, this low porosity, and hence the small amount of fluid contained in the samples, accounted for the negligible change in thermal conductivity for water-saturated versus vacuum-dried samples. This may also contribute to the lack of sensitivity to temperature [25—27].

8.6 Effects of Pressure

The thermal conductivity of geologic media may vary with confining pressure. The effect of pressure is relatively insignificant on thermal conductivity (3—25%), depending on the type of rocks. Change in thermal conductivity is attributed to closer intragranular contact due to compression of the sample. However, the degree/level of this effect is somewhat dependent on the composition and compressibility of the fluid trapped in the voids. Dell'hico, Captain, and Chansky concluded that pressure has a small impact on thermal conductivity of shales where the porosity is low. For the shale in the

Conasauga Group with an average porosity of 1.45%, the thermal conductivity at ambient temperature increased 2.1% when the pressure changes from 10 to 2.5 MPa [25–27].

8.7 Effects of Anisotropy

A number of research studies have reported significant differences between thermal conductivity, corresponding to heat flow in a direction perpendicular to the shale bedding plane and corresponding to heat flow in a direction parallel to the stratigraphic planes. In general, thermal conductivities parallel to the bedding planes are higher than those that are perpendicular to the bedding planes. For instance, Dell' Amico, Captain, and Chansky reported that the thermal conductivity values parallel to the bedding plane were 30% higher than those which are perpendicular to the bedding plane for a shale formation in the Conasauga Group [25–27].

9. PVT BEHAVIOR OF SHALE GAS MIXTURE

The reservoir fluid, oil or gas, may be very different from reservoir to reservoir. Based on the fluid composition, the property of the fluid varies considerably. The reservoir fluid is described graphically with phase diagrams as demonstrated in Fig. 2.12. A phase diagram displays the relationship between pressure and temperature ($P-T$ diagram), pressure and volume ($P-V$ diagram), or temperature and volume ($T-V$ diagram). For a single component system, the system will be at different phases including solid, liquid, and vapor under different pressures and temperatures [2,26,28].

The different phases are separated by solid–vapor; solid–liquid (melting curve), and liquid–vapor (vapor pressure curve) lines. The liquid–vapor line ends in a critical point. Beyond the critical point, the liquid and vapor phases are indistinguishable. For systems with two or more components the vapor/pressure curves of the system will lie between the vapor pressure curves of the single components. Instead of a single curve, a phase envelope can be drawn. Inside the phase envelope of the two phases, vapor and liquid, coexists in equilibrium. The bordering curves of the phase envelopes are called the bubble point line, where the first bubbles of gas start coming out of the liquid phase and dew point line, where the first drops of liquids start falling out of the vapor phase. From the PVT behavior, it can be observed, as long as the condensate and wet gas are in a gas solution, liquid drop out obstructing flow will not occur; however, as formation pressure declines, this could become a critical issue. In a black oil or volatile oil shale

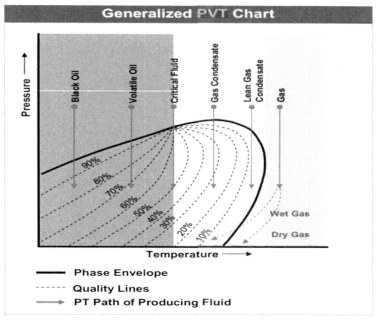

Figure 2.12 PVT behavior of fluids during production [29].

play, as formation pressure drops to the bubble point curve, gas comes out of solution, resulting in a decrease in gas oil ratio of the liquid phase and an increase in its viscosity, making it more difficult for the liquid to move out of the pore space. In addition, the free gas phase can move more freely through the pore throats of the rock, leaving behind the liquids [2,28].

The points with the maximum temperature and pressure on the phase envelope are known as cricondentherm and cricondenbar, respectively [2,28].

10. PETROLOGY AND GEOCHEMISTRY OF GAS SHALE

The successful and efficient evaluation of shale-gas reservoirs in the subsurface requires an adequate understanding of petroleum geology, geophysics, and geochemistry. In particular, various aspects of shale petrology and petroleum geochemistry provide fundamental information that is critical to finding and developing unconventional shale reservoirs. Shale-gas reservoirs vary from tight, low-permeability rocks to highly fractured rocks with variable bulk mineralogical composition, which controls the ductile versus brittle fabric of the shale (see Fig. 2.13). It is necessary to determine if natural gas stored in shales is microbial or thermogenic in

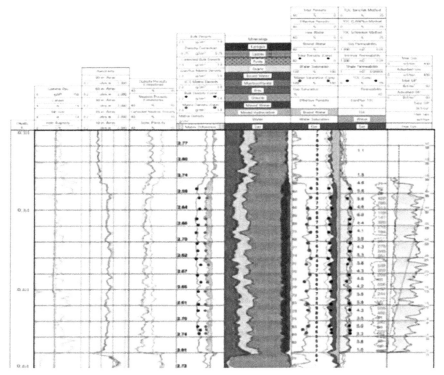
Figure 2.13 Formation properties.

origin in order to predict the likelihood of economic shale-gas production. Finally, it is critical to apply geochemical measurements and interpretations in the context of geologic characteristics to the decision-making process in shale-gas exploration and development.

Natural gas has to move through pore spacing in the shale to the well during extraction. However, pore spaces in shale are three orders less than those in conventional sandstone reservoirs which give a very low permeability [2,30]. Fig. 2.14 shows the nanoscale pores of shale [30].

10.1 Shale Lithology
10.1.1 Texture
Shale is considered to be a fissile rock, which means it can be broke down into thin layers parallel to the bedding plane because of the parallel orientation of clay flakes. Nonfissile rocks with the same composition and particles smaller than 0.06 mm are termed as mudstones. Rocks with less clay and similar composition are siltstones. Shale is a sedimentary rock [31].

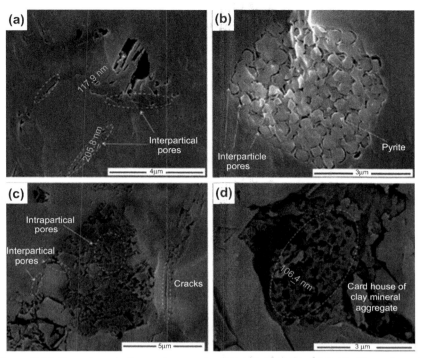

Figure 2.14 Scanning electron micrographs pictures of shales from Niutitang. Images (a–c) show interparticle pore spaces and (c–d) present intraparticle pore spaces [30].

10.1.2 Composition and Color

Shale rock is composed of quartz grains and clay minerals and the color is gray. Sometimes the addition of other minor constituents can alter the color of shale rock. When carbonaceous content is greater than 1%, shale rock appears to be black in color. Brown, red, and green colors indicate the presence of ferric oxide or iron oxide [31]. Clays form the major part of shale rocks. Clay minerals like montmorillonite, kaolinite, and illite can be present in shale rocks. Mudrocks and shale rocks contain approximately 95% organic matter [31].

11. ESTIMATION OF GAS IN PLACE VOLUME

Gas in place (GIP) is determined using the following equation:

$$\text{GIP} = 1359.7\ AhdG \tag{2.11}$$

where A refers to the drainage area; h is the reservoir thickness; d stands for the bulk density; and G refers to the total gas content.

Present day in situ GIP is a function of the geological factors which the reservoir has been subjected to, since deposition (reservoir "burping," uplift, faulting, and hydraulic stripping). Thus, accurate gas content data often cannot be calculated from knowledge of physical rock properties alone, but instead must be directly measured from freshly cut rock samples [32].

12. GEOLOGICAL DESCRIPTION OF GAS SHALE FORMATIONS

Table 2.9 shows Worldwide Marine Shales differentiated by geologic time period. In the table, black circles represent the number of occurrences for each age. Tectonics, geography, and climatic conditions affect the deposition of sediments which are organic-rich, such as marine sediments. These are thermally mature to an extent of converting kerogen into HC and thus

Table 2.9 Worldwide Marine Shales Differentiated by Geologic Time Period [32]

Million years ago	Period	North America	South America	Europe	Siberia and Central Asia	Africa	Australia and Asia
65	Quaternary and Tertiary			●●●●	●●		
135	Cretaceous	●●●●	●	●	●●	●●	
190	Jurassic	●		●●●●●● ●	●●		●
225	Triassic	●● ●		●●		●	
280	Permian	●		●●●●			
320	Pennsylvanian	●●●●			●●● ●●		
345	Mississippian	●●●●					
395	Devonian	●●●●●● ●●	●●●	●●●●		●●●	
435	Silurian	●●		●●●●●●●	●		
500	Ordovician	●●●●●		●●●●			
570	Cambrian	●		●●	●		
2,500	Proterozoic Archeozoic	●●●		●●	●	●	●●

have been a focus of exploration corporations. Shales from lake plains have been also explored but not as widely as the marine ones [32].

When shales form under the presence of high levels of organic matter and low amounts of O_2, they tend to be organic-rich. These geologic conditions were present during several geologic ages, not excluding the Devonian Period. Organic-rich shales from the pre-Cambrian Period to more recent times have been recognized. Yet, to be thermally mature, most gas shale plays emphasize on sediments from the Ordovician Period extending through the Pennsylvanian Period [14].

The geological description of shale gas formations is divided into two main areas as follows.

Proximal areas: A normal parasequence starts at a base in laminated very dark gray claystone, with sparse millimeter-thick silt beds and thin lags of skeletal phosphate. The thickness of the very fine sandstone or siltstone beds slowly increases to centimeter scale; these beds usually have planar parallel or present bedding and horizontal channels. Toward the surface, sandstone beds usually have erosive bases and fining toward the lower depth movements but are interbedded with mudstone.

The intensity of the horizontal channeling increases toward the lower depth and the encompassing mudstones become siltier and more extremely bioturbated with minor vertical channeling as well as dominantly horizontal channels. Sandstone beds become coarser at the top of the parasequence, and more common, with very fine overriding mudstone [32–36].

Sandstone rock tends to be directed by present lamination or extremely bioturbated with both horizontal and vertical channels, usually from the mudstone of the overlying parasequence. The grain size increases (from silt to about 0.125 mm) within the parasequences when the bed thickness increased [32–36].

Distal areas: Parasequences are considerably thinner in most distal areas, and comprise a greater ratio of laminated to parallel-bedded very dark gray mudstone and claystone than in proximal locations with a few centimeter-scale storm layers.

Lithologic signs still indicate an upward-shoaling tendency: percent siltstone, percent sandstone, the thickness of individual siltstone or sandstone bed sets, and maximum grain size. A considerable section of the silt- and sand-sized particles tends to be biogenic. These parasequences normally exhibit an increase toward the lower depth in skeletal phosphorus content, hydrogen index, TOC, and bioturbation and a slight decrease in biogenic silica abundance [32–36].

13. POROSITY AND PERMEABILITY: THEORY AND EXPERIMENTAL

Porosity and permeability are related properties of any rock or loose sediment. Most oil and gas has been produced from sandstones. These rocks are usually highly permeable with high porosity. Porosity and permeability are absolutely necessary to make a productive oil and gas well. The porosity consists of the tiny spaces in the rock that hold the oil or gas and permeability is a characteristic that allows oil and gas to flow through the rock.

The porosity of a rock is a measure of its ability to hold a fluid. Mathematically porosity is the open space in a rock divided by the total rock volume (solid + space or holes). Porosity is normally expressed as a percentage of the total rock which is taken up by spore space. For example, sandstone may have 8% porosity. This means 92% is solid rock and 8% is open space containing oil, gas, or water. Eight percent is about the minimum porosity that is required to make a decent oil well, though many poorer (and often noneconomic) wells are completed with less porosity. Even though sandstone is hard and appears very solid, it is really much like a sponge (a very hard, incompressible sponge). Between the grains of the sand enough space exists to trap fluids like oil or natural gas. The holes in sandstone are called porosity (originated from the word porous) [32−36].

The permeability of a rock is a measure of the resistance to the flow of a fluid through a rock, that rock has low permeability. If fluid passes through the rock easily, it is a highly permeable system. Permeability is a constant for a given porous medium and fluid. To separate the influence of fluid from that of the porous medium, the absolute permeability K is defined, which only describes the permeability of the porous medium [2,34].

$$K = k\frac{\eta}{\rho} \qquad (2.12)$$

in which K stands for the absolute permeability (L^2); k is the permeability, LT^{-1}; η represents the coefficient of dynamic viscosity of the fluid, TFL^{-2}; and ρ refers to the unit weight of the fluid, FL^{-3}. For clay-rich media, Eq. (2.1) strictly only applies to nonpolar fluids, since permeability is also affected by the valence and concentration of dissolved cations [2,34].

The permeability data in Fig. 2.15 range over six orders of magnitude, from about 2×10^{-22} m^2 to about 2×10^{-16} m^2. In a single porosity system, the permeability range decreases with porosity, from over four orders

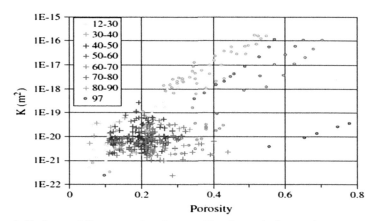

Figure 2.15 Permeability–porosity dataset. The legend shows the range of clay content of each band. Open circles and crosses are the data points of measured and modeled permeability, respectively [34].

of magnitude at a porosity range of 0.15–2. At lower porosities, the convergence of permeability reveals that mudstones seem to squeeze through the special collapse of those large pores which provide most of the fluid flow. The pore size distribution for lithologically diverse mudstones increases as porosity is reduced.

It shows that there is no simple and single relationship between porosity and permeability for mudstones. Furthermore, the inconstancy is mostly regulated by clay content; the higher the clay content, the lower the permeability at the identical porosity. The impact of clay content on the porosity–permeability relation is investigated clearly in Fig. 2.16. The data points in this figure were gathered from 58 samples of an 18 m-long core taken from the Gulf of Mexico. These samples have diverse clay contents (37–70%); however, they have a restricted range of porosity (0.20–0.24). The slight changes in porosity are regulated by clay content as estimated through different models of mechanical compaction for muds. Permeability values of mudstones correlate convincingly with clay content and change by an order of magnitude with changing the clay content. Earlier investigations have revealed that fine-grained classic sediment permeability is approximately correlated to void ratio ($\varepsilon = \phi/(1 - \phi)$) or porosity ($\phi$) through a log-linear function.

The reported datasets reveal that a linear relation between porosity and log permeability is acceptable over a limited range of porosity. Over the full

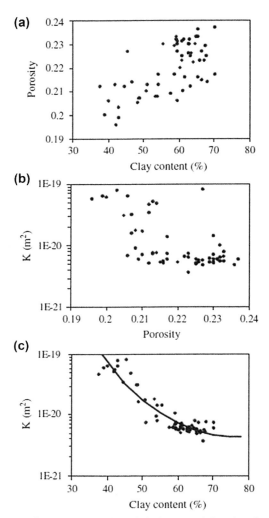

Figure 2.16 Porosity, clay content, and modeled permeability data for 58 samples from an 18 m core obtained from the Gulf of Mexico. (a) Porosity versus clay content, showing the dependence of porosity on clay content; (b) permeability versus porosity; (c) strong dependence of permeability on clay content and the fitting curve [34].

porosity range, the relationship between the logarithmic permeability and void ratio or porosity is well expressed by more complicated mathematical forms/expressions. The best fit of available data samples is illustrated in Fig. 2.17 [2,34].

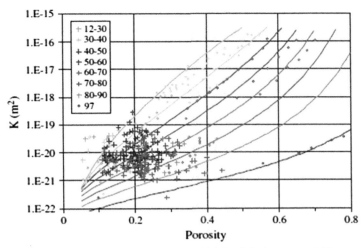

Figure 2.17 Comparison between measured/modeled permeability and our constructed relationship (curves are from the clay content constrained) [34].

In Fig. 2.17 (porosity—permeability relationship), the legend indicates the range of clay content for each band. Open circles and crosses represent data points of measured or modeled permeability respectively. Each curve characterizes the relationship at the middle value of clay contents of the band with the same color as follows [2,34]:

$$\ln(K) = a_k + b_k e + c_k e^{0.5} \tag{2.13}$$

where

$$a_k (\text{or } b_k \text{ or } c_k) = c_0 + c_1 \text{CF} + c_2 \text{CF}^{0.5} \tag{2.14}$$

in which, CF is the clay content in fraction, e is the void ratio, and a_k, b_k, c_k, c_0, c_1 and c_2 are the coefficients (m²). The influence of clay content on the permeability—porosity relationship is incorporated in the coefficients. At a fixed porosity or void ratio, the permeability—clay content equation thus takes the form of the relationship between the coefficients and clay content. The above equation shows that the correlation between clay content and coefficients is robust.

The relationship between permeability and porosity for fine-grained clastic sediments ("mudstones") is a key constitutive equation for modeling subsurface fluid flow and is fundamental to the quantification of a range of geological processes. For a given porosity, mudstone permeability varies over a range of two to five orders of magnitude. As mentioned before,

Table 2.10 Main Features of North American Gas Shales [33]

Formation	TOC [wt %]	Ro [%]	Porosity (total interc.) [%]	Matrix Permeability [mD]
Barnett	3.36	1.21	6.42	5.27 E-05
				5.46 E-06
				5.78 E-05
				3.26 E-05
				9.74 E-06
				3.53 E-05
				5.97 E-06
Marcellus	3.26	1.37	7.51	1.47 E-04
New Albany	4.89	0.73	5.06	1.28 E-06
Woodford	4.33	1.22	5.41	2.68 E-05

Substantial heterogeneity mainly in terms of porosity, TOC and Thermal Maturity

TOC range 1.83 – 4.89
Ro % range 0.73 – 1.92
Porosity range 4.23 – 8.99
Permeability range 1.28 E-06 – 3.53 E-04

the broad range can be explained by variations in lithology, which is defined simply and pragmatically by clay content (mass fraction of particles less than 2 μm in diameter). Using clay content as the quantitative lithology descriptor, a dataset (clay content range of 12–97%; porosity range of 0.04–0.78; six orders of magnitude permeability range) comprising 376 data points to derive a new bedding perpendicular permeability (K, m^2) – void ratio (e = porosity/(1–porosity)) relationship as a function of clay content (CF) through fitting an equation to our dataset as follows [2,34]:

$$\ln(K) = -69.59 - 26.79\text{CF} + 44.07\text{CF}^{0.5} + \left(-53.61 - 80.03\text{CF} + 132.78\text{CF}^{0.5}\right)e + \left(86.61 + 81.91\text{CF} - 163.61\text{CF}^{0.5}\right)e^{0.5}$$

(2.15)

The coefficient of regression (R^2) is 0.93. At a given porosity, inclusion of the quantitative lithological descriptor, clay content reduces the prediction range of permeability from two to five orders of magnitude to one order.

The ranges of porosity and permeability of North American Gas Shales is given in Table 2.10 [33].

14. POROSITY AND PERMEABILITY MEASUREMENTS: PRACTICAL METHODOLOGIES

Different techniques are used for porosity measurements whose description can be found in the literature. A part of the main techniques is listed below [34].

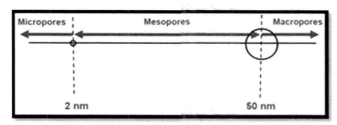

Figure 2.18 Size of pores (IUPAC Standard) [34].

1. Gas adsorption: This can measure only open pores and pore size >1 nm. It can determine pore structure diameter or micro- and mesopores.
2. Mercury porosimetry: This is similar to gas adsorption but measures pore size > 3 nm. It is good for meso- and macropores.
3. Helium porosimetry: This is an easy and established technique that can measure only open pores [34].

The sorting of pores by their size is demonstrated in Fig. 2.18.

Shale-gas rock volume contains a matrix made up of organic matter and inorganic minerals, as well as pore space between these parts, as shown in Fig. 2.19. The difference in the porosity values reported by various labs is clearly noticed. This is because of the way the term "porosity" is defined and employed. Several labs define a "total (dry) porosity," which is the pore space occupied by the mobile water, hydrocarbons, and irreducible water comprised of surface clay-bound water and capillary; while other experimentalists define a "humidity-dried" or "effective" porosity that excludes the pore space filled with the surface "clay-bound" water.

The "clay-bound" water measurements might not be precise owing to variable definitions or circumstances under which it is measured. For instance, it is a challenging task to convert a measurable "effective" porosity to a measurable "total" porosity. It is broadly acknowledged that the

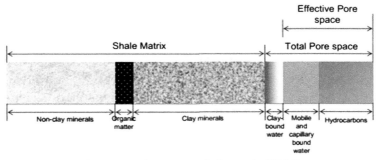

Figure 2.19 Porosity in shale matrix [35,39].

conventional reservoir core analysis has to be conducted at reservoir stress since rock properties determined in the absence of stress are pretty different from outputs at reservoir stress [35—39].

Porosity readings in shales are not easy because of the texture which is fine-grained. The small sizes of pores and ultra-low permeability also pose an issue. Nano—Darcy permeability values make traditional core measurement techniques extremely difficult. To overcome these problems, crushed rock methodology has been adopted by the upstream hydrocarbon industry for routine shale core analysis, referred to as the Gas Research Institute (GRI) technique [40]. This technique used for shale-gas formations provides fast results for porosity measurements.

A modified version of the saturation-immersion technique is utilized to determine total porosity in thermally mature mudrock lithologies such as gas shales that are characterized by low total porosity, a significant content of hydroscopic clay minerals, thermally mature organic matter, and a high amount of micro- and mesopore volumes [40].

When production from a reservoir is to be judged for economic viability, permeability of the gas shale is the most vital parameter to be considered [41].

Absolute permeability (referred to as permeability hereafter) of conventional rocks is measured on plug samples under reservoir stress using various methods such as the steady-state approach or pulse decay methods. Some of these methods have been employed with reasonable success on low-permeability rocks. For shale-gas reservoirs, the method commonly used is similar to that published by the Gas Research Institute where permeability of the rock matrix is measured using pressure decay on crushed rock samples. Like the case of porosity measurements, the permeability values measured on crushed rocks in the absence of reservoir stress can be quite different from the in situ matrix permeability magnitudes [27,35,36].

Similar to porosity measurements, it is concluded that a comparative study of matrix permeability measured by different laboratories using the pressure decay approach on crushed rock samples should be conducted, where the corresponding lab receives preserved sample splits of rock from the same depth intervals. The permeability values reported by different laboratories vary by two to three orders of magnitude. The reported values are as-received permeability and one of the sources of observed discrepancies may be differences in sample handling. Another primary difficulty in discerning the source of interlaboratory variation in permeability values is due to the absence of proper mathematical formulations which can be utilized in interpretation of the pressure decay response [27,36].

It should be noted that determination of permeability under one microDarcy is still considered a perplexing task. In a pore pressure oscillation method, the sample is first stabilized at certain pore pressure, then a small sinusoidal pressure wave is applied to the upstream side of the sample, and the pressure response at the downstream side is recorded. Permeability is deduced from the attenuation and phase shift of the downstream signal. Compared with other methods, this technique greatly reduces the pore pressure variation, increases the measurement sensitivity, and results in more accurate characterization [42].

A triple combo tool can also carry out the measurement of resistivity and porosity. The resistivity measurement of the shale that has the potential of gas is higher, compared to the shale having no gas potential. Porosity measurement of the gas-bearing shales also has distinct characteristics. Organic-rich shales exhibit more variability, higher density porosity, and lower neutron porosity. This designates the presence of gas in the shale. Lower neutron porosity may occur due to lower clay mineral contents in the shales [2,27,41].

The shales have higher bulk density in comparison with conventional reservoirs like sandstones or limestone because of the constituent material that plays an important role in the formation of shale. Kerogen has lower bulk density, which leads to a higher computed porosity. The grain density, derived from electron capture spectroscopy must be known in order to obtain the density porosity of the shale. Silicon, calcium, iron, sulfur, titanium, gadolinium, and potassium are the primary outputs of spectroscopy [2,27].

The spectroscopy data also provide information about the clay type, so that engineers can predict the sensitivity of the fracking fluid and understand the fracturing characteristics of the formation using the data. When the clay is in contact with water, it swells and hence inhibits gas production and a lot of operational issues (pressure increase, corrosion in production well) may arise. Smectite is the most common form of swelling clay. It also shows the rocks that are ductile. For long-term productivity of a shale gas well, acoustic measurements are very important as they supply the mechanical properties of anisotropic shale media. To enhance mechanical earth models and optimize drilling, the sonic scanner acoustic data are utilized. The mechanical properties may include bulk modulus, Poisson's ratio, Young's modulus, yield strength, shear modulus, and compressive strength, which are determined from compressional shear and Stoneley wave measurement [2,27].

When the difference between the vertically and horizontally measured Young's moduli becomes large, the closure stress will be higher for anisotropic media, compared to isotropic rocks. These anisotropic results are associated with the rock having higher clay volume. As the proppant is more embedded into ductile formation, it will be difficult to retain the conductivity of the fracture during production [2,27].

Another acoustic measurement that is beneficial in shale analysis is sonic porosity. Sonic porosity is much lower than neutron porosity in shales. This characteristic is a function of high clay-bound water which is common in shale. The high sonic porosity implies the gas potential within the pore space. When both the sonic and neutron porosity values are comparable, it means that shale may be oil-prone. To define the orientation and concentration, log analysts employ a wire line borehole image. It can be interpreted from these data whether the hole is open or not.

The measurement from these various tools can be combined in an integrated display like shale montage log provided by Schlumberger. The geologists can directly compare the quality of the rock by the formation properties that are presented using a single platform. Free and absorbed gas have the units of standard cubic foot per ton or scf/ton.

15. PORE SIZE DISTRIBUTION IN GAS SHALES

In gas shales, small pores (2—5 nm) dominate in number but the large pores (20—30 nm) dominate in their impact on permeability, based on the study conducted on BC shales (see Fig. 2.20) [43].

According to Fig. 2.21, conventional oil and gas reserves have higher permeability (e.g., $k \geq 1$ mD) and pore diameter (e.g., $d_{pore} \geq 1$ μm) than shale and tight gas formations. For instance, permeability and pore diameter are in the range of $1-10^{-3}$ μD and $10^{-1}-10^{-2}$ μm, respectively, for shale plays [44].

The presence of a porous system predominantly within the organic matter can be identified in SEM-FIB images [45]. Pores can make up to more than half of the volume of original organic matter. Hence, they are a considerable part of the fraction of total effective porosity in some gas shales (Fig. 2.22). It appears that the pores may have entirely different wettability, compared to the pores in the mineral matrix. The organic pores might be hydrocarbon wet and free gas could be found within them [31].

For further clarification about pore size distribution and surface area, the results for a real case study are briefly described here.

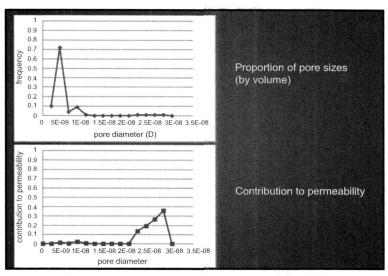

Figure 2.20 Pore size distribution in gas shales [43].

Core samples were drilled from Sichuan Basin. Eight samples were employed, SC1 to SC8, for characterization of original pore patterns [46]. CO_2 and N_2 adsorption isotherms were acquired at 77.3 and 273.15K. The working pressure was kept at 228 MPa [46].

Fig. 2.23 demonstrates gas adsorption against pressure for different samples. As expected, increasing pressure increases the amount of gas adsorbed.

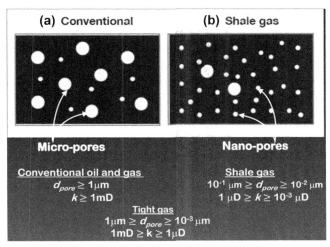

Figure 2.21 Porosity and permeability ranges for various oil and gas formations [44].

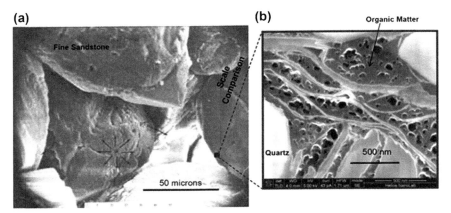

Figure 2.22 (a) Sandstone with fine grains imaged using an Scanning electron microscopy and (b) pore size in a Barnett Shale formation [45].

Isotherms obtained show a hysteresis loop formed by branches with a specific shape. For example, adsorption isotherms belonging to type 2 imply multilayer adsorption. At low pressures, gas adsorption increases and forms a monolayer adsorption, indicating a capillary condensation phenomenon [46]. Adsorption experiments also help to determine specific surface area and pore volume, as shown in Fig. 2.24.

16. CHALLENGES IN SHALE GAS CHARACTERIZATION

Nowadays, one of the main challenges in shale gas characterization is that it is not known how the adsorbed gas leads to higher production and recovery. There is no promising critical algorithm to deduce the existence of adsorbed gas from production data [47].

There are many problems being faced in the steps of shale analysis/characterization, such as [47,48]:

1. Sample crushing and grinding: This stage provides access to the local pore systems; however, the adsorption surface area is increased and exposed to oxygen.
2. Baking the kerogen and liquid HCs: The higher the extraction temperature in the lab, the higher the measured total porosity.
3. Measurements: Unusual measured gas curves show gas generation through various methods (bacterial, capillary evaporation in dual pore size, catalytic generation) [48].

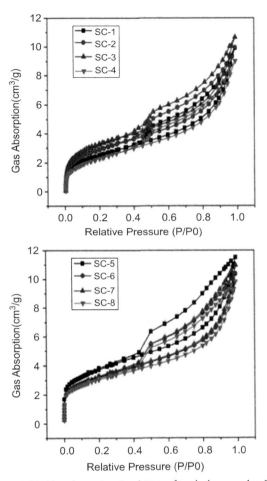

Figure 2.23 N$_2$ adsorption isotherms for shale samples [46].

It also has been found that results of porosity measurements vary, depending on investigation technique, as shown below [49]. The major challenge in the characterization process is the interdependency or interlinking of various parameters involved. As clear from Fig. 2.25, different characterization methodologies may attain different values/trends for gas shale properties that lead to major challenges in characterizing shaly formations. For further clarification, Fig. 2.26 highlights the complexity in the gas shale characterization models.

In general, the main reservoir characterization challenges for gas shales are as follows:

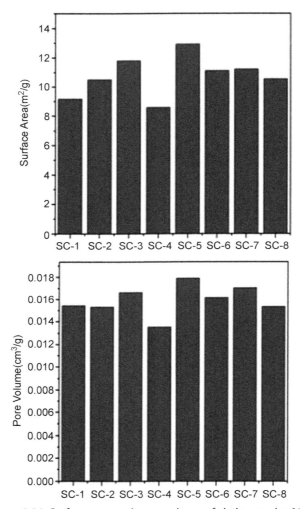

Figure 2.24 Surface area and pore volume of shale samples [46].

- No single log or core tool provides all information required for characterization of shale gas reservoirs.
- Conventional log methods are not able to give all characterization data needed for development of gas shales.

The challenges related to the development and production technologies of gas shales are also listed below:
- Minimize drilling costs;
- Optimize completion and fracturing design;
- Minimize environmental Impact.

Shale Gas Characteristics

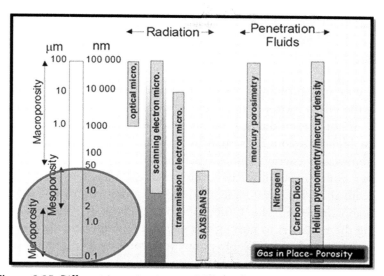

Figure 2.25 Different investigation methods to determine microporosity [49].

17. HOT TOPIC RESEARCH

Shale natural gas has high potential to appreciably boost the US security of energy supply, lower greenhouse gasemissions, and decrease the cost for end-users. Although gas production from shale reserves in the US has

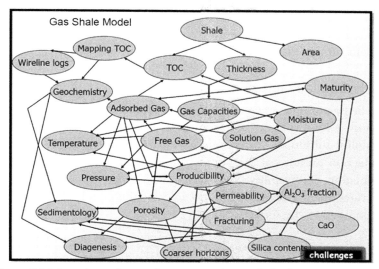

Figure 2.26 Interdependency of parameters in gas shale characterization [49].

been started for many decades, shale gas was not taken into account to be a considerable resource until the last decade when it was proved that hydraulic fracturing and new horizontal drilling and technologies eased the economic recovery.

The current contribution of shale gas is nearly 16% of natural gas production for the US. It is expected that this amount will experience momentous growth as this enormous gas supply is developed [2]. Natural gas can reinstate high-emission fuels such as coal and oil and facilitate various green/renewable energy sources such as solar, geothermal, and wind. However, concerns about the environmental risk, and safety prospects associated with shale gas development should be addressed before production can notably increase. The Office of Fossil Energy is concentrating on resolving these matters to guarantee a safe and environmentally sustainable resource of natural gas. The FE program is composed of the activities as follows.

17.1 Shale Reservoir Characterization

Gas-producing shales mainly consist of consolidated clay-sized particles with an elevated organic content. High temperatures and subsurface pressures alter the organic matter to gas and oil, which might travel to conventional hydrocarbon traps and/or remain within the shale formation. However, the clay content rigorously restricts fluid flow (e.g., water and gas) within the shales. It is, thus, crucial to comprehend the porosity, permeability, mineral and organic content, incidence of natural fractures, shale volumes, and thermal maturity to determine production potential. Suitable drilling and stimulation technologies required for commercial volumes and rates are dependent on these reservoir characteristics.

17.2 Hydraulic Fracturing Technology

Hydraulic fracturing is a practice in which a significant amount of water and sand, and a low volume of chemical additives are injected into low-permeability subsurface formations to increase the gas or oil flow rate. The injection pressure of the pumped fluid creates fractures that improve the formation permeability, and the sand or other coarse materials hold the fractures open. The majority of the injected fluid returns to the wellbore and is pumped to the surface.

The hydraulic fracturing technique has been employed for over 60 years in over one million wells [2]. Lately, public concerns about potential impacts on drinking water and other environmental damages have grown appreciably. As a result, Congress directed the Environmental Protection Agency (EPA) in 2010 to perform a study of this technology to further appreciate any possible influences of hydraulic fracturing on groundwater and drinking water. DOE/NETL is working closely with the EPA as it conducts the investigation and is also collaborating with the Department of the Interior to improve comprehension of these risks [31,44,49,50].

Hot topic research studies ongoing about characterization of shales and shale gas are generally in the following subjects:
- Implications of sequence stratigraphy, microfacies, and compositional variability for shale gas reservoir;
- Incorporating shale gas in gas to liquids technology;
- Characterizing the reservoir space quantitatively in new areas globally;
- Integrating the processing of shale gas production of C_2H_2;
- How the adsorption potential of shale gas is affected by the pore structure;
- Characterizing the effect of fracturing a gas shale;
- Extraction of organic matter from reservoir rock;
- Total porosity measurement in shale gas reservoirs by novel methods;
- Scale imaging of pore and organic structure in shale gas reservoirs;
- Characterizing the fracture properties of shale formations such as the Eagle Ford Shale, Texas;
- Digenetic modifications in shale gas reservoirs (e.g., the Eagle Ford Shale): implications for physical and chemical properties;
- Challenges of reservoir architecture, quantification, and characteristics of gas shales, such as thin-bedded reservoirs in the Plio-Pleistocene of the Columbus Basin, Offshore Trinidad.

The current shale gas projects are given below [50]:

(1) Colorado Group including First White Speckled Shale, Joli Fou, Shaftesbury, Fish Scales, Kaskapau, Blackstone, Second White Speckled Shale, Muskiki, Colorado Shale, Wapiabi, and Puskwaskau; (2) Fernie Formation including Fernie Shale, Pokerchip Shale, and Nordegg; (3) Muskwa Formation; (4) Duvernay Formation; (5) Exshaw Formation; (6) Lower Banff Formation; (7) Montney Formation; (8) Reirdon Formation; (9) Bantry Shale Member; and (10) Wilrich Formation.

The main engineering and research companies involved in shale gas development are listed below [50,51]:

- Apache
- Talisman Energy Inc.
- Total S.A.
- Ultra Petroleum
- Vale
- Waller LNG Services
- PetroChina
- PetroEdge Resources
- Mitsui
- Protege Energy
- Baker Hughes
- Naftogaz
- Noble Energy
- Petrohawk
- Petronas
- Southwestern Energy Company
- PTTEP
- Spectra Energy
- Statoil (STO)
- Warburg Pincus LLC
- Anadarko Petroleum Corporation
- Antero Resources
- BP
- Cabot Oil & Gas Corporation
- Carrizo Oil and Gas
- Hess Corporation
- Central Petroleum
- SM Energy
- Southern LNG Company
- Southern Union
- CE (Cambridge Energy)
- Cheniere
- Santos
- Chesapeake Energy Corporation
- Cove Energy
- Devon Energy Corporation
- Dominion Resources
- Exco Resources
- Excelerate Energy
- ExxonMobil Corporation
- Oregon LNG
- Pangea LNG
- Chesapeake Midstream Partners
- CONSOL Energy Inc.
- Pieridae Energy Canada Ltd
- Pioneer Natural Resources
- PKN Orlen
- Jordan Cover Energy Project
- PTTEP
- Chevron Corporation
- Korea Gas Corporation (kogas)
- Plains Exploration & Production Company (PXP)
- Atlas Energy
- BHP Billiton
- Progress Energy Canada Ltd
- Protege Energy
- Baker Hughes
- BC LNG Export Cooperative LLC
- BG Group
- Range Resources Corporation
- Reliance Industries Limited (RIL)
- Sasol
- Sempra
- Mitchell Energy
- Mitsubishi Corporation
- Shaanxi Yanchang Petroleum Group
- Shell Canada
- Repsol
- Rio Tinto
- East Resources
- Encana Corporation
- EOG Resources, Inc.
- EQT Corporation
- Royal Dutch Shell (Shell)
- San Leon Energy
- Sinopec
- Wintershall
- Woodside Petroleum

—cont'd
- Freeport LNG Development
- Gasfin Development
- Chevron Canada
- CNOOC
- CNPC
- Grenadier Energy Partners LLC
- Gulf LNG Liquefaction
- Haisla
- Japex
- KNOC—Korea National Oil Corporation
- Marathon Oil Corporation

REFERENCES

[1] Bustin AM, Bustin RM, Cui X. Importance of fabric on the production of gas shales. In: SPE unconventional reservoirs conference. Society of Petroleum Engineers; 2008.
[2] Speight JG. Shale gas production processes. Oxford (UK): Elsevier; 2013.
[3] Chopra S, Sharma RK, Marfurt KJ. Some current workflows in shale gas reservoir characterization. Focus; 2013.
[4] Sharma RK, Chopra S. Conventional approach for characterizing unconventional reservoirs. Focus; 2013.
[5] Cramer DD. Stimulating unconventional reservoirs: lessons learned successful practices areas for improvement. In: SPE unconventional reservoirs conference. Society of Petroleum Engineers; 2008.
[6] Daniels JL, Waters GA, Le Calvez JH, Bentley D, Lassek JT. Contacting more of the Barnett shale through an integration of real-time microseismic monitoring, petrophysics, and hydraulic fracture design. Society of Petroleum Engineers; 2007.
[7] Elgmati MM, Zhang H, Bai B, Flori RE, Qu Q. Submicron-pore characterization of shale gas plays. In: North American unconventional gas conference and exhibition. Society of Petroleum Engineers; 2011.
[8] Passey QR, Bohacs KM, Esch WL, Klimentidis R, Sinha S. From oil-prone source rock to gas-producing shale reservoir—geologic and petrophysical characterization of unconventional shale-gas reservoirs. Beijing, China. June 8, 2010.
[9] PGI. Assessment of shale gas and shale oil resources of the lower Paleozoic Baltic-Podlasie-Lublin basin in Poland. Polish Geological Survey; 2012.
[10] Slatt RM, Philp PR, Abousleiman Y, Singh P, Perez R, Portas R, Baruch ET. Pore-to-regional-scale integrated characterization workflow for unconventional gas shales. 2012.
[11] Abousleiman YN, Tran MH, Hoang S, Bobko CP, Ortega A, Ulm FJ. Geomechanics field and laboratory characterization of the Woodford Shale: the next gas play. In: SPE annual technical conference and exhibition. Society of Petroleum Engineers; January 2007.
[12] Weatherford. Shale gas/oil reservoir assessment. 2013.
[13] MAS. Physical characterization of shale. Micromeritics Analytical Services; 2013.
[14] Alexandar T, Baihly J, Chuck B, , et alToelle BE. Shale gas revolution. Oilfield Review 2011;23(3).

[15] Cluff B. Approaches to shale gas log evaluation. (Colorado, USA): Denver Section SPE Luncheon; 2011.
[16] Davis G. Petrophysics measurements: lithology, porosity, fluid, pressure and permeability. Society of Petrophysicists and Well Log Analysts; 2010.
[17] Passey QR, Creaney S, Kulla JB, Moretti FJ, Stroud JD. A practical model for organic richness from porosity and resistivity logs. AAPG Bulletin 1990;74:1777−94.
[18] Løseth H, Wensaas L, Gading M, Duffaut K, Springer M. Can hydrocarbon source rocks be identified on seismic data? Geology 2011;39:1167−70.
[19] Rickman R, Mullen M, Petre E, Grieser B, Kundert D. A practical use of shale petrophysics for stimulation design optimization: all shale plays are not clones of the Barnett Shale. In: Annual technical conference and exhibition. Society of Petroleum Engineers, SPE 11528; 2008.
[20] Treadgold G, Campbell B, McLain B, Sinclair S, Nicklin D. Eagle Ford shale prospecting with 3D seismic data within a tectonic and depositional system framework. The Leading Edge 2011;30:48−53.
[21] Zhang K, Zhang B, Kwiatkowski JT, Marfurt KJ. Seismic azimuthal impedance anisotropy in the Barnett Shale. In: 80th annual international meeting, SEG, expanded abstracts; 2010. p. 273−7.
[22] Bullin K, Krouskop P. Composition variety complicates processing plans for US shale gas. 2013.
[23] Bustin RM, Bustin A, Ross D, Chalmers G, , et alCui X. Shale gas opportunities and challenges. In: AAPG annual convention, San Antonio, Texas; 2009.
[24] Gardner GHF, Gardner LW, Gregory AR. Formation velocity and density—the diagnostic basics for stratigraphic traps. Geophysics 1974;39:770−80.
[25] Waite WF, Santamarina JC, Cortes DD, Dugan B, Espinoza DN, Germaine J, Jang J, Jung JWT, Kneafsey T, Shin H, Soga K, Winters WJ, Yun T-S. Physical properties of hydrate-bearing sediments. Rev Geophys 2009;47.
[26] Moridis GJ. Challenges, uncertainties and issues facing gas production from gas hydrate deposits. Lawrence Berkeley National Laboratory; 2011.
[27] Gilliam TM, Morgan IL. Shale: measurement of thermal properties. TN (USA): Oak Ridge National Lab; 1987.
[28] Dembicki H. Challenges to black oil production from shales. In: Geoscience technology workshop, hydrocarbon charge considerations in liquid-rich unconventional petroleum systems, Vancouver, BC, Canada; 2013.
[29] Emerging Oil Plays Canada 2012, http://www.emerging-shale-plays-canada2012.com/media/downloads/46-glenn-schmidt-manager-north-american-new-plays-talisman.pdf.
[30] Wang Y, Zhu Y, Chen S, Li W. Characteristics of the nanoscale pore structure in Northwestern Hunan Shale gas reservoirs using field emission scanning electron microscopy, high-pressure mercury intrusion, and gas adsorption. Energy & Fuels 2014;28(2):945−55.
[31] Wikipedia, http://en.wikipedia.org/wiki/Shale.
[32] Tourtelot HA. Black shale—its deposition and diagenesis. Clays and Clay Minerals 1979;27(5):313−21.
[33] Bonakdarpour M, Flanagan B, Holling C, Larson JW. The economic and employment contributions of shale gas in the United States. IHS Global Insight. America's Natural Gas Alliance; 2011.
[34] Labani M. Characterization of gas shale pore systems by analyzing low pressure nitrogen adsorption. Unconventional Gas Research Group. Curtin University; 2012.
[35] Worldwide Geochemistry. Review of data from the Elmworth energy Corp. 2008. Kennetcook #1 and #2 Wells Windsor Basin, Canada. p. 19.
[36] U.S. Office of Technology Assessment. An assessment of oil shale technologies. 1980.

[37] Dawson FM. Cross Canada check up unconventional gas emerging opportunities and status of activity. In: Paper presented at the CSUG technical Luncheon, Calgary, AB; 2010.

[38] Gillan C, Boone S, LeBlanc M, Picard R, Fox T. Applying computer based precision drill pipe rotation and oscillation to automate slide drilling steering control. In: Canadian unconventional resources conference. (Alberta, Canada): Society of Petroleum Engineers; 2011.

[39] Understanding Shale gas in Canada; Canadian Society for Unconventional Gas (CSUG) brochure.

[40] Kuila U. Measurement and interpretation of porosity and pore-size distribution in mudrocks: the hole story of shales. Colorado School of Mines; 2013.

[41] Pemberton SG, Gingras MK. Classification and characterizations of biogenically enhanced permeability. AAPG Bulletin 2005;89(11):1493–517.

[42] Wang Y, Knabe RJ. Permeability characterization on tight gas samples using pore pressure oscillation method. Petrophysics-SPWLA-Journal of Formation Evaluation and Reservoir Description 2011;52(6):437.

[43] Harris NB, Dong T. Porosity and pore sizes in the Horn River shales. Earth and Atmospheric Sciences, University of Alberta; 2012.

[44] Society of Petroleum Engineers, http://www.spe.org/dl/docs/2012/ozkan.pdf.

[45] Klimentidis R, Lazar OR, Bohacs KM, Esch WL, Pedersen P. Integrated petrography of mudstones. In: AAPG 2010 annual convention, New Orleans, Louisiana; 2010.

[46] Jun-yi L, et al. Nano-pore structure characterization of shales using gas adsorption and mercury intrusion techniques. Journal of Chemical and Pharmaceutical Research 2014;6(4):850–7.

[47] Lewis AM. Production data analysis of shale gas reservoirs. Louisiana State University and Agricultural and Mechanical College; 2007.

[48] Vasilache MA. Fast and economic gas isotherm measurements using small shale samples. Journal of Petroleum Technology 2010;44:1184–90.

[49] Maiullari G. Gas shale reservoir: characterization and modelling play shale scenario on wells data base. San Donato Milanese (Italy): ENI Corporate University; 2011.

[50] Alberta Geological Survey, http://www.ags.gov.ab.ca/energy/shale-gas/shale-gas-projects.html.

[51] Visiongain. The 20 leading companies in shale gas 2013-competitive landscape analysis. 2014.

CHAPTER THREE

Exploration and Drilling in Shale Gas and Oil Reserves

1. INTRODUCTION

In the past, hydrocarbons trapped in shale and tight reservoirs have been uneconomical to produce, as the return on investment using traditional drilling and production methods was too low. As a result of the advances in drilling technology, exploration and drilling in these reservoirs has increased, unlocking untapped resources. These technologies include horizontal drilling, hydraulic fracturing, and advanced drilling fluids, among others. Increases in shale oil and gas production have been crucial in allowing Canada, as well as many other countries, to offset the decline in conventional hydrocarbon reservoirs. This chapter will outline some of the new and emerging technologies in shale oil and gas exploration as well as their impact on operations in the energy industry today.

2. EXPLORATION TECHNIQUES

There are several exploration techniques employed in the oil and gas industry for exploring for shale oil and gas. One of these techniques is seismic surveying. Seismic surveying uses large machinery at the surface that creates a vibration sending seismic waves down through the earth. These seismic waves reverberate differently off the different ground layers and are recorded by geophones at the surface. Based on how the waves are reflected back and how long they take to return a 2D or 3D model of the earth can be created. This process can be seen in Fig. 3.1.

A third seismic model can also be created and this is a 4D model. The 4D model incorporates seismic data or core samples taken over an extended period of time. This allows the changes in the rock formation over time to be examined. Using these seismic images potential traps and oil reservoirs can be determined based on the rock formations and rock types.

Another method of exploration, which is generally carried out after seismic surveying, is geophysical well logs. This is the method of actually drilling into the earth's surface and removing a core sample which is then

Shale Oil and Gas Handbook
ISBN: 978-0-12-802100-2
http://dx.doi.org/10.1016/B978-0-12-802100-2.00003-4

© 2017 Elsevier Inc.
All rights reserved.

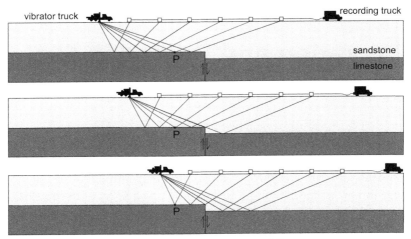

Figure 3.1 Seismic imaging [1].

further analyzed in a lab. Analyzing core samples in a lab allows the prospector to identify any oil or gas trapped in the core sample as well as the key potential rock properties such as porosity, permeability, and wettability. This process could then further lead to the identification of a valuable reservoir [2].

3. ADVANTAGES AND DISADVANTAGES OF EXPLORATION TECHNIQUES

Each exploration technique has some advantages and disadvantages. For the 2D seismic imaging this analyses a small slice of the earth which shows the rock formations and different rock types. This method is less expensive than the 3D and 4D seismic although it does not show as much information as the other seismic imaging.

3D seismic images display much more detail about the land as they show the rock types and formations over an area. The level of detail and the distribution over an area of land make this a better option than the 2D seismic, although due to the level of analysis required for the imaging it is much more expensive. 3D seismic imaging can cost upwards of $100,000 or more per square mile [1]. Fig. 3.2 displays a 3D seismic image.

The third seismic image is 4D seismic and while it displays more information than the 3D seismic, there are a lot of drawbacks to this method. The issue with 4D seismic is it requires seismic data to be taken over extended

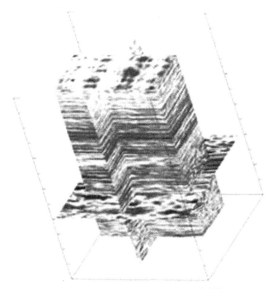

Figure 3.2 3D seismic image [1].

periods of time to compare the images and determine the changes in the rock formations. This can get very expensive and may not show anything more valuable than the 2D or 3D image.

Finally, core sampling is an effective method of exploration as it allows hands-on lab tests to confirm that there are in fact hydrocarbons in place in the ground. The biggest drawback to this method is that you must first know where to drill for the core sample to have a chance of being in a potential, valuable reservoir, this is where the seismic imaging becomes useful.

3.1 Exploration and Equipment Stages

The major stages of exploration are given below [1].

Stage 1: Recognition of the gas resources
- Land acquisition, safe seismic and authorizations of drilling location, land utilization agreements.
- Primary geochemical and geophysical investigations in targeted areas

Stage 2: Initial evaluation of drilling.
- Seismic surveys to obtain the geological specifications of gas-bearing formations such as formation discontinuities or faults that may affect the potential of the reservoir.
- Primary vertical drilling to estimate the characteristics of shale gas resource so that generally, core samples are taken.

Stage 3: Pilot project of drilling
- Initial drilling of horizontal well(s) in order to distinguish reservoir characteristics and optimize completion methods (which may include multistage fracturing process).
- Consecutive vertical wells drilling in additional regions of potential shale gas zones.
- Primary production tests.

Stage 4: Pilot production trails
- Multihorizontal wells drilling in a single layer as part of a full-size pilot test.
- Optimization of the completion techniques including drilling and multistage fracking through microseismic tests.
- Pilot production tests.
- Field developments through designing and accomplishment of pipeline flow systems.

Stage 5: Commercial development
- The commercial decision to launch.
- Government permission for construction of gas plants, pipelines, and drilling.

There is generally some equipment that is employed for shale oil and gas reservoir exploration processes, such as seismic vibrator, geophone, and onshore drilling rig.

3.1.1 Seismic Vibrator

The seismic vibrator is used in the seismic surveying and is the device that actually creates the seismic waves that are sent through the earth. These vibrators are usually made as attachments for a truck. This makes them very mobile and allows for frequency adjustments to ensure they can reach resolution requirements [3]. An image of a seismic vibrator truck is shown in Fig. 3.3.

Figure 3.3 Nomad 90 seismic vibrator [4].

3.1.2 Geophones

Geophones are the devices used to pick up the reverberated seismic waves from seismic surveying. These devices use magnets and a coil of copper wire. When the geophone is moved, due to a seismic wave, the coils of wire cut the magnetic field. From this, the velocity and direction of the seismic wave can be determined.

3.1.3 Onshore Drilling Rig

An onshore drilling rig is required once a potential reservoir is identified. An onshore drilling rig drills deep into the earth's surface to a desired depth and recovers a core sample of the reservoir rock. These drill rigs can come in many different sizes depending on the rock types it needs to drill through and the depth it needs to drill to. The components of a conventional onshore drill rig are shown in Fig. 3.4.

4. CHALLENGES AND RISKS IN SHALE GAS AND OIL FORMATIONS

There are several risks involved with shale oil and gas. Some of these include high-pressure reservoirs, ground water contamination, and water shortages. These shale wells can reach up to 13,500 psi. This high pressure requires a large amount of drilling mud and chemical additives to ensure that the well does not encounter a blowout.

One of the other major risks encountered is ground water contamination. When drilling at high pressures it is easy for a gas reservoir to begin to leak or migrate upwards once it is fractured. Once the well begins to leak or gas starts to migrate it is possible for the gas to reach a groundwater source. If oil or gas reaches a ground water source, then it will contaminate the ground water in the area.

Another major risk of shale oil and gas recovery is water shortages. Just one of these wells could use up to millions of gallons of water to help mitigate the pressure encountered. Once the water or fluid is injected into a well the drill chemicals contaminate it. With the world's water supply already being stretched thin, using large amounts of water for drilling fluids in every well can heavily increase the risk of creating a water shortage [6].

5. TYPICAL EXPLORATION COSTS

Exploration costs can get very expensive for an area before an economical find is discovered. 3D seismic is one of the more common

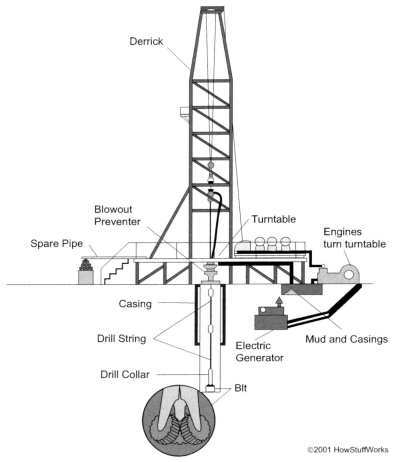

Figure 3.4 Components of an onshore drilling rig [5].

seismic models used when exploring for oil and gas. These seismic images can get very expensive if a large area is to be examined. 3D seismic images can cost upwards of $100,000 per square mile surveyed. Once they are collected even more money is required to analyze the images and create a model. Companies can spend anywhere from $1 million to upwards of $40 million to create a seismic model of an area [7].

The next step of exploration for shale reserves is to drill a preliminary well for core samples. These wells can take anywhere from 14 to 35 days to drill and at a cost of approximately $150–200 per foot. The overall cost of these wells can be upwards of $3 million just for one. However, one exploration well will not likely provide the information required for the area. On average 10—20 wells may need to be drilled before the exploration is complete in some areas and the findings for each well would have to be analyzed and reported [8].

6. SURFACE MINING

Oil shale is a type of rock that consists of organic carbon and minerals. It is generally a sedimentary rock that contains solid kerogen. Surface mining is one of the main methods of recovering this oil shale. Surface mining is when the areas of kerogen rock are stripped away using heavy equipment and carried to a refinery as shown in Fig. 3.5. Blasting or cutting is used to break up the rock surface, and then the rocks are crushed and trucked away. This method of recovery is intrusive to the environment leaving large holes where the rock was removed and taken away [9].

7. UNDERGROUND MINING

Underground mining is another method of recovering oil shale. This method is also called the room-and-pillar method. This method is conducted underground leaving a much smaller impact on the environment. In this method all of the material is removed in a horizontal plane underground creating an array of rooms and pillars. The pillars provide the support to the underground rooms and prevent mine collapse. It is important in this method of mining that the pillar size is sufficient to support the mine. In general practice the pillar is generally the same size as the room. An example of an underground room-and-pillar mine is displayed in Fig. 3.6 [10].

Figure 3.5 Example of surface mining shale oil.

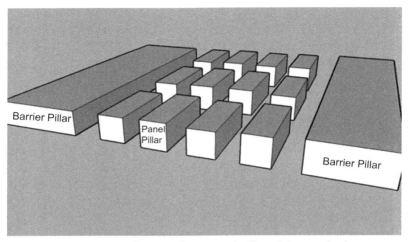

Figure 3.6 Underground room-and-pillar mining method.

8. DRILLING TECHNOLOGIES

8.1 Procedures, Technologies, and Equipment

Once an oil or shale gas reserve has been identified and proven to be economical through exploration techniques, the next step in the recovery process is to drill down into the reservoir to begin producing the hydrocarbons.

Before commencing the drilling procedure, the appropriate infrastructure must be set in place. Preparations will depend on the type of environment (onshore versus offshore wells), but there will need to be a type of water source or well, a means to dispose of drilling fluids and waste, and a level area to set-up the drilling rig platform. Drilling also involves input from a multidisciplinary team of drillers, maintenance personnel, engineers, geologists, and reservoir specialists [11].

Along with infrastructure and personnel, necessary equipment must be used to drill a well. This equipment may be supplied from many types of drilling rigs, including drilling ships, jack-ups, land rigs, semisubmersibles, among others depending of the environment. The largest of the drilling equipment needed is a mast or derrick that is laid over the substructure and holds/connects all the equipment in line with the drill hole [12]. Some of the major components include large diesel engines, electrical generators and motors to drive the mechanical system. A rotary table is used to rotate the drill string, which is held in place by a swivel or large handle that can support the heavy weight and make a tight seal on the hole while drilling. A Kelly bushing (a four- or six-sided pipe) is implemented to transfer the rotary motion to the rotary table and drill sting. All of the rotating equipment is driven by a top drive, located at the top of the derrick. The drill string consists of thick piping sections around 30 ft long, connected by large-diameter drill collars. A drill bit is located at the end of the string and is first entered into the hole, some drilling bits are shown in Fig. 3.7. The bits consist of sharp edges that grind and cut up the rock, and can come in many shapes or materials—including tungsten, steel, and diamond. Near the drill bit, there is often well logging or data-tracking equipment installed in the drill string which helps the engineers at the surface have a better understanding of what is going on in the well [11].

There is also safety equipment installed at the top of the well to protect the workers and environment in case of an emergency. A blow-out preventer (BOP) is placed at the top of every well which is specially designed to stop hydrocarbons from coming up to surface in case of a blowout.

Once the drilling plan has been initiated with the appropriate personnel, infrastructure, and equipment in place, the drilling may begin. The drill string and drill bit are lowered onto the top of the well, with the weight of the drill string and the rotating motion cutting into the rock and descending deeper into the well. Every 30 ft, a new section of drill pipe is added onto the drill string. Fig. 3.8 depicts onshore drillers helping stack drill string sections that are to be descended into the well.

There are three main stages involved in well drilling—running the drill bit into the ground until it reaches the depth of the targeted zone, running a smaller-diameter casing into the drilled hole, then cementing the casing in place. This process is repeated multiple times, using a smaller-diameter drill bit each run, until the well is drilled as deep into the ground as intended to

Figure 3.7 Types of drilling bits [13].

Figure 3.8 Onshore personnel stacking drill string to be lowered into a well.

reach the reservoir [14]. The sizes and types of casing implemented in well design are further elaborated on in Section 3.5.

9. HORIZONTAL, VERTICAL, AND DIRECTIONAL DRILLING

At the start of petroleum recovery, and oftentimes when drilling exploration wells, the well is drilled directly vertical from the derrick. However, since the 1920s, directional drilling has become an integral part of petroleum production [15].

Directional drilling may be defined as controlling the direction, angle, and deviation from the vertical path of the wellbore to reach a specific underground target or location. This type of drilling may be completed for a multitude of reasons.

Primarily, nonvertical wells are drilled to hit targets that cannot be reached by vertical drilling. This may be the case if a reservoir is located beneath an important area where drilling is impossible or forbidden. Secondly, directional wells can be used to minimize the surface footprint of the drilling operation. Horizontal well paths can reach areas within a reservoir with only one well drilled, whereas vertical drilling may take many different drilled wells to recover the same amount of hydrocarbons. This action may also increase the pay zone within the target rock unit. Directional drills may be completed to intersect and relieve pressure from an "out-of-control" well [16]. Directional drilling is also often used to sidetrack around underground obstructions or to drill through a steeply inclined fault.

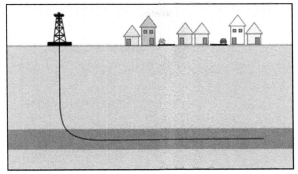

Figure 3.9 Horizontal well drilling [16].

The three primary types of directional wells include horizontal wells, multilateral wells, and extended-reach wells. Horizontal wells, shown in Fig. 3.9, are high-angle wells with an inclination greater than 80o. Horizontal wells play perhaps the most important role in shale formations. The low-permeability rock seen in these types of reservoirs contains significant amounts of gas, which are challenging to recover from tiny pore spaces. To stimulate productivity in these areas, companies often drill horizontally through the rock unit, then use hydraulic fracturing to produce artificial permeability [16].

Multilateral wells (Fig. 3.10) are a new type of drilling technology in which several wellbore branches radiate from the same main wellbore. Thus, directional drilling must be used to reach the different target points. Extended-reach wells are another type of horizontal drilling where the measured depth and the true vertical depth have a ratio greater than at least 2:1 [17].

Directional drilling can be completed by simply pointing the drill bit in the correct direction. However, more complex methods of directional drilling can include downhole steerable mud motors that are located in the bottom hole assembly (BHA) near the bit. These motors use push or pull methods with drilling fluid against the wellbore to bend the drill string in the correct orientation [15].

9.1 Advantages and Disadvantages of Drilling Techniques

As described in the previous section, one obvious disadvantage to drilling vertical wells is the inability to reach as many targets as directional drilling. Directional drilling offers benefits such as increased productivity and pay zone of the reservoir, decreased costs associated with the same amount of productivity (it may take one horizontal well to recover a substantial amount

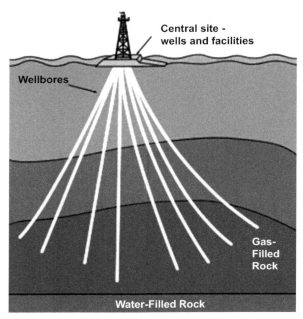

Figure 3.10 Multilateral or multiple wellbore drilling [15].

of hydrocarbons in a reservoir that would take multiple vertical wells), and lower environmental impact. Directional drilling also offers advantages of being able to drill into more difficult faults, to drill around obstructions, and to relieve uncontrolled wells [17].

However, directional drilling also has many disadvantages. Multilateral wells, a type of directionally drilled wellbore, can be particularly challenging when trying to get sufficient flow through one of the laterals to lift fluids and clean the wellbore. This technology is also fairly new and technically challenging. Lack of experience in this area can be detrimental to a petroleum company [17]. Directional drilling also requires additional personnel to operate the downhole mud steerable motors and extra logging or data-acquiring techniques to ensure the bit is steered closer to the target. Directional drilling also poses additional safety risks by having to ensure two wells do not cross over one another. This risk is significantly reduced with straight-vertical drilling.

9.2 Drilling Fluids

While drilling, fluid or "mud" is pumped through the drill bit. Liquid drilling fluids, also referred to as drilling mud, consist of a fluid base (water, petroleum, or synthetic compounds), and other chemical additives to aid in the drilling of boreholes into the earth. The three main drilling muds utilized in drilling applications are water-based mud (WBM), oil-based mud (OBM), and synthetic-based mud (SBM). Depending on the cost, environmental impacts, and technical requirements of the drilling operation, drilling mud selection is a key component in success of the operation [18].

When utilizing a rotary drill bit, as seen in Fig. 3.11, mud is pumped down through the pipe, exits out of holes in the drill bit and returns to the surface through the annulus (the space between the drill pipe and the wall of the borehole). As the drill bit rotates, rock fragments or cuttings break off into the drilling mud and the borehole deepens. One of the main functions of the mud is to carry these cuttings to the surface and to avoid fragment buildup in the borehole [18]. Other functions of drilling mud include maintaining borehole stability, cooling and lubrication of the drill bit, controlling formation pressures, and transmitting hydraulic energy to drilling equipment.

WBM are the most commonly used drilling fluids in engineering applications, known for their cost-effectiveness and versatility. WBM includes a fluid base of fresh water, seawater, or brine mixed with clays and other chemical additives. On most onshore locations, WBM can be disposed

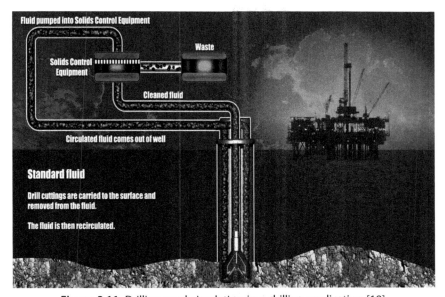

Figure 3.11 Drilling mud circulation in a drilling application [19].

within the environment depending on what chemical additives are included in its composition [20]. OBM was developed for challenging drilling operations where WBMs consistently did not provide sufficient results. These challenging drilling situations include reactive shales, deep wells, and horizontal and extended-reach wells. OBMs excelled in these operations because of their enhanced lubricity, shale inhibition, cleaning abilities, and ability to withstand greater heat without breaking down. To produce the same effects of OBM without the harmful environmental impacts, engineers developed SBMs, which share the desirable drilling properties of OBMs but are free of polynuclear aromatic hydrocarbons, have lower toxicity, faster biodegradability, and lower bioaccumulation potential. Due to high costs, SBMs are typically recycled or re-injected, rather than disposed into the environment [21].

Drilling muds are important fluids that are engineered to perform essential functions during the construction of a well. The most basic and important function of drilling muds includes carrying the rock fragments or cuttings created by the drill bit to the surface. The mud's ability to carry the cuttings to the surface depends on the cutting size and shape, as well as the mud's density and speed of flow traveling up the well [22]. It is important, as engineers, to sustain the mud's capability to execute this task in order to avoid lost circulation and "stuck pipe" situations.

Drilling fluids are also a key component in maintaining control of the well. Hydrostatic pressure exerted by the mud when released onto the drill bit provides an offset to the formation pressure that would otherwise force formation fluids into the borehole, thus losing control of the well (Fig. 3.12). These fluids also maintain the wellbore stability, lubricate the bit, and transmit hydraulic energy to the drill bit [22].

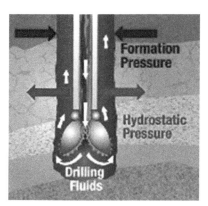

Figure 3.12 Hydrostatic pressure of drilling mud [19].

9.3 Drilling Risks and Challenges

Drilling into shale gas and oil reserves is a technically challenging duty that poses many risks. While the petroleum industry faces many challenges related to social impact, environment, and economic concerns each day, the following subsections explain typical risks overcome by engineers specifically related to drilling. All of these risks pose a threat to working personnel and the environment if not handled properly.

9.4 Fluid Loss

One challenge in drilling is fluid loss, or loss of circulation. This occurs when the drilling fluid perforates into the formation, and the well is uncontrolled. Fig. 3.13 displays how partial loss occurs when only some mud returns to the surface, but total loss is when no mud is retrieved at the surface because it is all being pushed into the formation.

This problem occurs when formations are inherently fractured or when excessive downhole pressure (due to high mud weight, improper hole cleaning, or high-pressure gas) induces fractures. The issue may also be caused by casing set too high, or the existence of improper drilling conditions [23].

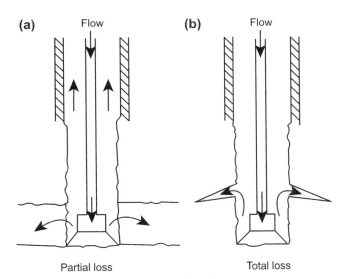

Figure 3.13 Fluid loss of circulation when drilling [23].

Though oftentimes impossible to prevent completely due to high-permeability or fractured zones, loss circulation may be improved significantly or partially avoided by multiple means. Maintaining proper mud weight to ensure the correct weight on bit, minimize annular pressure, and clean the hole adequately is major. It is also worthwhile to study where weaker formations exist, and set casing in place to protect the wellbore [23].

There are also preventive tests that can be conducted to limit the possibility of loss of circulation. Leak off tests (LOTs) are conducted by closing in the well and pressuring up the open hole below the last casing string. The point where the pressure drops off indicates the strength of the wellbore. Additionally, formation integrity tests (FITs) may also be conducted. These tests determine whether the wellbore can tolerate the maximum mud weight anticipated when drilling. Should either the LOT or FIT display inadequate well formation, cement is typically run down the well as a remediation effort [23].

9.5 Borehole Instability

Borehole instability includes the closing or collapse of the wellbore, fracturing in the wellbore walls, as well as hole enlargement. Instability may occur when formation foreign fluids are introduced. To prevent the flow of formation fluids into the hole, the drilling mud must exert a greater pressure than that of the fluids in the porous rocks that are penetrated by the bit. To prevent this from occurring, the mud properties and chemistry must be properly maintained. Mud weight must be carefully selected based on parameters, hydraulics must be utilized to control the equivalent circulating density of the fluid, and the type of drilling mud should be compatible with the borehole and formation being drilled [24].

9.6 Stuck Pipe

One major problem when drilling is sticking of the drill string, also known as the stuck pipe. This problem costs the oil industry hundreds of millions of dollars each year, and occurs in 15% of oil wells [25].

Mechanically, stuck pipe may also occur for two reasons; primarily, due to improper hole cleaning, also perceived as excessive drilled cuttings accumulation in the annulus fluid. Because of this, it is common practice to circulate the fluid several times before pulling the drill string to the surface. Increases in torque, drag, and circulating pressure are all indicators of excessive cuttings. Secondly, stuck pipe may occur due to borehole closing. This occurs when the mud weight is too low, which can lead to the collapse of the wellbore. To free mechanically stuck pipe, drilling engineers should try to lower the equivalent circulating density (if stuck due to cuttings accumulation) or increase the mud weight (if the wellbore is collapsed) [26].

9.7 Typical Drilling Costs

According to Pioneer Natural Resources, drilling a well can take approximately 60 days and $15 million to complete on land. Based on reports from the Canadian Association of Petroleum Producers, the cost of drilling an offshore well in Atlantic Canada usually takes from 3—4 months and costs upwards of $150—200 million per well [27].

Thus, drilling costs and time vary greatly on the environment and type of well. Some wells drilled in offshore Newfoundland have taken as short as 30 days and others close to 365 days, totaling close to a billion dollars depending on daily fees and rig rental rates.

Horizontal drilling, as often seen particularly in shale reserves, is expensive. It can cost up to three times as much per foot as drilling a vertical well. However, this increase in cost is justified by the increased productivity from the well [16].

10. HYDRAULIC FRACTURING

10.1 What is Hydraulic Fracturing?

Also known as fracking, hydraulic fracturing is a well stimulation technique where oil and gas recovery is enhanced by means of creating fractures in rock formations using high-pressure injection of fracturing fluids (as shown in Fig. 3.14). The propagation of fractures in the reservoir increases the

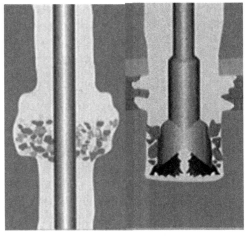

Figure 3.14 Stuck pipe: Cuttings accumulation (left) and wellbore collapse (right) [25].

overall permeability of the formation, which allows hydrocarbons to flow more freely, thus enhancing production. This method is optimal for shale reservoirs with low permeability's (Darcy factor is less than 10 mD) [28].

Generally, hydraulic fracturing consists of three major phases. The first phase involves the injection of fracturing fluid into the formation at a high enough rate, keeping in mind the fluid rheological properties, to overcome the compressive stresses of the earth and tensile strength of the formation rock allowing for fracture to occur. The second phase as it is shown in Fig. 3.15 involves the continuation of fluid injection for further crack propagation, resulting in an increase in the length and width of the cracks. Lastly, the third phase of hydraulic fracturing consists of the addition of proppant into the injection fluid to fill the propagated cracks and prevent the fractures from closing when pressure is reduced. The addition of proppant also increases the permeability of the formation and resulting oil flow for production [29].

10.2 Equipment

The fracturing treatment process requires specialized equipment and materials that can vary based on the type of well being drilled. Commonly stored in tanks or containers at the well site, surface equipment can consist of multiple pumping units, blending units, fracturing fluid storage tanks, chemical storage tanks, proppant supplies, ancillary equipment, and control monitoring units (Fig. 3.16) [30].

Well management and monitoring is a key factor in the hydraulic fracturing process. Control units located onsite of the fracturing operation monitor and record the rate and pressure at which the fracturing fluid is pumped into the wellbore, fluid additive rates, and proppant concentrations to ensure safe operation as well as provide critical data for treatment optimization [31].

Figure 3.15 Crack propagation within rock formation [29].

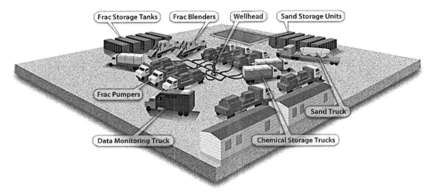

Figure 3.16 Hydraulic fracturing equipment.

10.3 Theory

Hydraulic fracture occurs when the fluid pressure within the shale formation exceeds the minimum principle stress (σ_h) and the tensile strength of the rock. The continuation of pumping the hydraulic fluid at increasing pressures causes the formation fracture to propagate in the direction of least resistance. Theoretically, the condition for fracture occurs when:

$$P_{\text{frac}} = 2\sigma_h - p_f + T_0$$

where P_{frac}, maximum well pressure or fracture initiation, T_0, tensile strength, σ_h, minimum horizontal stress, and p_f, fluid pressure.

During the first and second pump cycles of the hydraulic fracturing process, important characteristics such as the critical fracturing pressure and tensile strength can be determined from the well pressure response, as shown in Fig. 3.17.

The linear section observed during the first pump of the hydraulic fracturing treatment represents the elastic deformation of the system as the fluid compression in the borehole. At the peak of the response, fracture initiates and creates a vertical fracture. After fracture, the well pressure drops, resulting in unstable fracture growth as the fracture propagation rate is higher than the fluid injection rate. However, the continuation of pumping fluid results in restabilizing of the fracture propagation which is represented by the constant well pressure, as seen in the second pump response. During the second pump, the tensile strength becomes zero as the fracture has already occurred. The difference between the first and second peaks represents the ideal measure for tensile strength of the formation [32].

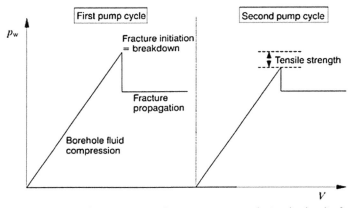

Figure 3.17 Theoretical first and second pump response during hydraulic fracturing.

With the continuous injection of fracturing fluids, the resistance to flow within the rock formation increases, resulting in the wellbore pressure also increasing until it exceeds the breakdown pressure of the formation. The breakdown pressure of the formation is the sum of the in situ stresses and the tensile stress of the rock. Once a fracture is created, the extension of the crack occurs at a pressure known as the fracture-propagation pressure and can be equated using the following parameters:

Fracture Propagation Pressure = In Situ Stress + Net Pressure Drop
+ Near Wellbore Pressure Drop

where the net pressure drop is equal to the decrease in pressure down the fracture as a result of the fracturing fluid, along with any increase in pressure due to tip effects. Near-wellbore pressure drop can be a combination of the pressure drop due of the viscous fluid flowing through the crack and/or the pressure drop as a result of tortuosity (convoluted pathways) between the wellbore and crack [33].

In situ stresses describe the confined pressures acting on the rock formation underground and can be broken into three principal stresses as shown in Fig. 3.18 where σ_1 represents the vertical stress (σ_v), σ_2 is the minimum horizontal stress (σ_h), and σ_3 is the maximum horizontal stress (σ_H). The control pressure required to create and propagate a fracture is dictated by the magnitude and direction of the principal stresses outlined in Fig. 3.18.

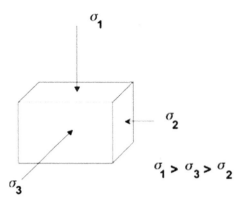

Figure 3.18 In situ stresses—principle stress profile.

During hydraulic fracturing a crack will propagate perpendicular to the minimum horizontal stress creating a vertical fracture. The minimum horizontal stress can be determined using the following equation:

$$\sigma_{min} = \frac{\nu}{1-\nu}(\sigma_1 - \alpha p_p)\alpha p_p + \sigma_{ext}$$

where ν, Poisson's ratio, σ_1, overburden pressure, α, Biot's constant, p_p, pore pressure, and σ_{ext}, tectonic stress.

Based on the theoretical analysis for hydraulic fracturing, it is evident that the fracturing fluid properties, as well as rock formation properties, are very important in the formation of the fracture and propagation.

10.4 Fracturing Fluids and Additives

Fracturing fluids containing proppants and additives are pumped into the well at high pressures during treatment to create fractures. The main functions for fracturing fluids include extending fractures, transporting proppant, and providing lubrication for the fractioning process. Depending on the rock formation and properties, different fluid bases can be utilized to provide optimal performance such as water, foam, oil, acid, alcohol, emulsion, and liquefied gases such as carbon dioxide [28]. For shale formations, "slickwater" treatments are used with low-viscosity fluids and low propping agents pumped at high rates to create narrow, complex fractures. Typically, water-based fluid composition is comprised mainly of water, with minimal proppants and a significantly lower portion of additives, as shown in Fig. 3.19 [34].

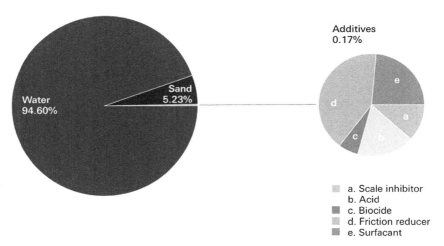

Figure 3.19 Typical fracturing fluid composition.

Similar to the fluid base, additives for hydraulic fracturing also depend on the conditions of the well and enhance the viscosity, friction, and formation compatibility, as well as provide fluid-loss control for the fracturing fluid [33]. Common additive classes as well as their functions include:
1. Biocide—for avoiding growth of bacteria or other fauna;
2. Buffer—pH control;
3. Breaker—reducing viscosity or enhanced fluid retrieval;
4. Corrosion inhibitor—protect casing and equipment;
5. Crosslinker—support gel formation and increase viscosity for carrying proppant downhole;
6. Friction reducer—creates laminar flow;
7. Gelling agent—support gel formation, viscosity for carrying proppant downhole, and ideal proppant carriage;
8. Scale inhibitor—avoid precipitates from mineral scalings on casing or wellhead;
9. Surfactant—emulsification and salinity tolerance.

10.5 Fracturing Proppant

Proppants are a key factor of the hydraulic fracturing process. As shown in Fig. 3.20, proppants are small particles suspended within the fracturing fluid base. They are utilized for holding fractures open after the treatment as well as forming conduits for fluid flow into the wellbore [35].

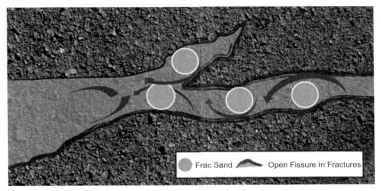

Figure 3.20 Proppant agent in rock formation.

Ideal proppants are strong, resistant to crushing and corrosion, have a low density, and are readily available at a low cost. Interstitial spacing of proppant particles is also an important characteristic to consider, as spacing should be sufficiently large to allow fluid flow but must maintain the mechanical strength necessary to withstand closure stresses.

The most common propping agents include silica sand, treated sand, and ceramics with sizes ranging from 106 μm to 2.36 mm. Resin-coated sand (RCS), is the most commonly used treated sand and is optimal for compressive rock formations. RCS has a lower density and higher strength properties than regular sand but is also more expensive. Ceramic proppant agents include sintered-bauxite, intermediate-strength proppant, and lightweight proppant. These proppants are optimal for stimulating deep wells (greater than 8,000 ft), where in situ stresses are very high [36].

10.6 Fluid Rheology

Rheology is the science of fluid flow and deformation. Fracturing fluids are non-Newtonian fluids, meaning their rheology is viscous in nature and varies based on applied stress. Fracturing fluid properties that dictate fluid rheology are also affected by shear rate within the wellbore, chemical additive concentrations, and proppant types, as well as temperature [29]. Fracturing fluid properties are also important when determining optimal proppant selection, as the base fluid must have rheological characteristics for supporting the transport, suspension, and distribution of proppants.

Rheology is a fundamental component of hydraulic fracturing when predicting fracture growth and geometry. This is essential for optimal execution of the well stimulation technique as accurate measurements for crack propagation provide specific details on the requirements for designing and executing treatment. Failure in fluid selection due to rheology can result in unsuccessful treatment of the reservoir and therefore reduced oil production.

10.7 Fracture Treatment Design and Optimization

To design the fracture treatment for a shale formation, important data must be taken such as the in situ stress profile and permeability to determine the optimum fracture fluid and proppant selection. A key factor in the optimization of the treatment design is determining the effect of fracture length, conductivity on the productivity and recovery factor from the well. Using the obtained production data and reservoir characteristics, a hydraulic fracture model can be developed to find the optimum fracture length and conductivity at a minimum cost.

The fracturing fluid selection for the treatment design is selected based on the following factors:
1. Reservoir temperature;
2. Reservoir pressure;
3. The expected value of fracture half-length;
4. Water sensitivity.

As most oil reservoirs contain water, a water-based fracturing fluid is used as part of the fracture treatment. Whereas acid-based fracturing fluids are commonly used in carbonate reservoirs and oil-based fluids are optimal in oil reservoirs, it is proven that water-based fluids are not successful.

Similar to fracturing fluids, proppant selection is a key factor to optimal fracture treatment design. To determine this, the maximum effective stress on the proppant should be evaluated. Typically, for maximum effective stress less than 6,000 psi, a sand proppant agent should be used. If the maximum effective stress ranges from 6000 to 12,000 psi, a ceramic or intermediate-strength proppant agent should be used pending the reservoir temperature. Additionally, if the effective stress is greater than 12,000 psi, a high-strength bauxite proppant should to be used [37].

10.8 Fracture Modeling and Simulators

Fracturing simulators are used to provide detailed schematics of fracture geometries and distributions, which are used in the optimization of fracture treatments. The first mathematical 2D model developed by Howard and Fast in the 1950s, was used to design fracture treatments where the fracture width was assumed to be constant to determine the fracture area on the basis of the fluid leak-off characteristics of the formation. With the progression of technology, more in-depth 2D fracture propagation models were developed to find fracture geometry more accurately with reasonable success. Today, 3D fracture propagation simulators using high-powered computers are used in the oil and gas industry to determine realistic fracture geometry and dimensions. These models use accurate data describing the layers of the fracturing formation as well as the layers above and below the zone of interest [37].

The first key component of an efficient, fully 3D, hydraulic fracture simulator is the geometric representation of the model. Representation implies computer storage and visualization of the model topology and geometry. Simulation of crack growth is more complicated than many other applications of computational mechanics because the geometry and topology of the structure evolve during the simulation. For this reason, a geometric description of the body that is independent of any mesh needs to be maintained and updated as part of the simulation process. The geometry database should contain an explicit description of the solid model including the crack.

The three most widely used solid modeling techniques, boundary representation (B-rep), constructive solid geometry (CSG), and parametric analytical patches (PAP), are capable of representing uncracked geometries [37a–c].

Fracturing simulators, such as "MFracSuite" by Baker Hughes, provide real-time data of the fracture geometry as well as wellbore heat transfer, proppant transport, and perforation erosion. Fracturing simulators evaluate the most productive zones for hydrocarbon recovery [38].

10.9 Fracture Characteristics

Fractures were characterized by density or intensity of fracturing, which is compared to rock type, physical parameters (depth, altitude, and topographic setting), and hydraulic conductivity measurements. The intensity of fracturing was computed as the distance between all fractures that were observed in the index wells—regardless of the fracture orientation, mode of fracturing, or rock type. This estimate of fracture intensity is referred to as the "interfracture spacing."

Fractured-crystalline aquifer systems are extremely heterogeneous and complex [39].

Hydraulic fractures generally propagate in the horizontal or vertical orientation based on the stress directions and preexisting planes of weakness

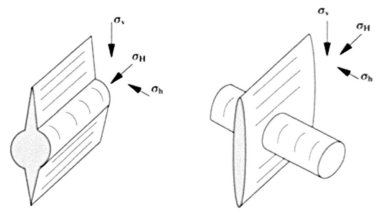

Figure 3.21 Fracture orientations—parallel to the borehole (left) and perpendicular to the borehole (right) [32].

(or naturally occurring fractures) within the formation [39]. The fracture will propagate in the direction based on principal stress location in relation to the borehole:
1. σ_H is parallel with the borehole: $P_{frac} = 3\sigma_h - \sigma_v - p_f + T_0$
2. σ_h is parallel with the borehole: $P_{frac} = 3\sigma_H - \sigma_v - p_f + T_0$

As shown in Fig. 3.21 and the above cases, the first case (1) the fracture will be parallel with the borehole because the smallest in situ stress is normal to the borehole. In the second case (2), the fracture will be normal to the borehole because the smallest in situ stress is parallel to the borehole [32].

Fracture conductivity is also a critical characteristic of fractures and determines the flow capacity through the fracture and is a function of the width, permeability, and length of the fracture. The conductivity of the fracture will decrease during the life of the well due to increasing stress from the formation on the propping agents, corrosion, propping embedment into the formation, and proppant crushing.

Using the fracture geometry and characteristics, well spacing and development strategies can be implemented to recover maximum hydrocarbons from the reservoir.

10.10 Fracturing Rock Properties

Fracturing is typically required in two scenarios; reservoirs that have very low permeability or reservoirs that have been producing for a long time where the operator is looking to extend production in order to produce the remaining residual oil from the small pore spaces. Fracturing is not a drilling process, it is conducted after the final reservoir section has been drilled to enhance oil or gas production from the well. During exploration drilling, certain drilling tools can be used to log rock properties in the wellbore to help production and reservoir engineers to understand the reservoir properties better. This information is then used to determine if the fracturing process is required to achieve optimal production.

The fracturing process is an unconventional oil and gas production method. The term conventional simply refers to oil and gas that flows naturally from the reservoir to the surface. However, unconventional production refers to production from tight formations with low permeability or production methods that require further intervention other than simply drilling a production well into a reservoir.

Rock formations that are said to have tight oil or tight gas are sometimes referred to as shale oil or shale gas reservoirs. A tight oil or gas reservoir refers to a formation having permeability between 0.1 mD and 0.001 mD, meaning that they are relatively impermeable [40]. In some tight shale reservoirs, the permeability can be even less ranging from 0.001 mD to 0.0001 mD [40]. Essentially, this very low permeability means that the pore spaces holding fluid in the formation are very poorly connected and, therefore, cannot easily flow to a production well. These types of formations require a fracturing process to be implemented in order to produce the hydrocarbons.

There are three major types of unconventional hydrocarbon reservoirs that could require hydraulic fracturing [41]:

1. *Tight oil and tight gas sands*—These sand formations could be oil- or gas-bearing, however their fine grains provide little to no permeability. If there are no naturally occurring fractures in the formation, then they will almost always require fracturing to release their hydrocarbons.
2. *Shale gas*—Shale gas can be produced from shale rock where the natural gas has become trapped inside the fine-grained shale material. Fracturing processes can release the gas from the tight shale formation.
3. *Coal bed methane*—This refers to the natural gas that has become trapped inside coal formations. Fracturing is required to release the gas from the coal rock.

10.11 Fracturing Processes

After the well is drilled and several hole sections are completed with steel casing and cement, the final reservoir section is reached. If fracturing is required, the reservoir section likely consists mainly of tight shale rock. In the final reservoir section, the last piece of steel casing is cemented in place and then small holes are perforated through the casing to gain access to the shale formation. The end of the reservoir section is referred to as the toe and the beginning of the reservoir section is often called the heel [42]. Fracturing is often completed in stages starting from the toe of the well and working back to the heel.

Special completions equipment is placed inside of the casing down in the wellbore. This equipment is set up to facilitate various pressure-up stages in order to fracture the rock. An initial cleanout process uses circulation of an acidic fluid to clear the casing of any cement debris that may cause blockages and interrupt the fracturing process [43]. Following a cleanout pumping stage, the fracture fluid is pumped into the well at high pressure. Each fracturing stage requires the surface equipment to pump the fracking fluid at very high pressures such that it creates fractures in the rock formation and opens pathways for the reservoir fluids to flow. The proppant is pumped along with the fracking fluid, usually water, to hold open the fracture locations [42]. The result of fracturing in a shale formation is shown in Fig. 3.22.

The hydrodynamic dispersion in fractured crystalline rocks (such as granites), where the connected fracture networks dominate groundwater flow and contaminant transport, is an important issue for environmentally sensitive underground engineering projects. Generally speaking, two basic mechanisms are involved in this transport process.

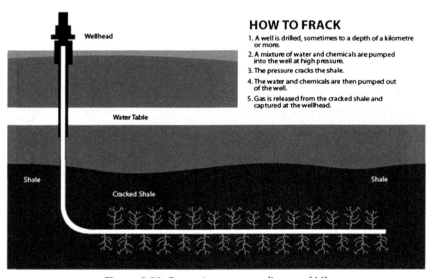

Figure 3.22 Fracturing process diagram [44].

One is molecular diffusion. This is caused by the random motion of molecular species, which is independent of whether the fluid is moving or at rest, and contributes little at long travel distances or with high fluid flow velocities.

The second mechanism is macroscopic dispersion. This is due to the differences in the fluid flow velocity fields in the fracture networks (channeling), caused mainly by the variations in trace lengths, orientations, and apertures of the fractures. Actually, the separation between the two mechanisms is rather artificial, as they are essentially mixed together.

10.12 Risk Evaluation in Fracturing Process

Geologic, product price, and mechanical risks are the main risks that should be always evaluated by the well operator [1].

Sensitivity analysis of reservoir models and fracture propagation models help to identify the data uncertainties. The main issue with hydraulic fracturing processes is that all stages of treatment are done once, implying a significant amount of money is spent over a short time period, while the production rate and cumulative recovery are very far from the desired extent. Several factors, such as the reservoir reaction to fluid injection and mechanical problems with the surface facilities and wells, may result in the treatments failing [1–3].

It has been proved based on economical analysis that almost the whole costs and a part of the revenue should be spent to assess the reservoir and mechanical risks. For instance, the treatment of five fractures might not be successful if even 80% of the expected revenue and 100% of the expected expenses are assigned to the corresponding operations/activities to determine the optimal length of the fractures. Fig. 3.23 demonstrates variation of preferred fracture length in terms of the economical analysis [1–3].

After the fracture treatment is optimally designed, it is very important that the service company and well operate work together to perform effective quality control before, during, and after the treatment operation so that the optimum treatment is pumped into the well in a precise manner [1–3].

10.13 Challenges and Risks of Fracturing Formations

There are several benefits to fracturing as a method of oil and gas production. It allows access to shale oil and gas reserves that would not flow naturally,

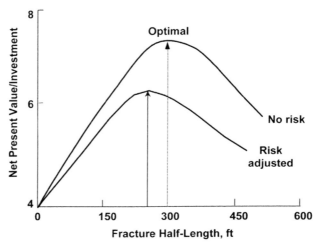

Figure 3.23 Net present value versus fracture half-length.

thus increasing the profit a company can make from the sale of hydrocarbons, while also increasing the amount of resources available for fuel and other products. It could also help extend the life of an oilfield, thus increasing the amount of jobs in the economy. However, with increased benefit also comes increased risk, and in this case the risks are both environmental and economic.

In an article published by the National Geographic in 2013, several negative effects were highlighted with regards to fracturing in the oil and gas industry. Among some of the risks were several environmental impacts such as increased emissions from pumping equipment and trucking services to run the fracturing operations, increased flare emissions from produced gas that must be burned due to operational constraints, and unintentional emission or spills of fracturing chemicals used to enhance the fracturing process [45]. In other cases, fracturing rock in shallow wells or in wells near a ground water supply could cause serious health and environmental risks as well. Fracturing in wells that are closer to the surface could eventually lead to gas bubbling from the well up through the fractured cap rock and into the atmosphere. Similarly, if a well is fractured near the ground water level, there is a potential for gases to pollute the water supply [46]. This could lead to serious health conditions for residents living in nearby areas and can be very hard to detect without constant monitoring of nearby water supplies.

11. EXPLORATION WELLS
11.1 Well Design and Construction

Exploration wells are mainly used for the purpose of data collection and prospecting for new hydrocarbon resources. However, every well drilled requires a design and construction plan. Well design is much more difficult for exploration wells than development wells. Exploration wells are often drilled in areas where the rock strength, pore pressures, and fluid types are unknown. Due to the immense uncertainty it can be very difficult to plan ahead and create a well design.

Well design often begins with a directional plan. Once the final target location for the well has been decided, engineers use all the available data they have with regards to the reservoir and rock properties to determine the well structure. This includes data such as seismic surveys that have been shot in the area, or pressure and fluid gradient data from other wells that may have been drilled in the nearby areas. The pressure data are important because they help an engineer determine the size of the hole that can be drilled in the area, and thus, the casing shoe location.

When an exploration well is drilled various hole section diameters are used based on the pore pressure present in the rock. As the well is drilled deeper into the earth, the pore pressure increases and thus the fracture gradient of the rock is higher. Since a heavier drilling fluid must be used in deeper hole sections, the upper, previously drilled sections, must be isolated from the heavy fluid [47]. To isolate each hole section a steel casing pipe is installed in the well bore and cemented in place. The cement helps to isolate fluids from the rock from flowing up to the surface and also improves the structural integrity of the casing pipe.

In general, there are typically four or five main hole sections drilled in every well. The first two sections are the conductor and surface sections. The conductor section has the largest diameter hole (often 30—36′) and the shallowest depth; it acts as the foundation for the well. The surface section is also a relatively large diameter, and extends slightly further than the conductor hole. Next, the intermediate section is one of the largest sections, which is drilled into the beginning of the reservoir section. The intermediate casing shoe is often placed just inside the reservoir cap rock. The final hole section is the production or reservoir section. This hole section is drilled through the reservoir which is desired to be produced.

Fig. 3.24 shows a vertical schematic of a typical well design with casing and casing cement shown. Exploration wells may be vertical or directional,

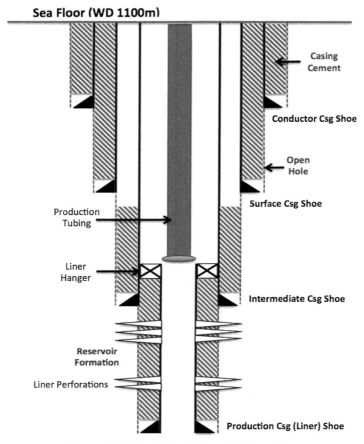
Figure 3.24 Typical vertical well schematic.

however in a development well the reservoir section may extend further into the reservoir to increase production or injection capability. For exploration wells, the purpose of drilling into the reservoir may only be to determine the reservoir properties for future development wells in the area.

11.2 Casing and Perforating Wells

Casing and perforating a production or injection well is a common completion strategy used to gain access to the reservoir rock when there is casing in place in the reservoir hole section. This is a quick and effective method to gain access to the reservoir if performed correctly. Once the casing is installed and cemented in place, perforating guns must be run into the well and placed inside the reservoir section casing. A charge is activated from the surface that fires the perforating guns. Hard perforating bullets are released from the guns and penetrate through the casing into the reservoir rock [48]. This opens a flow path for the fluids to enter into the wellbore.

For exploration wells, it is more common to log the reservoir data while drilling the reservoir section rather than after the section is cased and perforated. However, if for some reason it was desired to produce from the exploration well for a set amount of time to predict pressure depletion or flow rates, this method could be used to initiate flow.

11.3 Well Completion Equipment

There is a huge amount of completion equipment and technology available in today's oil and gas industry. As technology continues to develop, new and more efficient products and tools are being developed to increase production and reduce well construction and production costs.

Besides the perforating equipment used to open the well to the reservoir, the most common and versatile piece of completions equipment is a packer. A packer is simply a sealing component that fits inside the casing to isolate the reservoir fluids from upper casing strings and the production tubing as shown in Fig. 3.25. This is a key element that is designed and selected to withstand specific pressure and temperature ratings in the well.

Figure 3.25 Production packer [49].

Other than the production packer, downhole completion equipment could include items such as pressure and temperature gages to monitor the well during production, production screens to prevent sand or particles from entering into the production tubing, and chemical injection and gas lift valves to enhance production and increase the life of the well.

11.4 Typical Well Completion Costs in Shale Reservoirs

Well completions costs range depending on the location and environment in which the well is located. Offshore wells are typically more expensive than wells which are drilled on land simply because of the type of operation that is required. Offshore rig rental rates are significantly higher than land rig rental rates and the logistics costs to procure equipment for offshore operations also increase the completions costs significantly.

According to an online source referencing land well costs in Western Canada, the cost for completions and casing of the well could be between 1–2 million dollars [50]. This cost could range depending on the depth of the well of course, as a deeper well would require more casing and more completions equipment. However, for offshore wells the cost of drilling and completions is much more substantial. According to an article posted about the offshore oil and gas business in Atlantic Canada, drilling and completing an offshore well could take between 3–4 months and could cost between 150 and 200 million dollars [27]. It is assumed that the majority of this cost estimate is attributed to the daily rental rate for an offshore drilling rig, however a substantial portion would include the cost of the completions equipment and the logistical costs associated with such a complex operation.

11.5 Borehole Instability in Shale Formations

Borehole instability is mainly caused by the mechanical failure of rocks from internal or external stresses, erosion of wellbore due to fluid pumping rates and chemicals in the drilling fluid that interact and breakdown the rock layers [51]. All of these issues result in a breakdown of the rock around the wellbore, which can create issues while drilling, such as stuck pipe, collapsed hole, circulation failure, and inability to log the rock properties. These issues are even more apparent when drilling through shale formations due to the properties of the shale rock.

Drilling through shale presents many challenges. Shale is mainly comprised of highly cemented siltstone, which is very fragile when stress is applied in certain directions. When drilling through shale vibrations from the drill string and drilling equipment can induce failure in the shale rock causing it to essentially crumble and cave in around the drill string [52]. Other issues, such as drilling with a heavy mud weight, could cause the shale to become unstable as well. Also, some reservoirs may already have naturally fractured shale zones. In this case it is relatively impossible to avoid borehole instability in this zone since the shale has already become fragile from existing fractures in the rock.

When conducting exploration drilling it is extremely difficult to prepare for drilling through unstable rock. Since the rock properties are often estimated as unknown before the drilling operation begins, a complex problem could quickly arise due to unstable boreholes. This risk is very apparent if the goal of a production or exploration well is to access a shale oil or gas formation. These operations can become extremely costly should a well have to be sidetracked due to a lost hole or if equipment becomes stuck in the hole and has to be left behind.

11.6 Equations Related to Exploration Drilling

While exploration drilling is a very operations-based undertaking, there are theoretical equations that form the basis of well planning and control. The mathematical equations used during drilling are vast, with many requiring computational software due to their complexity. The following are some common equations used regularly by drilling engineers.

The equations associated with hydraulic fracturing are discussed in the previous sections.

11.7 Normalized Rate of Penetration

Normalized rate of penetration (NROP) is the speed at which a bit may move through the formation, resulting from various drilling parameters. This allows drilling speed to be optimized by varying the factors shown in the equation below [53].

$$NROP = ROPx \frac{(W_n - M)}{(W_o - M)} x \left(\frac{N_n}{N_o}\right)^r x \frac{(Pb_n x Q_n)}{(Pb_o x Q_o)}$$

where ROP, observed rate of penetration; W_n, normal bit weight; W_o, observed bit weight; M, formation threshold weight; N_n, normal rotary speed; N_o, observed rotary speed; r, Rotary exponet; Pb_n, normal bit pressure drop; Pb_o, observed pressure drop; Q_n, normal circulatrion rate; and Q_o, observed circulation rate.

11.8 Rate of Penetration—Bourgoyne and Young

The most comprehensive model for calculating rate of penetration (ROP) in drilling is the Bourgoyne and Young model, developed in 1986. It is a function of four different factors per the below equation [53].

$$ROP = (f_1)(f_2)(f_5)(f_6)$$

Factors 1 and 2 are fairly simple to calculate using formation data as seen in the below equations. The correction factor is obtained through regression analysis of the oil field data obtained during early evaluation [53].

$$f_1 = K$$

where K, drillability.

$$f_2 = e^{2.303a_2(10000-D)}$$

where D, vertical depth (ft); a_2, depth correction exponent.

The third factor (f_5), is known as the weight on bit factor, and is calculated per the following equation [53].

$$f_5 = \left[\frac{\left(\frac{W}{d_b}\right) - \left(\frac{W}{d_b}\right)_t}{4 - \left(\frac{W}{d_b}\right)_t} \right]^{a_5} = [W]^{a_5}$$

where W, weight on bit (1000 lbf); d_b, bit diameter [in]; $(W/d_b)_t$, threshold bit weight per inch of bit diameter at which the bit begins to drill, 1000 lbf/in; a_5, weight on bit exponent.

The final factor in calculating ROP is the rotary speed factor, formed from the rotary speed and the rotary speed exponent per the equation below [53].

$$f_6 = \left(\frac{N}{60}\right)^{a_6} = [N]^{a_6}$$

where N, rotary speed (rpm); a_6, rotary speed exponent.

Once the above four factors are combined, their product forms the ROP. As seen below, this can be simplified into one equation using the components of the four individual equations.

$$ROP = (f_1)(f_2)(f_5)(f_6) = (K) \times (W)^{a_5} \times (N)^{a_6} \times \left(e^{a_2(10000-D)} \right)$$

11.9 Abandonment and Reclamation of Exploration Wells

The abandonment and reclamation of an exploration well is an important engineering undertaking and can often account for up to 25% of the total cost associated with a well [54]. This can be done either using a drilling rig or a dedicated plug and abandonment (P&A) vessel. The P&A of a well begins with the removal of all completions equipment such as downhole equipment, packers, and injection mandrels where applicable. In the case of an exploration well, this is usually fairly minimal as the purpose of the well was to obtain data and fluid/rock samples and it was never actually used for production. Following removal of completion equipment, any subsea equipment that was used in the completion of the well is removed to avoid impacting marine wildlife or vessel routes. After this, the casing of the well must be pulled and the wellhead cut using specialty tools.

Once the well completion has been successfully removed, pulled to shore, and disposed of properly, the well is filled with cement that prevents the flow of reservoir fluids into the sea. It is critical that this step is performed correctly to ensure that the cement is cured fully and there are no leaks. In an exploration well this is of particular concern as the reservoir pressure is still very high and the oil saturation is at the initial level, making a spill even more potent. In a production well the pressure would have declined naturally over the production life and the reservoir fluid would have been reduced to a water cut of approximately 95% [54,55]. To minimize the curing time of the cement in exploration wells, a concentrated sand slurry is pumped into the well along with the cement during P&A [54].

11.10 Research and Development of New Exploration and Drilling Techniques

In an attempt to increase recovery and make more hydrocarbon reservoirs economical, there is a constant race to develop and commercialize new drilling technologies and methods. There are several techniques that are currently undergoing further research and development in order to optimize them for regular field use. These techniques include zipper fracturing and carbon dioxide injection.

Zipper fracturing uses two wells drilled very near one another, hydraulically fractured at the same time. This cracks the rock more deeply and effectively, as shown in Fig. 3.26. In some onshore test wells, this has led to doubled volumes of recovered oil and gas [55].

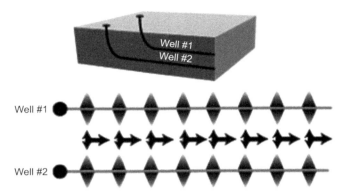

Figure 3.26 Diagram of zipper fracturing [55].

Carbon dioxide injection is executed using two horizontally drilled wells, organized in a staggered line drive. The well on the top injects supercritical carbon dioxide and the well on the bottom produces oil. Normally a gas, when pressurized beyond the critical point carbon dioxide, is a supercritical fluid meaning it has the density of a liquid with the viscosity of a gas. This makes it an excellent solvent which extracts the methane and light components from the reservoir fluids, allowing the oil to flow into the lower well for collection [56]. This technique is still being researched but it is believed that it will impact the oil and gas industry significantly.

The new direction of research in exploration and drilling of shales is briefly described below.

11.10.1 Technological Evolution

Nowadays, all efforts are being made to design wells with small diameter into exact targeted areas/locations in the unconventional resources for optimum recovery through the development of efficient technologies for lateral and horizontal drilling. To attain this goal, a large portion of the formation containing hydrocarbons should be in the drainage area of the wells. The reservoirs should also experience stimulation strategies for permeability improvement. This is achievable only if long lateral horizontal wells are drilled so that they spread over a wide area of the hydrocarbon layer [3–5].

11.10.2 Transverse Fractures

The most complex hydraulic fracture network arises from a horizontal wellbore which is perpendicular to the maximum in situ stress. The main reason for this complexity is the presence or/and propagation of a series of transverse fractures along the horizontal lateral. It is clear that the effective reservoir permeability and consequently production rate experience an increase because of a complex fracture network in the porous formation [1—4].

Engineering and research activities in the shale reservoir field are developing new strategies in diagnostic technologies and discovery of the sweet spots in real time over drilling operations. Beside exploring efficient techniques for control of fracture navigation and orientation, the optimized techniques for drilling and completion and the effective link between the generated data and reservoir characteristics at various circumstances are the future targets of the research and engineering centers [1—4].

REFERENCES

[1] McFarland J. How do seismic surveys work. Oil and Gas Lawyer Blog; 2009.
[2] Dybkowska K. Shale oil exploration and production methods. [Online]. Available from: http://infolupki.pgi.gov.pl/en/technologies/shale-oil-exploration-and-production-methods.
[3] PetroWiki. [Online]. Available from: http://petrowiki.org/Seismic_data_acquisition_equipment.
[4] Wikipedia. April 5, 2016 [Online]. Available from: https://en.wikipedia.org/wiki/Seismic_vibrator.
[5] Martinelli A. Fracking versus conventional oil drilling: an investor's guide. Energy and Capital; 2014.
[6] Botkin DD. The dangers of gas drilling. [Online]. Available from: http://www.desmogblog.com/fracking-the-future/danger.html.
[7] Hill KB. A seismic oil and gas primer. [Online]. Available from: http://www.loga.la/flash/HS/kevinhillLSUS.pdf.
[8] RA Associates. The current costs of drilling a shale well. April 7, 2016 [Online]. Available from: http://www.roseassoc.com/the-current-costs-for-drilling-a-shale-well.
[9] Adams M. Wirtgen. 2016 [Online]. Available from: http://www.wirtgen.de/en/news-media/press-releases/article_detail.2500.php.
[10] Room and pillar. November 4, 2014 [Online].
[11] Freudenrich C, Strickland J. How oil drilling works. April 12, 2001 [Online]. Available from: http://science.howstuffworks.com/environmental/energy/oil-drilling3.htm.
[12] Energy F. Equipment used for purposes of oil extraction. 2016 [Online]. Available from: http://www.flowtechenergy.com/oilfield-equipment/drilling-equipment/.
[13] HRDC. Drilling and well completions. 2016 [Online]. Available from: http://www.petroleumonline.com/content/overview.asp?mod=4.
[14] Zion Oil & Gas. The excruciating difficulty of drilling for oil — 51 steps. 2016 [Online]. Available from: https://www.zionoil.com/updates/excruciating-difficulty-of-drilling-for-oil-in-51-steps/.
[15] RigZone, How does directional drilling work? [Online]. Available from: http://www.rigzone.com/training/insight.asp?insight_id=295.
[16] King H. Directional and horizontal drilling in oil and gas wells. [Online]. Available from: http://geology.com/articles/horizontal-drilling/..
[17] PetroWiki. Directional drilling. June 26, 2015 [Online]. Available from: http://petrowiki.org/Directional_drilling#Types_of_directional_wells.

[18] Neff JM. Composition, environmental fates, and biological effects of water based drilling muds and cuttings discharged to the marine environment. Battelle; 2005 [Online]. Available from: http://www.perf.org/images/Archive_Drilling_Mud.pdf.
[19] 3M oil and gas. November 3, 2011 [Online]. Available from: http://i.ytimg.com/vi/CDK771L5glU/maxresdefault.jpg.
[20] AES Drilling Fluids. Drilling fluids. AES Drilling Fluids, LLC; 2012 [Online]. Available from: http://www.aesfluids.com/drilling_fluids.html.
[21] Drilling waste management information system. [Online]. Available from: http://web.ead.anl.gov/dwm/techdesc/lower/.
[22] Williamson D. Drilling fluid basics. 2013 [Online]. Available from: http://www.slb.com/resources/publications/oilfield_review/~/media/Files/resources/oilfield_review/ors13/spr13/defining_fluids.ashx.
[23] Society of Petroleum Engineers. Lost circulation. June 30, 2015 [Online]. Available from: http://petrowiki.org/Lost_circulation.
[24] Society of Petroleum Engineers. Predicting wellbore stability. December 5, 2014 [Online]. Available from: http://petrowiki.org/Predicting_wellbore_stability.
[25] Schlumberger. Stuck pipe: causes, detection and prevention. October 1991 [Online]. Available from: http://www.slb.com/~/media/Files/resources/oilfield_review/ors91/oct91/3_causes.pdf.
[26] Society of Petroleum Engineers. Stuck pipe. January 2015 [Online]. Available from: http://petrowiki.org/Stuck_pipe.
[27] Canadian Association of Petroleum Producers. Offshore oil and gas life cycle. 2015 [Online]. Available from: http://atlanticcanadaoffshore.ca/offshore-oil-gas-lifecycle/.
[28] Gandossi L, Von Estorff U. An overview of hydraulic fracturing and other formation stimulation technologies for shale gas production. Publications Office of the European Union; 2015.
[29] Edy KO, Saasen A, Hodne H. Rheological properties of fracturing fluids.
[30] CSFU Gas. Understanding hydraulic fracturing. [Online]. Available from: http://www.csug.ca/images/CSUG_publications/CSUG_HydraulicFrac_Brochure.pdf.
[31] Hydraulic fracturing: the process: BC Oil & Gas Commission. [Online]. Available from: http://fracfocus.ca/hydraulic-fracturing-how-it-works/hydraulic-fracturing-process.
[32] Fjaer E, Holt R, Horsrud P, Raaen A, Risnes R. Petroleum related rock mechanics. 2nd ed. Elsevier B.V.; 2008.
[33] Fracturing fluids and additives. [Online]. Available from: http://petrowiki.org/Fracturing_fluids_and_additives.
[34] Fracturing fluids: types, usage, disclosure. [Online]. Available from: http://www.shale-gas-information-platform.org/categories/water-protection/the-basics/fracturing-fluids.html.
[35] Schlumberger oilfield glossary. [Online]. Available from: http://www.glossary.oilfield.slb.com/en/Terms/p/proppant.aspx.
[36] Proppant: the greatest oilfield innovation of the 21st Century. [Online]. Available from: http://info.drillinginfo.com/proppant-the-greatest-oilfield-innovation/.
[37] Fracture treatment design. [Online]. Available from: http://petrowiki.org/Fracture_treatment_design.
[37a] Mantyla M. An introduction to solid modeling. 1988.
[37b] Mortenson ME. Geometric modeling. New York: John Wiley & Sons; 1985.
[37c] Hoffmann CM. Geometric and solid modeling: an introduction. Morgan Kaufmann Publishers Inc.; 1989.
[38] Hydraulic fracturing: increase oil and gas recovery. [Online]. Available from: http://www.bakerhughes.com/products-and-services/reservoir-development-services/reservoir-software/hydraulic-fracturing.

[39] The source for hydraulic fracture characterization. 2005 [Online]. Available from: https://www.slb.com/~/media/Files/resources/oilfield_review/ors05/win05/04_the_source_for_hydraulic.pdf.

[40] GO Canada. Natural resources Canada. 2016 [Online]. Available from: http://www.nrcan.gc.ca/energy/sources/shale-tight-resources/17675.

[41] United State Environmental Protection Agency. In: The process of hydraulic fracturing, 10; 2015 [Online]. Available from: https://www.epa.gov/hydraulicfracturing/process-hydraulic-fracturing.

[42] Halliburton. Interactive fracturing 101. 2016 [Online]. Available: http://www.halliburton.com/public/projects/pubsdata/Hydraulic_Fracturing/disclosures/interactive.html.

[43] Registry CD. FracFocus. 2016 [Online]. Available from: https://fracfocus.org/hydraulic-fracturing-how-it-works/hydraulic-fracturing-process.

[44] LINBC. BC LNG Info. 2016 [Online]. Available: http://bclnginfo.com/learn-more/environment/hydraulic-fracturing-fracking/.

[45] Nunez C. The great energy challenge. November 11, 2013 [Online]. Available from: http://environment.nationalgeographic.com/environment/energy/great-energy-challenge/big-energy-question/how-has-fracking-changed-our-future/.

[46] O'Day S. Top environmental concerns in fracking. March 19, 2012 [Online]. Available from: http://www.oilgasmonitor.com/top-environmental-concerns-fracking/.

[47] Society of Petroleum Engineers. Casing design. June 25, 2015 [Online]. Available from: http://petrowiki.org/Casing_design.

[48] Society of Petroleum Engineers. Perforating. June 29, 2015 [Online]. Available from: http://petrowiki.org/Perforating.

[49] Weatherford. Injection production packers. 2016 [Online]. Available from: http://www.weatherford.com/en/products-services/well-construction/zonal-isolation/inflatable-packers/injection-production-packers.

[50] Petroleum Services Association of Canada. 2015 Well costs study. March 30, 2015.

[51] Society of Petroleum Engineers. Borehole instability. June 26, 2015 [Online]. Available from: http://petrowiki.org/Borehole_instability.

[52] Bol G. Borehole stability in shales06. Society of Petroleum Engineers; 1994 [Online]. Available from: https://www.onepetro.org/download/journal-paper/SPE-24975-PA?id=journal-paper%2FSPE-24975-PA.

[53] Solberg SM. Improved drilling process through the determination of hardness and lithology boundaries. Norwegian University of Science and Technology; 2012.

[54] Fjelde KK, Vralstad T, Raksagati S, Moeinikia F, Saasen A. Plug and abandonment of offshore exploration wells. In: Offshore technology Conference, Houston (Texas, USA); 2013.

[55] Badiali M. 2 new drilling techniques that will shatter US oil expectations. 2014 [Online]. Available from: http://dailyreckoning.com/2-new-drilling-techniques-that-will-shatter-us-oil-expectations/.

[56] Messer AE. New technique both enhances oil recovery and sequesters carbon dioxide. 2015 [Online]. Available from: http://dailyreckoning.com/2-new-drilling-techniques-that-will-shatter-us-oil-expectations/.

CHAPTER FOUR

Shale Gas Production Technologies

1. INTRODUCTION

"The rise of shale gas is shaping up to be the biggest shift in energy in generations" [1].

Shale is a classification of sedimentary rock known as "mudstones." It is a rock formed from the compaction of clay-size mineral particles (mud) and silt. Shale is "fissile" and "laminated" meaning that it is comprised of many thin layers that readily split along the laminations. Shale gas is natural gas that is trapped within these shale formations.

Black shale contains organic material that can breakdown to form oil and natural gas. The black color is obtained through the tiny bits of organic matter trapped within the mud that the shale is formed from. Under the pressure and temperature of the rock formation, this material within the mud is turned into oil and natural gas.

Similarly to other formations, the hydrocarbons (due to low density) migrate upwards within the formation until trapped by a cap rock such as sandstone. If this happens, it creates a conventional reservoir that can be drilled to create a production well. However, many shale reservoirs are classified as unconventional. This essentially means the reservoir has low permeability and requires stimulation for production.

Shale gas has significant obstacles to overcome in terms of production. Since there is a quantity of gas trapped within pore spaces of rock, it is sometimes inaccessible. In recent years, the combination of horizontal drilling and hydraulic fracturing methods has allowed worldwide access to large volumes of shale gas that were previously uneconomical to produce. The production of unconventional shale gas reservoirs has created new life in the North American natural gas industry [2].

In Texas, drillers learned that fluid could be pumped into a production well with enough pressure to fracture pore spaces, and increase the permeability of the well. This was a revolutionary discovery in unconventional gas production.

In the years following this discovery, directional drilling also became a useful tool. A "pay zone" is the portion of reservoirs that will generate valuable hydrocarbons. Since these pay zones are typically much wider than they are in vertical length, it was desired to have the ability to control where the well was drilled. With the technology that allowed drillers to drill into a reservoir, and then turn the well 90 degrees, a much larger portion of the unconventional well was made accessible.

This combination of technology revolutionized the shale gas industry and made way for a significant number of large natural gas fields [3].

The growth of unusual shale gas manufacturing will boost a sense of energy safety in the main marketplace; however it also fetches a difficult set of confronts at worldwide and limited stages as discussed below:

- While changing the electricity production to natural gas from other foundations can have a severe impact on the climate.
- Utilization of large amounts of poisonous chemicals and water used in the hydraulic fracking process not only becomes the reason for contamination but also a source of threat to the drinking water.
- The massive use of chemicals, associated emissions, and truck traffic has considerable impacts on the atmosphere and ecology.
- A number of communal, civilizing, and financial costs for the local arises from the different issues.
- A lot of confronts are involved with in-commission companies relating to scale and the numerous operatives and service providers working in a single area, raising concerns for coordination and the expectation and organization of dangers, including mishaps and occupational health exposures.

2. PRODUCTION OF SHALE GAS

The process through which the financial formation of natural gas from shale reserves has been achieved. It has totally changed the energy market of the United States (US) in terms of shale gas and expanded the interest of shale gas usage in power generation and transportation. However, besides this, hydraulic fracturing has major impacts on the environment which have to be covered.

The permeability of shale is low as compared to conventional natural gas sources that limit gas and water flow. Natural gas is located in the distinct and unconnected pores of the shale play. Hydraulic fracturing is the process that allows the gas to flow via connecting of these pores. The steps include in shale gas production are as follows.

2.1 Construction of Road and Well Pad

The area required by a well that provides a steady base for a drilling rig; water storage tanks; loading and unloading areas for water trucks, piping, pumping and control trucks. Cleaning plus leveling of several acres of land is involved for preparation of a pad. The size of the well pad depends upon the distance downward of the well and amount of wells drilled.

2.2 Drilling

Shale gas reserves are found at different depths. Most reserves are situated at rock bottom of 6000 feet and can be comparatively thin. The gas extraction from a skinny film of rock requires horizontal drilling as shown in Fig. 4.1. This is achieved by drilling perpendicularly downward. Utilizing heavy-duty industrial drill bit, a typical well drilled in several stages starting with a large-diameter drill bit and then a small-diameter drill bit as the drilling is advanced. After drilling each portion of the well, a nested steel protective casing is inserted that will protect the groundwater and maintain the integrity of the well.

2.3 Casing and Perforating

During the drilling procedure, the drilling is closed several times and different diameters steel strengthened pipe covering is installed that not only secures the fresh water under the earth, but also maintains the integrity

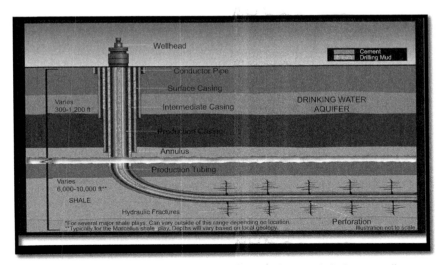

Figure 4.1 Horizontally drilled, hydraulically fractured shale gas well.

of the wellbore. This is also called surface casing. Cement is carried down into the empty space sandwiched between the covering and immediate mineral structure. This completely isolates the well from the deepest private or municipal water wells. A blowout preventer is installed after the surface casing has been cemented. The blowout preventer is a series of high-pressure safety valves and seals attached to the top of the casing to control well pressure and prevent surface releases. Next, a small drilling assembly is passed down through the surface casing. At the bottom of the casing, the bit drill continues its journey in the natural gas target area as deep as 8000 ft below the surface. The drilling method employed below the surface casing uses drilling mud which is a nonhazardous mixture. The drilling mud is used for the following purposes:
- Transporting the cuttings from the bottom of the hole to the surface;
- Cooling of the drill small piece and lubrication of the drill string;
- Supporting the walls of the hole to prevent it from caving;
- Exerting hydrostatic pressure to overbalance the pressure of the formation and thus prevent flow of formation fluids into the well;
- Ornamental drilling by jetting action through the bit nozzles.

A few hundred meters above the target shale, drilling assembly comes to a stop. The entire string is retracted to the surface to adjust the drilling assembly and install a special drilling tool. This tool allows the drill bit to gradually turn until the horizontal plan is reached (see Fig. 4.1). The remainder of the well is drilled in this horizontal plan while in contact with the gas-producing shale. Drilling continues horizontally through the shale at links greater than 4000 ft from the point where it entered the formation. Once drilling is completed, the equipment is retracted to the surface. Then the small-diameter casing called the production casing is installed throughout the total length of the well. The production casing is cemented and secured in place by pumping cement down through the end of the casing depending on the regional geological conditions. The cement is pumped around the outside casing wall to approximately 2500 ft above the producing shale formation or to the surface. The cement creates a seal to ensure that the formation fluids can only be produced within the production casing. After each layer of the casing is installed, the well is pressure-tested to ensure integrity for continued drilling. A cross-section of the well below the surface reveals many shielding layers; conductor covering, surface covering, drilling sludge, manufacturing casing, and then production tubing, through which the produced gas and water will flow. With seven layers of protection, horizontal drilling offers many advantages when compared to vertical drilling.

The covering adjacent to the horizontal part of the well through the shale configuration is then perforated using minute explosives to permit the stream of hydraulic fracturing fluids out of the well into the shale and the ultimate flow of natural gas out of the shale into the well. Since horizontal wells contact more of the gas-producing shale, few wells are needed to optimally develop a gas field. Multiple wells can be drilled from the same pad side, e.g., development of a 1280-acre track of land using conventional vertical techniques could require as many as 32 vertical wells with each having its own pad side. However, one multiwall pad side with a horizontal well can affectively recover the same natural gas reserves from the 1280-acre track of land. That reduces the overall surface disturbance by 90%.

2.4 Hydraulic Fracturing and Completion

Even the well covering is perforated, only a small amount of natural gas will flow liberally into the well from the shale. The network of the fracture must be shaped within the shale deposits which permit the gas to run away from the unconnected spots and the natural fractures where the gas is fascinated within the rock. In this process, the hydraulic fluid is used in a huge quantity, characteristically millions of gallons that contain a large amount of water and cement which is pumped at a high force in to the well. The composition of the well is clearly shown in Fig. 4.2.

The other remaining fracking fluid is composed of different chemicals that are used for different purposes. Some of the chemicals along with their

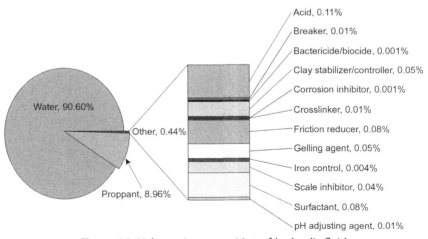

Figure 4.2 Volumetric composition of hydraulic fluid.

types and purposes are as shown in Table 4.1. These chemicals not only increase the property of the fluid but also prevent the growth of organisms and blockage of the shale fracture, decay and scale preventers that protects the reliability of the well. The perforation gun is inserted in the well and an electric spark is passed through the gun via a connecting wire, which creates a hole through the casing, cement, and the targeted shale rock. These perforated holes connect the reservoir and the wellbore.

The gun which is perforated is removed and grounded for the further step. The hydraulic fluid, as discussed above, under controlled conditions (see Table 4.1) is fed into deep underground reservoir formations. This helps

Table 4.1 Volumetric Composition of Hydraulic Fracturing Fluid

Product Category	Main Ingredient	Purpose	Other Common Uses
Water	Approximately 99.5% water and sand	Expand fracture and deliver sand	Landscapping and manufacturing
Sand		Allows the fractures to remain open so the gas can escape	Drinking water filtration, play sand, concrete, and brick mortar
Other	Approximately 0.5%		
Acid	Hydrochloric acid or muriatic acid	Helps dissolve minerals and initiates cracks in the rock	Swimming pool chemical and cleaner
Antibacterial agent	Glutaraldehyde	Eliminates bacteria in the water that produces corrosive byproducts	Disinfection, sterilizer for medical and dental equipment
Breaker	Ammonium persulfate	Allows a delayed breakdown of the gel	Used in hair coloring, as a disinfectant, and in the manufacture of common household plastics
Corrosion inhibitor	N,N-dimethyl formamide	Prevents the corrosion of the pipe	Used in pharmaceuticals, acrylic fibers and plastics
Crosslinker	Borate salts	Maintains fluid viscosity as temperature increases	Used in laundry detergents, hand soaps, and cosmetics

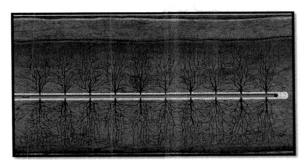

Figure 4.3 A view of a single perforated region.

to improve the performance of the stimulation. This stimulation fluid is pumped in under high pressure and out through the perforations that were noted earlier.

This process forms fractures in gas and oil reservoir rock as depicted in Fig. 4.3. The sand in the fracture fluid remains in the fracture of the rock and keeps them open when the pump pressure is released. This permits the last intent oil and natural gas to pass to the well hole. This first drive section is then isolated with a particularly calculated plug and the perforated guns are used to perforate the next stage.

2.5 Production, Abandonment, and Reclamation

During the production process, the recoverable gas is collected in a production well network through small-diameter gathering pipelines. The production lifetime of shale gas wells is not fully developed because of the initial stages of its development. It is generally observed that the shale gas wells experience quicker decline than conventional natural gas production. It has been estimated that almost half of the well's lifetime production, or estimate recovery, occurs within the first 5 years in north—central Arkansas. If the well is unable to give its production at an economical rate, then the wellhead is removed from the location and the wellbore is filled with cement to avoid the leakage of gas into the air. The surface is reclaimed, and the site is abandoned to the land's surface holder.

3. ROCK PROPERTIES

Shale is a broad term to describe various rock compositions, such as clay, quartz, and feldspar, all possessing the same physical characteristics. Shale, otherwise known as mudstones, has exceptionally fine-grained particles, with a diameter of less than 4 μm, but may contain silt-sized particles (up to 62.5 μm) [4]. Although the composition of shale varies it is difficult to determine the change with the naked eye due to the small size of these particles.

Shale gas reservoirs are known as "unconventional reservoirs," this means that in order to achieve a profitable production rate the reservoir will require additional stimulation [5]. Permeability is defined as the ease of flow within a reservoir, therefore low permeability means that the gas does not flow easily within the reservoir. Shale gas reservoirs have low permeability, therefore they will require stimulation, such as fracturing, to ensure a good flow of gas [6].

Porosity is the amount of free space within the reservoir rock meaning that it can contain the organic-rich material (hydrocarbons) or water. Since shale has such a small diameter it is difficult to determine the porosity of these reservoirs. These small pores can hold surface water from the capillary forces [4]. Due to the difficulty of measuring porosity in these reservoirs there is a recommendation to consider bulk volume gas (BVG) as opposed to porosity [4].

3.1 Effect on Production

Due to the many different compositions that make up shale, the reservoir properties change based on the type of shale within the reservoir. For example, if they contain 50% quartz or carbonate the rock tends to be more brittle and responds well to well stimulation [4].

Barnett and Eagleford Shale are the two largest plays found in Texas [7]. In Fig. 4.4, the compositions of the shale found at both these plays, as well as clay-rich gas-bearing mudrock, are compared in a tertiary diagram. This diagram is important because it shows how much the composition of shale varies; when comparing only three plays the entire tertiary diagram is almost completely covered. A significant observation noted in the journal article "From Oil-Prone Source Rock to Gas-Producing Shale Reservoir—Geologic and Petrophysical Characterization of Unconventional Shale Gas Reservoirs," [4] is that the current plays in production tend to stay below the 50% total clay line. This is due to the better stimulation that can be achieved when quartz or total carbonate is 50% and better stimulation directly affects the production rate of the reservoir.

Shale Gas Production Technologies

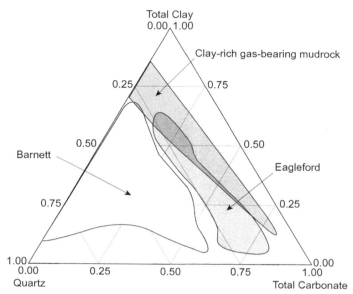

Figure 4.4 Mineral composition in various reservoirs [4].

4. PRODUCTION METHODS

For unconventional reservoirs such as shale gas, production is not as simple as drilling a hole and waiting for the rise of petroleum. These reservoirs require careful and precise manipulation in order to successfully extract the desired amount of fluid. Fig. 4.5 shows the typical shale production layout. As seen here, the well is vertically drilled until it is near the depth of the reservoir, and then the trajectory is manipulated such that the well will travel horizontally along the length of the pay zone. Hydraulic fracturing will then take place, creating many branches, increasing permeability, and allowing for linear gas flow into the production well. This is a revolutionary technology in unconventional reservoir production. Shale production methods involve drilling deep wells; swallow wells cannot achieve the desired production flow rate [8].

4.1 Artificial Well Stimulation

In shale reservoirs, the natural formation permeability is inadequate and does not provide sufficient petroleum production. The production rate would

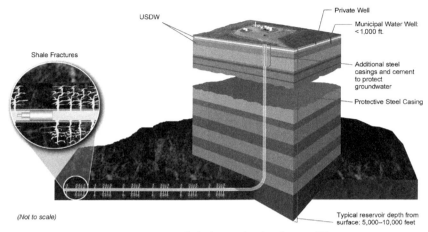

Figure 4.5 Typical shale production layout [9].

not be high enough to recover the capital investment required to drill and complete wells [10].

In this case, there are methods of artificial well stimulation that are widely used to increase the permeability of a reservoir. There are two main types of stimulation, hydraulic fracturing and matrix acidizing. When a formation has an average effective permeability of 1 mD or less, the requirement for stimulation is justified.

In conventional petroleum reservoirs, the flow toward the well bore is mainly radial. Low permeability, as discussed above, would be a major constraint in fluid flow through the porous medium. This is when hydraulic fracturing comes into play. A highly conductive fracture is placed along the length of the gas reservoir, changing the flow from radial to the wellbore to linearly along the fracture [10]. This is illustrated in Fig. 4.6.

Hydraulic fracturing within the tight reservoir is most commonly horizontal. Starting from the end of the wellbore, a hydraulic fluid is pumped into the reservoir until enough pressure is achieved to initiate and propagate a formation fracture. A "proppant" is normally placed in the fracturing fluid to help keep the fracture aperture open after the pressure of fracturing is released [10].

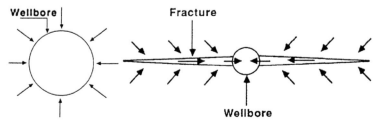

Figure 4.6 Difference in flow to fractured and unfractured wells [10].

The placement of this proppant is governed by the properties of the fracturing fluid. Some fluids carry the proppant along the length of the fracture, whereas some allow the proppant to drop through the fluid after the fluid has transitioned down the well [10]. An example of a proppant used in a reservoir is sand.

There are two main types of fracturing fluids, namely slicks and gels. Slicks have a dynamic viscosity less than water and can better flow back from fractures. Gels have a dynamic viscosity greater than water and for this reason can carry the proppant further into the fractures [10]. A recent technological innovation is the combination of these fluids so that the gels can carry proppant deep into the well, enzymes are injected to break the gel down, and then it becomes slicks for better production.

Surfactants are a type of chemical that is used to modify the surface chemistry of the fracture fluid [10]. This is sometimes required as some water-based fracture fluid can lock hydrophobic rock surfaces and interfere with overall flow. For this reason, some natural gas liquids can be used. This is advantageous as there is chemical compatibility with the hydrocarbons and the fluid will be released as a gas during the production phase.

The pressure at the well surface is built-up until the breakthrough of the formation is developed and a fracture occurs. Fig. 4.7 is an example of a pressure—time curve of the fracturing process. After the peak, the decline in pressure represents where the proppant is being added to the well.

As the high injection rate of fluid is continued, the fracture progresses along the well becoming proportionally wider closer to the well. A typical well fracture requires 500—1000 m^3 of fluid. The fracturing process is closely monitored commonly using "microseismic monitoring," which is tracking fracture initiation and growth [10]. Potential spills of fracturing fluid must be minimized, and if there is a spill it must be monitored and contained to ensure there is no environmental damage.

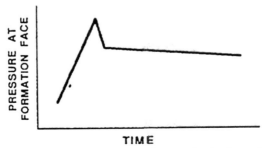

Figure 4.7 Pressure versus time plot for hydraulic fracturing [10].

Although it is not common for shale reservoirs, matrix acidizing is a common way to improve permeability in a well [10]. The acid is injected into the reservoir therefore enlarging pore spaces and increasing the ability of fluid flow. This method is commonly used to fix well damage that has occurred due to the drilling and completion phases of a petroleum site. A preferred well stimulation method used is a combination of both acidizing and fracturing, whereas an acid such as hydrogen chloride is used as the fracturing fluid. However, this is for formations with significant solubility to acid, which is more likely to be sandstone than shale [10].

4.2 Directional Drilling

Most underground wells that are drilled for water, oil, or natural gas are vertical. This means the length of the well is the true vertical depth. The necessity to be able to control where a well is drilled was determined, and directional drilling was established. Directional drilling is a technological breakthrough that has revolutionized shale gas production.

There are many applications of directional drilling, with the applications pertaining to petroleum production being [11]:

- *Hit targets that cannot be reached by vertical drilling*: this allows reservoirs under structures to be accessed without interfering with the location of the structure itself;
- *Drain a significant area from a single location*: In reducing the surface footprint of petroleum production, multiple wells can be drilled from one location, allowing for a neat arrangement of well pads and ease of product gathering;
- *Increase the length of the* "pay zone": This is an important concept in shale gas production. By drilling through horizontally orientated reservoirs, the length of the well that is in the reservoir can be drastically increased. This is especially significant in hydraulic fracturing whereas the fracture can be spread along the length of the pay zone (area where petroleum is present);
- *Seal/relieve an existing well*: If there are problems with an existing well, another well can be drilled into it to either relieve some of the pressure to the surface or seal the existing well;
- *Nonexcavating areas*: In areas where the ground cannot be dug up, similar to drilling under cities, utilities can be installed using directional drilling.

As mentioned above, the most important benefit of directional drilling has been in shale gas plays. The combination of hydraulic fracturing and directional drilling is vital in the process of creating more pore spaces in a shale formation, thereby increasing permeability and fluid flow through the porous medium.

In areas where a vertical well would only produce a small amount of gas, these methods have created a vast network of production wells that are sustaining the shale gas industry.

With the horizontal well through the reservoir rock, the fracture is extended for a significant length, propped open with sand as discussed earlier, and then completed and produced [11].

5. DRILLING METHODOLOGY

Directional drilling commences at the surface as a vertical well. This drilling will commence until the drill front is approximately 100 m above the target. At this point, there is a hydraulic motor attached between the drill pipe and the drill bit. This motor can alter the direction of the drill bit without affecting the pipe that leads up to the surface. Furthermore, once the well is being drilled at a certain angle, many additional instruments are placed down the hole to help navigate and determine where the drill bit should go. This information is then communicated to the surface and then to the motor, which will control the direction of the bit.

Combining horizontal drilling with hydraulic fracturing can cause the drilling and completion stages of a well to cost upwards of three times as much as a vertical well. However, with the improved production rate and total recovery, this capital cost is easily justified with revenue. As stated earlier, many wells today would not be accessible or feasible without these methods [11]. Fig. 4.8 illustrates the concept of directional drilling.

6. OPTIMIZATION

With the recent coming of shale gas production, the optimization of this production is ongoing. New drilling methods have allowed for well configurations that improve the recovery rate and overall recovery of a gas reservoir. By having a well pad centralized around a number of wells, shale gas production is more organized, efficient, and opens doors for a number of possibilities.

Artificial lift is most commonly used in vertical wells. This can be with the use of an electronic submersible pump (ESP), or by using water or gas injection. In oil reservoirs, water and gas injection (or a combination of both) is

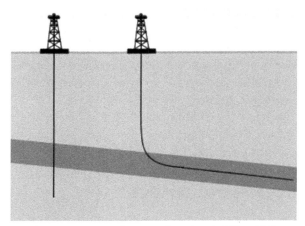

Figure 4.8 Vertical versus directional drilling [11].

used to maintain reservoir pressure and improve the overall recovery flow of petroleum products. However, in a horizontal shale gas reservoir, the implementation of pumps faces the challenge that it must bend and travel along a horizontal trajectory with the possibility of the liquid build up being in several places, and the overall concept of having the pump operating on its side.

A new development in subsurface pump placements has enabled engineers to reach these directional wells. By removing the liquid from a shale gas reservoir, the overall flow of gas out of the well is improved.

This enhanced recovery combined with the well fracturing has enabled gas production rates to reach full potential. Optimization of the process will always be performed.

7. LIMITATIONS

A major limitation in the production of gas shale is the impact it has on the environment and public health. With the majority of shale gas reservoirs requiring the use of hydraulic fracturing, its effects have been a topic for countless studies.

Hydraulic fracturing brings with it many concerns, such as the potential to affect air quality, contaminate groundwater/drinking water, the migration of gases/fracture fluids to the surface, and the improper handling of wastewater. The most significant concern with hydraulic fracturing is the fluid used. This fluid contains many harmful chemicals, which if released into the environment can harm not only numerous species but the public as well.

Whether it is onshore or offshore, the environmental and societal implications remain largely the same. Many hydraulic fracturing projects require approval of society before they are commenced; therefore this limits the production of shale gas in many areas. For example, shale gas production in Nova Scotia, which is in Atlantic Canada, was stopped in 2011 because the provincial government would not allow hydraulic fracturing [8].

Another limitation for the production of natural gas is the nature of the fluid. Hydrogen sulfide is a harmful gas that is commonly present in many gas reservoirs. When this gas is present, equipment and piping must be designed in order to handle it, along with the processing facilities or the market that will buy the gas, making it economically feasible, especially when it is already in a remote location.

8. GAS LIQUID SEPARATION

Gas liquid separation is often referred to as the rising of gas vapors and settling of liquid droplets in a vessel where the liquid exits from the bottom and the gas exits from the top [12]. The equation for terminal velocity of liquid droplets explains the separation of liquid from the vapor phase.

$$V_t = \sqrt{\frac{4gD_p(\rho_L - \rho_G)}{3\rho_G C'}}$$

where V_t is terminal velocity so that a liquid drop can settle out of the gas, g is the force of gravity, D_p is the droplet diameter, ρ_L and ρ_G are liquid and gas density respectively, and C' is the drag coefficient.

The separation of gas and liquid is an extremely important process in the oil and gas industry. In order for operations to run smoothly, gas and liquid separation must occur. The separation is vital in protecting processing equipment and achieving product specifications [13]. If the liquid contaminants are properly removed from the gas stream, the amount of downtime and maintenance required for process equipment, such as compressor and turbines, is decreased significantly [14].

There are different methods used to achieve gas liquid separation. However, one of the most commonly used technologies is a gravity separator, where the force of gravity and difference in densities control the separation [14].

8.1 Separator Design

A gravity separator used for gas liquid separation is commonly referred to as a knock-out drum and is typically orientated vertically. These types of

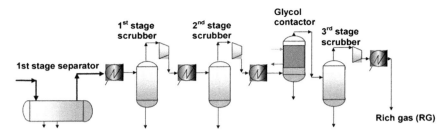

Figure 4.9 Separation stages [13].

separators are frequently used as the first-stage scrubber for gas liquid separation because they are more efficient with bulk or larger liquid particles. These separators are generally used when the internals are required to have minimum complexity [14]. The vessel is designed based on the inlet flow and desired separation efficiency. It is common for separators to be designed in stages to achieve the required separation as demonstrated in Fig. 4.9 where "scrubbers" represent the knock-out vessels [15].

The gravity separator is able to achieve better separation when the gas velocity is low or the vessel is large [14]. Some selection criteria for gas liquid separators include duty, orientation, and components [16]. They are often designed based on the required residence time for the liquid phase to coalesce out of the gas but sized with the capability of operating within a large range of flows. They are commonly designed in functional zones known as, inlet, flow distribution, gravity separation, and outlet [15]. Some of the major factors considered for separator design are required performance, overall flow, fluid composition, required emulsion, retention time, and alarms. These parameters are used to conduct theoretical calculations to estimate vessel volume. The diameter is sized from the cross-sectional area calculated from the settling theory of each phase [17]. The settling theory involves varying equations and calculations depending on the phase and vessel orientation. When designing a gas liquid separator, many parameters must be considered as well as the desired application to ensure that proper performance can be achieved.

8.2 Material Selection

Material selection is another consideration that must be made prior to fabricating a gas liquid separator. The material choice depends significantly on the operating conditions, the composition of components in the fluid, and the surrounding environment. Selecting the proper material for the manufacturing of process equipment is really important in the prevention of material failure. Failures in material, such as brittle fracture and mechanical fatigue, can lead to extreme and catastrophic incidents where equipment can be damage and personnel safety can be compromised [18].

The properties of a material are considered to ensure that it can withstand the conditions of the process without failure. Some properties that are taken into account as per routine are tensile and yield strength, ductility, toughness, and hardness. A material's chemical and physical properties are also used to determine how corrosion-resistant it is [18]. Corrosion is a major cause of material failure.

Typical materials selected for gas liquid separators include carbon steel, 304 stainless steel, and 316 stainless steel, where the process environment is not extremely corrosive. In a corrosive environment (like offshore where there are chlorides present from the seawater) or when certain chemicals are used in the process, a more robust material, such as high-alloyed stainless steel or Hastelloy, a nickel alloy, should be selected [19].

9. CORROSION ISSUES

Corrosion is commonly known as degradation that occurs in material due to process fluid or chemicals. Corrosion can occur over time or without any warning (such as chloride stress corrosion cracking) and therefore selecting a corrosion-resistant material is extremely important [20]. Different examples of corrosion cracking can be seen in Fig. 4.10. Corrosive environments include alkaline environments, chlorine, water, etc. Some examples of corrosion materials include carbon dioxide and hydrogen sulfide [21]. Corrosion reactions are determined so that they can be mitigated appropriately. Some alloys are treated to provide increased corrosion resistance. Heat

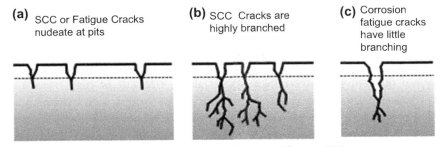

Figure 4.10 Corrosion cracking and fatigue [20].

treatment such as annealing and quenching are used to offer stress relief and increase the ductility of the material. The hot working increases a material's resistance to corrosion [18].

10. TRANSPORTATION AND STORAGE

After the production and refinement of the natural gas it is important to consider the transportation and storage of these combustible materials. This section discusses the most common ways to transport and store natural gas.

10.1 Transportation

Once shale gas is produced from the reservoir, another phase in the overall process begins at the wellhead. Natural gas is largely moved from place to place by means of a pipeline. In other cases, such as offshore gas production, it may be desired to have the gas liquefied upon production. A combination of extreme pressurization and cooling can accomplish this [22].

Most commonly on land, it is feasible to have pipelines built to accommodate the transport of natural gas from the production site to the location of high demand. There is a vast network of pipelines spanning North America, which provide highways for natural gas all over the continent [22].

The types of gas pipelines in the US include gathering, interstate/intrastate, and distribution. Gathering, as specified by the name will route natural gas to the processing facility that is responsible for processes such as dehydration, sweetening, condensate removal, etc. Gathering pipelines may require additional service such as a specialized sour-handling pipe or a pigging system [22]. A pigging system may be required to push heavier hydrocarbons through a syncline in a pipe for example (i.e., if it was traveling through a valley). Here a plug-like tool is made to the same diameter as the pipe, which will travel through the low-lying area being pushed by the pressure of the gas behind it and pushing with it anything that may have built up in the line [22].

Interstate/intrastate pipelines are the longest section of pipe, which are required for transportation to the area of high demand (urban centers). Required here, and potentially required on gathering pipelines, are compressor stations [22]. Since the natural gas is flowing by the pressure gradient, the natural drop in pressure, which occurs along a length of pipe, must be compensated for. Compressor stations are a type of "pit-stop" for natural gas that will compress it using a turbine, engine, or motor and provide a boost to the fluid to keep it moving along. They are typically located at anywhere between 40- and 100-mile (64—160 km) intervals along a pipeline [22].

Distribution piping will provide natural gas to consumers, the final step in the process.

10.2 Storage

If required, natural gas can be stored for periods of time prior to distribution piping. This can be attributed to the fact that natural gas is a seasonal fuel, as more will be consumed in the winter months [23]. This has been the traditional outlook, however, natural gas is now being used as an electricity fuel so it has become used more frequently in the summer months as well.

Nonetheless, there is a requirement for storage of this light hydrocarbon mixture. The storage facilities are located near markets that are not in close proximity to producing facilities. In the early 1950s, seasonal demand became evident and this led to the development of underground storage facilities.

There are three main types of underground storage facilities; depleted reservoir, aquifer, and salt cavern. These storage facilities are reconditioned underground, and natural gas is injected, building the pressure and essentially creating a storage vessel.

Depleted reservoirs are natural gas reservoirs that have already reached the end of their lives. With the natural gas removed from the ground, a storage vessel with the properties that make it capable of holding natural gas (evidently) is created. In 1915, the first underground storage facility was created [23]. With a depleted reservoir determined to have the appropriate porosity and permeability, it can become a storage vessel. There is a portion of this gas that is physically unrecoverable. With a depleted reservoir, the injection of this gas is not necessary as cushion gas because it is already present in the reservoir.

Aquifers are underground, permeable, and porous rock formations holding water. A less desired method of underground storage is in aquifers. The development of an aquifer is much more cost-intensive as the water must be drained and the reservoir conditioned prior to injection of gas. Also, the formation properties must be thoroughly tested to ensure that it is capable of handling the fluid. With these types of storage facilities, aboveground processing utilities must be put into place. These types of storage methods will only be used where there is no close access to depleted reservoirs [23].

Similarly to an aquifer, salt caverns may be used to store natural gas. These will require similar preconditioning methods, but the physical properties of the formation will be much more suitable to hold natural gas. However, these caverns are much smaller and much less common [23].

11. MATHEMATICAL FORMULAS FOR TRANSPORT FLOW

Mathematical formulas and equations are frequently used to model reservoirs and describe the flow conditions of the reservoir fluid. These equations are a result of mass, momentum, and energy conservation laws. Darcy's law is derived from the momentum equation and demonstrates that the relationship between the velocity of the fluid and the pressure gradient is linear [24].

Shale gas is a compressible fluid and has specific mathematical equations for transport flow depending on the flow type (linear or radial). When the flow conditions within a reservoir are at steady state and Darcy's law is valid, as the flow continues along the path the mass flow rate is constant. Pressure and density decrease as the flow proceeds while volumetric gas flow rate increases [25].

11.1 Linear Flow

Assuming that shale gas is a nonideal gas, the density for linear flow can be denoted as:

$$\rho = \frac{P\,M_w}{ZRT}$$

where P is pressure and M_w is molecular weight, Z is compressibility factor, R is the gas constant, and T is temperature. From this, for all x between 0 and L (see Fig. 4.11), the density multiplied by the volumetric flow rate (ρq) is constant and when the cross-sectional area is constant then so is the density multiplied by the velocity (ρv) also constant [25].

Considering that at steady-state conditions $\rho q = \rho_1 q_1 = \rho_2 q_2$ and assuming molecular weight, compressibility, viscosity, and temperature are constant, the substitution of the density and Darcy's flow rate equations generates the following [25]:

$$P_2 q_2 = -P\frac{K\,A}{\mu}\frac{dp}{dx}$$

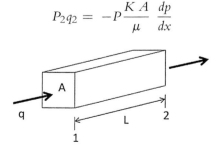

Figure 4.11 Linear, single-phase, compressible flow [25].

Then integrating from P_1 to P_2 and $x = 0$ to $x = L$, the following is obtained:

$$P_2 q_2 = \frac{KA}{\mu} \frac{P_1^2 - P_2^2}{2L}$$

The equation can be rearranged and $P_m = (P_1 - P_2)/2$ can be substituted to produce the final transport flow equation for linear, single-phase shale gas [25]:

$$q_2 = \frac{KA}{\mu} \frac{P_m(P_1 - P_2)}{LP_2}$$

This equation can be used to mathematically model reservoir flow and predict its behavior.

In addition to the formula above, the flow rate can also be related to standard or surface conditions using the following equation:

$$q_2 \frac{P_2}{Z_2 T_2} = q_{sc} \frac{P_{sc}}{T_{sc}}$$

where P_{sc} is equal to 1 atm and T_{sc} is $60°F$.

11.2 Radial Flow

Again, assuming that the gas is ideal, the density can be denoted the using the same equation for radial flow. Fig. 4.12 demonstrates radial flow within a reservoir where the subscripts w and e represent well and drainage respectively.

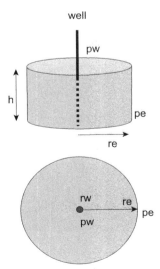

Figure 4.12 Radial, single-phase, compressible flow [25].

Similar to linear flow, steps can be followed to derive a transport equation for radial flow. Under radial conditions, q is equal to velocity multiplied by area, where area is equal to $2\pi rh$. The following equation represents the transport equation for radial, single-phase, compressible flow of shale gas:

$$q_w = \frac{\pi K h}{P_w \mu \ln\left(\frac{r_e}{r_w}\right)} \left(P_e^2 - P_w^2\right)$$

Like when the flow is linear, radial flow can also be related to the flow rate at standard conditions. The equation below is very similar to the liner flow relation except in this formula, the subscript is w and stands for well.

$$q_w \frac{P_w}{Z_w T_w} = q_{sc} \frac{P_{sc}}{T_{sc}}$$

12. SHALE GAS PRODUCTION

Shale gas is one of the many reservoirs that are made up of natural gas. Natural gas is used for many commodities in day-to-day life such as residential use and electric power [26] (see Fig. 4.13 for the historical usage of natural gas in the US). Originally this unconventional reservoir was not an economically feasible way to produce this resourceful gas. In recent years there have been advancements in technology that have allowed for shale gas production to become a profitable endeavor [8].

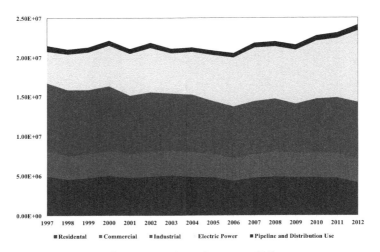

Figure 4.13 Natural gas uses [26].

13. RESEARCH AND DEVELOPMENT

Shale gas can be found in shallow or deep wells. The US has been utilizing shallow wells (1000—3000 ft) for many years; these reservoirs were produced using vertical wells and minimal fracturing, thus making them economically feasible [27]. In recent years, the US has developed a new technology that allows deep wells to be effectively stimulated, thus creating an enhanced flow of products. This technology involves the combination of hydraulic fracturing and horizontal drilling [8]. The Barnett play, the first deep shale play, located in the Forth Worth Basin (North Texas) led the pathway to this new technology. Barnett introduced the use of horizontal wells with hydraulic fracturing in 2002 [27]. The first deep well drilled by Barnett was approximately 7000—8000 ft deep [27].

13.1 Current Status

This new technology has allowed the US to become the main producer of shale gas reservoirs, even though it does not possess the largest technically recoverable shale gas reserves. The top five countries with the greatest amount of technically recoverable shale gas are China, Argentina, Algeria, the US, and Canada [8]. As per Fig. 4.14, in 2014, North America was the primary source of shale gas. Over half of the natural gas produced by the US and approximately 15% of Canada's production was shale-based. The only other country producing shale gas at this time was China, with less than 1% of its natural gas production being from shale [28].

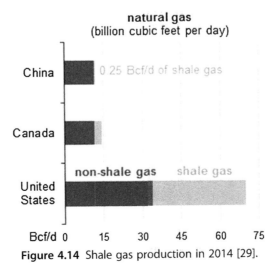

Figure 4.14 Shale gas production in 2014 [29].

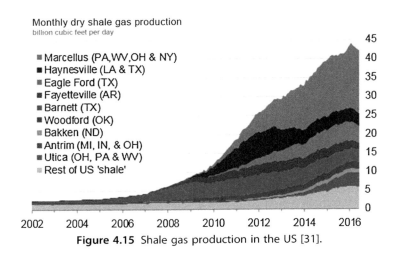

Figure 4.15 Shale gas production in the US [31].

Within the US there are many different companies that are producing shale gas (see Fig. 4.15 for the comparisons). Marcellus, which is located in Pennsylvania and West Virginia, is the largest producer at approximately 16.28 billion cubic feet (bcf) per day in 2015 [30]. This play expanded its production by 10.3 bcf per day in just over 3 years; this is an increase of approximately 162% [30].

However, with the constant increase in production the North American market continues to be overflooded with natural gas. Due to the immense amount of natural gas available the price naturally drops. For many years the lowering price did not affect the production rate of shale gas, however it was seen in September 2015 that the production of shale gas began to slow [32].

14. FUTURE PROSPECTS

Although the production of shale gas in North America has slowed, the future of shale gas production is still promising. Before shale gas became an economically feasible resource, Canada and the US were beginning to become worried about the supply of natural gas. Conventional reservoirs were starting to produce less, thus there was no foreseen future for natural gas exploration. "Unconventional gas production is forecast to increase from 42% of total US gas production in 2007 to 64% in 2020" [33].

Fig. 4.16 shows the number of wells, both oil and gas, that have been drilled within the US with respect to region by 2010. It sums up to a total of more than four million wells with approximately half being stimulated by

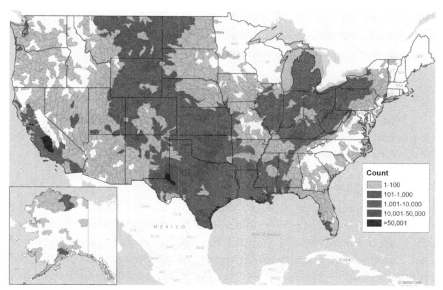

Figure 4.16 Oil and gas wells drilled in US [9].

utilizing hydraulic fracturing. In recent years 95% of wells drilled have been stimulated using hydraulic fracturing [9]. With this being said, as the technology advances the production systems advance as well, thus there is no foreseeable future that does not involve shale gas production.

No country has been able to reciprocate the success that the US has achieved with this new technology [34]. Some countries, such as Mexico, have considered utilizing this advanced software, however there are always challenges that must be faced. In 2015 a preliminary study was performed on the shale gas that lies in Mexico. Within the study it was found that most of the recoverable shale oil and gas is located in the underdeveloped area of Mexico. This poses a challenge because in order to produce the shale oil and gas basic structures would have to be built before production structures can begin to be constructed [34].

China, the country with the largest recoverable shale gas reservoirs, has also tried to mimic the success that the US has achieved with this new production method. However many challenges have been faced over the past few years. The two main challenges are economics and geology. The current economy in China is overflooded with natural gas, therefore production of shale gas has slowed since conventional reservoirs are cheaper and obtain the same product. Shell, one of the investors in a field in China, has recently pulled out because of the "challenging geology and mixed drilling results" [35]. These factors have forced China to change the desired expected production of shale gas in 2020 to a third of the original estimate [35].

15. ECONOMIC CONSIDERATIONS

The economic considerations for this chapter entail the current market pricing, and breakeven gas price. These two factors are very important for the future of shale gas production.

15.1 Pricing of Natural Gas

The recent evolution of shale gas production has the ability to affect the price of natural gas. Shale gas production was always considered an unattainable resource due to the low permeability of the reservoirs, thus it has never been considered as a natural gas resource for each respective country. For example, Canada's natural gas production peaked in 2006 after which it then began to decline [8]. The decline in natural gas production was expected to continue until the new technology, discussed above, was developed. The new technology offered an opportunity for Canada to continue to produce natural gas in an economy that, without this source of natural gas, would require the import of liquefied natural gas.

With this increase in the availability of natural gas the price has been drastically affected. From 2008 to 2015 the price of natural gas has been reduced by 75% [8], this trend can be seen in Fig. 4.17. This has both positive and

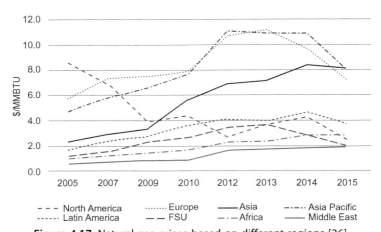

Figure 4.17 Natural gas prices based on different regions [36].

negative repercussions for the Canadian economy. The residents of Canada saved an immense amount of money due to the low gas prices, however the low gas prices have also caused the Canadian market to drill less wells since the profit is not as large. Although fewer wells are currently being drilled, the opportunity to export overseas and turn a higher profit may someday be utilized [8]. Fig. 4.17 also shows that Asia and Europe have much higher gas prices than North America; this can be attributed to the low availability of natural gas in these regions.

This drastic difference in pricing shows the availability of shale gas in each respective region. North America is currently experiencing an overflow of natural gas availability, thus the price has dropped immensely. Economic analysts, such as Art Berman, do not believe this is a sustainable economy. For many years the drop in price has not affected the production increases in shale gas, until September 2015 [32]. In September 2015 there was a slight drop in production rate (-1.2 bcf per day), although this was only a small drop it is significant because it was the first decrease in production in a while.

15.2 Breakeven Gas Price for Marcellus

As per Art Berman [32], the increase in the price of natural gas is inevitable. The companies that are producing natural gas from unconventional reservoirs cannot afford to produce gas at these low prices. Table 4.2 shows the breakeven oil price for the largest shale play in the US, Marcellus.

"Spot prices at most market locations rose this report week (Wednesday, July 20, to Wednesday, July 27). The Henry Hub spot price

Table 4.2 Marcellus Breakeven Gas Price

Marcellus	Wells	EUR (bcf)	B/E Price
Anadarko	241	6.17	$4.25
Cabot	280	9.36	$3.42
Chesapeake	575	7.20	$3.91
Chevron	199	4.93	$4.89
EQT	220	9.42	$3.41
Range	643	3.85	$5.75
Shell	305	3.20	$6.56
Southwestern	238	5.81	$4.73
Tailsman	354	4.31	$5.33
Average for COG, CHK, and EQT.			$3.58

CHK, Chesapeake.
Drilling Info and Labyrinth Consulting Services, Inc.

rose 8¢ from $2.72 per million British thermal units (MMBtu) last Wednesday to $2.80/MMBtu yesterday" [31]. This price is still below the current breakeven gas price for Marcellus, therefore with the current price of natural gas the largest play in the US is not making any money; in fact it is losing money (see Table 4.2).

For the North American economy this is a great concern. If the price of natural gas does not begin to rise soon the shale gas plays will have to produce less. If the production of shale gas slows they will not be offsetting the reduction of conventional gas production thus a depletion of the natural gas resource will begin. Thus the rise of natural gas prices is one that must occur for the economy to continue to thrive.

16. CONCLUSIONS

The ability to develop and economically produce from shale gas reservoirs has breathed new life into the natural gas industry. With the rise of hydraulic fracturing and directional drilling technologies, countless reservoirs that were once deemed as not economically feasible are now being developed and produced.

Shale formations are considered unconventional reservoirs due to the inability of fluid to flow through them. Porosity is the measure of the pore spaces within a rock, and permeability is a property that describes the interconnection of these pore spaces. Even though shale has reasonable porosity, the properties of this sedimentary formation cause it to have a low permeability value, classifying it as an unconventional reservoir.

With the first hydraulic fracturing occurrence taking place in the late 1940s, this technology has been further developed and perfected into the safer, more environmentally friendly method that is used today. With ongoing studies into the environmental effects of hydraulic fracturing, there are still many opinions on it, with some implying that it should not be performed at all.

The first implementation of the directional drilling was performed in the 1920s, however the real potential of this technology was further developed many decades after this. Directional drilling became a powerful tool in oil and gas production, and pertaining to shale reservoirs it was combined with formation fracturing to create vast opportunities for production.

Shale gas production methods, combined with enhanced recovery technology, state-of-the-art processing facilities and gas storage and transportation methods create an economically viable industry that has development prospects for many years into the future. The North American natural gas industry has provided a reliable domestic economic boost and this will continue for many years.

REFERENCES

[1] Zakaria F. The shale gas revolutional, Toronto Star; 30.03.2012.
[2] Geology.com. Shale. [Online]. Available: http://geology.com/rocks/shale.shtml; 01.01.2016.
[3] Administration EI Energy In Brief. [Online]. Available: http://geology.com/energy/shale-gas/; 01.01.2010.
[4] Passey QR, Bohacs K, Esch WL, Klimentidis R, Sinha S. From oil-prone source rock to gas-producing shale reservoir - geologic and petrophysical characterization of unconventional shale gas reservoirs. In: International oil and gas conference and exhibition in China, Beijing; 2010.
[5] Halliburton. Unconventional resevoir wells. 2016 [Online]. Available: http://www.halliburton.com/en-US/ps/cementing/cementing-solutions/unconventional-reservoir-wells/default.page?node-id=hfqela4g.
[6] Cipolla CL, Lolon EP, Erdle JC, Rubin B. Resevoir modelling in shale-gas Resevoirs. SPE Resevoir Evaluation and Engineering 2010;13(04):638—53.
[7] Courthouse Direct. What you need to know about Eagle Ford and Barnett shale. Courthousedirect.com Inc; July 29, 2013 [Online]. Available: http://info.courthousedirect.com/blog/bid/314206/What-You-Need-to-Know-About-Eagle-Ford-and-Barnett-Shale.
[8] Parliment of Canada. Shale gas in Canada. January 30, 2014 [Online]. Available: http://www.lop.parl.gc.ca/content/lop/ResearchPublications/2014-08-e.htm#a8.
[9] U. S. D. o. Energy. Natural Gas from Shale: Questions and Answers. [Online]. Available: http://energy.gov/sites/prod/files/2013/04/f0/how_is_shale_gas_produced.pdf; 01.01.2011.
[10] A. A. o. P. Geologists. Stimulation. [Online]. Available: http://wiki.aapg.org/Stimulation 7.07.2016.
[11] Geology.com. Geoscience News and Information—Geology.com. [Online]. Available: http://geology.com/articles/horizontal-drilling/; 01.01.2016.
[12] Enggcyclopedia. Gas liquid separation. 2015 [Online]. Available: http://www.enggcyclopedia.com/2011/05/gas-liquid-separation/.
[13] Statoil. Gas/Liquid separation. October 18, 2010 [Online]. Available: http://www.ipt.ntnu.no/~jsg/undervisning/naturgass/lysark/LysarkRusten2010.pdf.
[14] Corporation PALL. Liquid/gas separation technology. 2016 [Online]. Available: http://www.pall.com/main/fuels-and-chemicals/liquid-gas-separation-technology-5205.page?.
[15] Society of Petroleum Engineers. Oil and gas separators. PetroWiki; July 6, 2015 [Online]. Available: http://petrowiki.org/Oil_and_gas_separators.
[16] Shell, "Gas/Liquid Separators - Type Selection and Design Rules," December 2007. [Online]. Available: http://razifar.com/cariboost_files/Gas-Liquid_20Separators_20-_20Type_20Selection_20and_20Design_20Rules.pdf.
[17] Society of Petroleum Engineers. Separator sizing. PertroWiki; July 6, 2015 [Online]. Available: http://petrowiki.org/Separator_sizing#Separator_design_basics.
[18] Center for Chemical Process Safety. Guidelines for engineering design for process safety. New York: American Institution for Chemical Engineers; 1993.
[19] EATON Powering Business Worldwide. Gas liquid separators. 2016 [Online]. Available: http://www.eaton.com/Eaton/ProductsServices/Filtration/GasLiquidSeparators/GasLiquidSeparators/index.htm.

[20] Pearce J. Stress corrosion cracking—Metallic corrosion. AZO Materials; February 22, 2001 [Online]. Available: http://www.azom.com/article.aspx?ArticleID=102.
[21] Society of Petroleum Engineers. Corrosion problems in production. PetroWiki; January 19, 2016 [Online]. Available: http://petrowiki.org/Corrosion_problems_in_production.
[22] Naturalgas.org. The transportation of natural gas. [Online]. Available: http://naturalgas.org/naturalgas/transport/; 20.9.2013.
[23] Naturalgas.org. Storage of Natural Gas. [Online]. Available: http://naturalgas.org/naturalgas/storage/; 20.9.2013.
[24] Society for Industrial and Applied Mathematics. Flow and transport equations. 2006 [Online]. Available: https://www.siam.org/books/textbooks/cs02sample.pdf.
[25] Zendehboudi S. Reservoir analysis & fluid flow in porous media. May 2016 [Online]. Available: https://online.mun.ca/d2l/le/content/218361/viewContent/1968392/View?ou=218361.
[26] Union of Concerned Scientists. Uses of Natural Gas. [Online]. Available: http://www.ucsusa.org/clean_energy/our-energy-choices/coal-and-other-fossil-fuels/uses-of-natural-gas.html#.V5ypHLgrK00.
[27] Kuuskraa VA. Case study #1. Barnett shale: the start of the shale gas revolution. April 6, 2010 [Online]. Available: http://www.adv-res.com/pdf/Kuuskraa_Case_Study_1_Barnett_Shale_China_Workshop_APR_2010.pdf.
[28] U.S.Energy Information Administration. North America leads the world in production of shale gas. U.S. Department of Energy; October 23, 2013 [Online]. Available: http://www.eia.gov/todayinenergy/detail.cfm?id=13491.
[29] U.S. Energy Information Administration. Shale gas and tight oil are commercially produced in just four countries. U.S.Department of Energy; February 15, 2015 [Online]. Available: http://www.eia.gov/todayinenergy/detail.cfm?id=19991.
[30] Brackett W. How marcellus & utica compare to other shale basins. August 27, 2015 [Online]. Available: http://extension.psu.edu/natural-resources/natural-gas/webinars/how-marcellus-and-utica-production-compares-to-other-shale-basins/how-marcellus-and-utica-production-compares-to-other-shale-basins-powerpoint.
[31] U.S. Energy Information Administration. Natural gas weekly update. U.S. Department of Energy; July 28, 2016 [Online]. Available: http://www.eia.gov/naturalgas/weekly/.
[32] Berman A. Natural gas price increase inevitable in 2016. February 21, 2016 [Online]. Available: http://www.artberman.com/natural-gas-price-increase-inevitable-in-2016/.
[33] A. P. Institute. Facts about shale gas. [Online]. Available: http://www.api.org/oil-and-natural-gas/wells-to-consumer/exploration-and-production/hydraulic-fracturing/facts-about-shale-gas; 01.01.2016.
[34] Tunstall T. Prospective shall oil and gas in Mexico. Shale Oil and Gas Magazine; September 28, 2015 [Online]. Available: http://shalemag.com/shale-oil-gas-mexico/.
[35] Guo A. BP taking a bet on China's shale gas while shell backs out. Bloomberg; April 1, 2016 [Online]. Available: http://www.bloomberg.com/news/articles/2016-04-01/bp-taking-a-bet-on-china-s-shale-gas-while-shell-backs-out.
[36] International Gas Union. Wholesale Gas Price Survey—2016 Edition. May 2016 [Online]. Available: http://www.igu.org/sites/default/files/node-news_item-field_file/IGU_WholeSaleGasPrice_Survey0509_2016.pdf.

CHAPTER FIVE

Shale Gas Processing

1. INTRODUCTION

Shale gas is natural gas, one of several forms of unconventional gas (also known as methane or CH_4). Shale gas is trapped within shale formations with low permeability, which is fine-grained sedimentary rock which acts as its source as well as reservoir. The shale rock acts as both the storage material and creator of the gas through the decomposition of organic matter (see Fig. 5.1). Therefore, the technologies/techniques used at one well may not translate into success at another shale gas location [1,2].

Due to the significant high cost of production, there has not been much interest in exploration of shale reserves in the world. However, over the past few years shale oil and gas reserves have become more important, especially because of the development of new production technologies, high cost of petroleum in the market, and geopolitical benefits.

A key aspect of releasing natural gas from shale rock formations is the combination of two established technologies—horizontal drilling and hydraulic fracturing, where water with chemical additives is injected under high pressure underground to crack the rocks and expedite release of the trapped oil and gas. The fracturing fluid is a mixture of 90% water, 9% propping agent, and less than 1% of functional additives containing pH adjusting agent, biocides, clay stabilizer, scale and corrosion inhibitors, gelling agent, friction reducer, surfactants, and acids. On a rig platform, a blender mixes the fluid with propping agent (generally sand) and other additives, and supplies the fracture fluid slurry with the help of high-pressure pumps to wells. After completing the process of injecting the high-pressure mixture, the well is ready to produce natural oil and gas (depending on the type of the basin) which is later transported through several pipelines to the production facilities and ultimately to the market.

The required infrastructure in the development of shale gas is most similar to the conventional gas collection and processing infrastructure. However, the main difference is in the type of gas pollutants, components, and resulting quantity of produced water recovered from the wells. During hydraulic fracturing, a large quantity of water is utilized. From a financial perspective, transportation and refining of shale gas cost as much as

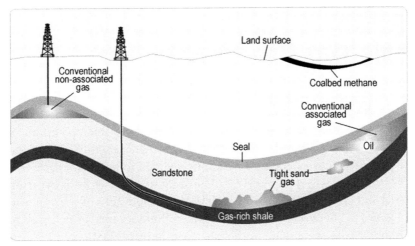

Figure 5.1 Schematic geology of natural gas resources [2a].

conventional gas. The supply chain for shale gas includes wells, a collection network, pretreatment, natural gas liquid (NGL) extraction and fractionation facilities, gas and liquid transporting pipelines, and storage facilities.

2. SHALE GAS PROCESSING: BACKGROUND

The production of shale gas with high commercial value involves several stages from treating the shale formations for extraction to refinement in gas processing plants. The first stage is ground extractions which utilize the following steps consecutively, horizontal drilling, hydraulic fracturing, and multistage fracing [3].

In the horizontal drilling step, a well is first drilled vertically to a specific depth, usually about 6000 ft, where it meets the shale gas formation [3]. The drill bit changes direction to increase the angle of drilling until eventually the wellbore becomes in horizontal contact with the shale gas reservoir [3]. After horizontal contact is made, the well is extended to a lateral depth of approximately 1000–30,000 ft [3]. This allows for maximization of the surface area of the wellbore that is in contact with the shale gas formation [3].

Following the completion of the drilling operation, a steel casing is introduced into the well and is consolidated in place by injected cement around the casing [3]. A special tube-perforating gun inserts holes all around the casing to allow intimate contact between the wellbore and the shale reservoir [3].

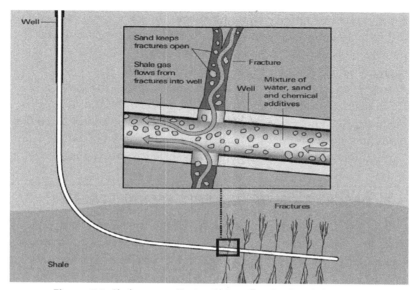

Figure 5.2 Shale gas well stimulation. *Graphic by Al Granberg.*

The second step in extraction is hydraulic fracturing which fractures the shale formation to increase its permeability [3]. This involves the injection of a fluid under pressure (usually water but sometimes gas) that contains fine sand grains as well as chemical additives [3]. The pressurized fluid will produce new fractures as well as increase the size of existing ones [3]. The purpose of the sand grain suspension is to hold the fractures open once the pressurized fluid is withdrawn from the wellbore [3]. As a result, gas in the shale continuously flows to the wellbore. The procedures illustrated above are shown in Fig. 5.2.

The horizontal well is then separated into segments where each fracturing event is isolated from the adjacent segment [3]. This process is called multistage fracing and it maximizes the efficiency by which gas is recovered from the formation [3]. The dividers are then removed allowing the natural gas to flow through the wellbore and eventually to the surface.

A gas processing facility is usually constructed on the site where shale gas production is occurring. The key components of a gas processing plant are [4]:

- Condenser unit—to condense off free water and other condensable components from the gas;
- Dehydration unit—to remove the free water from the condensate;

- Amine unit—to remove contaminants in the gas such as H_2S and CO_2 (other techniques such as molecular sieving can be employed to remove moisture and these contaminant gases);
- Mercury removal unit—to remove mercury from gas;
- Nitrogen rejection unit—to remove nitrogen from the gas;
- Demethanizer unit—to separate natural gas from NGLs;
- Fractionators' unit—to separate the NGLs into fractions of the different types present in the mixture (e.g., ethane, propane, butane)

Fig. 5.3 demonstrates different steps in shale gas processing.

3. DESCRIPTION OF GAS PROCESSING STAGES

The first level of treatment usually begins at the wellhead. Gas condensates and free water are separated at the wellhead using mechanical separators. Extracted condensate and free water are directed to separate storage tanks and the gas flows to a collection system. Along the path, contaminants such as carbon dioxide and hydrogen sulfide are removed at a gas processing plant by using a physical solvent called amine solution. This preliminary processing is important as CO_2 and H_2S are highly corrosive, and the risk of corrosion in pipelines is significantly high, which can cause consequences like line eruption, leakage, and explosion.

The treated gas using an amine solution later must be dehydrated to pipeline quality. Dehydration can be accomplished through absorption or adsorption methods. The process of separation and dehydration, and associated costs to these processes, are similar for both types of gases (conventional and shale). If there is a significant amount of nitrogen present, it should be extracted using a cryogenic plant which includes supercooling equipment in order to meet the minimum heating value required by the pipeline.

If the gas does not meet the required pipeline quality, such as dew point, then it is typically processed at a refrigeration unit. In refrigeration units, the gas is cooled and the NGLs precipitate out, removing more than 90% of the propane and about 40% of the ethane. Other heavier components in the mixture are also almost entirely removed. The temperature drop causes the ethane and other heavier hydrocarbons to condense, removing a considerable amount of the natural gas liquids and 90–95% of ethane.

The fractionation process is utilized because each hydrocarbon component has a unique boiling point. During the time periods that NGL prices are low, extraction will be minimized as it does not make economic

Shale Gas Processing

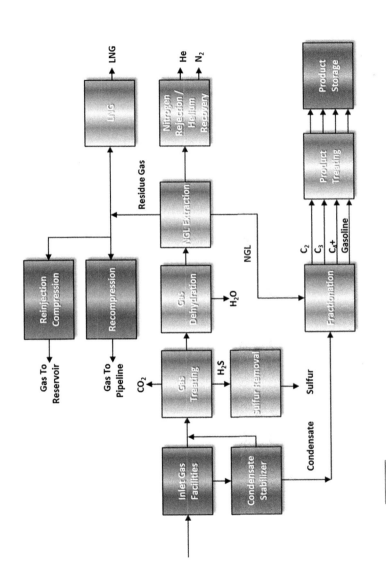

Figure 5.3 Shale gas processing steps [5].

justification. Some gas liquids, and particularly ethane, will be left in the gas because they may not be economic to separate.

Fig. 5.4 shows a typical processing flow diagram for shale gas. The process is very similar to the conventional gas processing operation.

4. HYDRATE FORMATION AND INHIBITION

Under elevated pressures and low temperatures, water molecules can condense to form five to six membered rings which join with adjacent rings to produce crystal lattice structures known as hydrates [7]. The formation of these hydrates is undesirable, since gas molecules can be trapped within these ice crystals. This results in a significantly decreased yield during shale gas extraction [7]. In addition, hydrate may plug pipelines due to their tendency to agglomerate together and adherence to the wall of the pipe [7]. It is important to understand the chemistry of these hydrates in order to prevent their formation.

There are a number of promoters that induce the formation of hydrates. These promoters are classified as either primary promoters or secondary promoters.

The presence of free water and the necessary physical conditions, such as high pressures and low temperatures, are considered primary promoters [7]. The composition of the surrounding mixture will also impact the physical properties of hydrates [7].

The disturbance of the mixture in the shale formation through application of high pressures at regular intervals and high-velocity injections can lead to the formation of hydrates [7]. These factors which induce hydrate formation through agitation are classified as secondary promoters [7]. Introduction of preformed hydrates from external sources into a shale gas reservoir can also accelerate the formation of additional hydrate crystals [7].

Hydrate formation can be prevented by the utilization of different types of inhibitors or inhibition techniques. One way to reduce hydrate formation is to remove the free water present which will take away the precursors necessary to produce these hydrates [7]. Although this method is effective, it is very costly and unpractical to utilize the large volumes of gas produced from the shale formations.

An alternative to inhibition by dehydration is to inject chemical inhibitors to slow down or stop hydrate formation [7]. The use of chemical inhibitors is more practical and economical. Chemical inhibitors can be

Shale Gas Processing

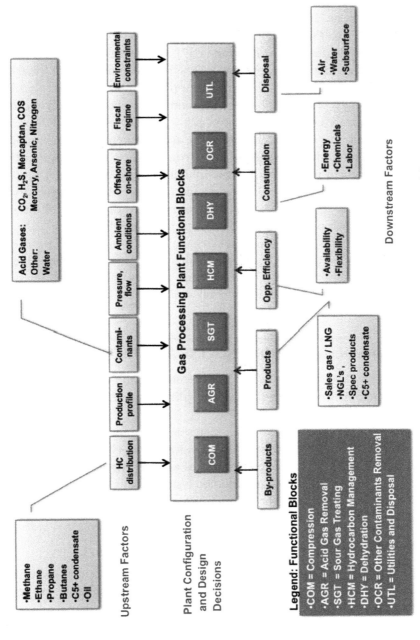

Figure 5.4 Typical flow scheme of shale gas processing [6].

classified under three broad categories such as thermodynamic inhibitors, kinetic inhibitors, and antiagglomerants [7].

Thermodynamic inhibitors alter the physical properties under which hydrates are able to form. Their most common effect is to lower the temperatures for the hydrate formation process [7]. This will significantly lower the amount of gas trapped in ice crystals. The most common class of thermodynamic inhibitors used is glycols, specifically monoethylene glycol and diethylene glycol [7]. Methanol is sometimes used but is usually more expensive, compared to glycol inhibitors [7].

Kinetic inhibitors work by increasing the activation energy required for the reaction that produces hydrate [7]. As a consequence, the rate at which hydrates are formed is reduced. Kinetic inhibitors are also known as low-dosage hydrate inhibitors because the dosage of these inhibitors necessary to prevent hydrate formation is very low, compared to the dosage of thermodynamic inhibitors required to have about the same effect [7].

Antiagglomerants are chemical species which have no impact on the thermodynamics or the kinetics of the hydrate reaction but prevent the coagulation of the hydration cages with one another [7]. Antiagglomerants are usually used along with kinetic inhibitors to deliver optimum results.

5. GAS DEHYDRATION PROCESS AND TECHNOLOGIES

The dehydration of natural gas is the removal of the water that is associated with natural gases in vapor form. Dehydration prevents the formation of gas hydrates and reduces corrosion.

There are different gas dehydration methods encountered in gas processing units with their own advantages and disadvantages. The most common methods of dehydration used for natural gas are triethylene glycol (TEG) dehydration units and solid bed dehydration systems. Most gas sales contracts specify a maximum value for the amount of water vapor allowable in gas. A typical value is $-15°C$ at 70 bars [8].

5.1 TEG Dehydration System: General Description

The gas stream with free water leaving the condenser unit is transferred to a dehydration unit [9]. The free water is separated and removed from the gas stream. Glycol chemical species are generally used in dehydration units, particularly TEG [9]. Fig. 5.5 illustrates the basic schematic of a TEG dehydration system.

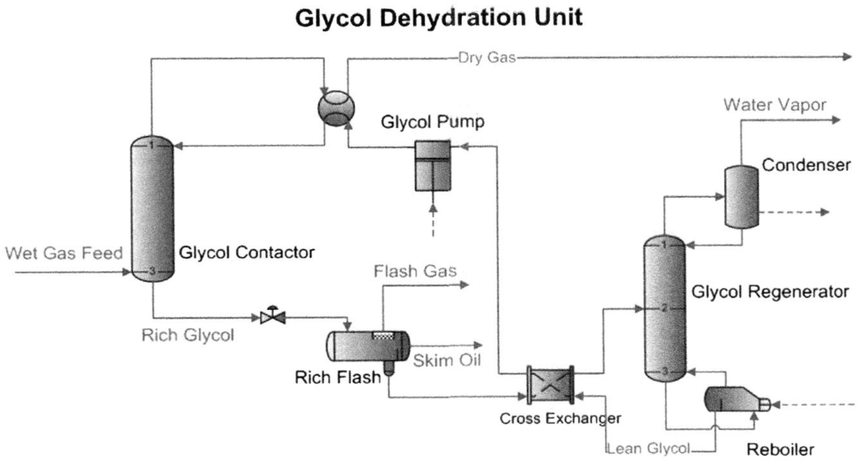

Glycol Dehydration Unit

Figure 5.5 Dehydration system utilizing glycol [9].

First, the wet gas enters the bottom of the glycol contact tower where it will flow upstream [9]. From the top of the tower, lean glycol (pure glycol) is injected and is forced to flow downstream countercurrent to the wet gas [9].

Upon contact between the glycol and the wet gas, the glycol absorbs the moisture and other contaminants including methane, volatile organic compounds, and other hazardous air pollution [9]. The gas that has been stripped of the water and contaminants is then released at the top of the tower to the next stage of treatment.

The rich glycol (glycol overburdened with moisture and contaminant) is transferred to a flash vessel. This serves to strip away any gaseous or liquid hydrocarbons from the glycol [9]. The rich glycol is then pumped to a glycol reboiler/regenerator. Most of the water and contaminants are removed and the glycol is restored to its original purity [9]. The lean glycol is recirculated back to the glycol contact tower [9].

A major issue with glycol dehydration systems is foaming. A considerable amount of the glycol can be lost as a result of foaming [10]. This in turn causes a reduction in the treatment of wet gases in the glycol contactor [10]. Foam formation prevents effective contact between the TEG and the wet inlet gas. The foaming process can be inhibited by injecting monoethanolamine into the contactor [10]. Usually, foam is produced when either the concentration of the lean TEG entering the contact tower is low or the temperature difference is too large between the lean TEG and the wet gas injected [10]. Another issue is the methane emission from these units. A large amount of methane may be emitted to the atmosphere through the recirculation of the glycol [11].

Methane emissions from traditional glycol units are a major concern while performing gas dehydration. A new innovative technology called zero emission dehydrators utilizes different technologies in order to prevent emissions [11]. Instead of using pressurized gases to pump the shale gas through the circuit, electric-controlled pumps are used which essentially produce zero emissions [11]. Moreover, the glycol regeneration step in the zero emission dehydrator does not utilize a gas stripper but a water exhauster instead [11].

5.2 Gas Dehydration: Traditional and New Technologies

Water vapor present in natural gas streams has many adverse effects potentially causing slug formation, hydrate formation, and corrosion in pipelines and downstream equipment. To prevent hydrate formation, the removal of water or dehydration is imperative. This is also important as it avoids other phenomena such as corrosion and plugging in processing plants and transmission systems. One efficient strategy for preventing hydrate formation is the use of hydrate inhibitors which may be injected in the gas stream to deter hydrates from forming. Several gas conditioning steps precedes dehydration that includes; namely separation/removal of acid gas (if available), and separation of free hydrocarbons in the liquid form.

Recovery of Hydrocarbon NGLs or/and Hydrocarbon Dew Pointing are few other methods that could be used for gas conditioning prior to gas sales. Other process alternatives to achieve dehydration are discussed below [12].

5.2.1 Dehydration Process Alternatives

There are five methods which can be utilized to reduce the water vapor of a gas content. Almost all the methods are applicable at the process pressure of the gas as listed below [12];
- Absorption with a deliquescing solid (e.g., calcium chloride)
- Adsorption with solid desiccants such as mole sieve, silica gel, and alumina
- Absorption using liquid desiccants that may include methanol and glycol
- Cooling to lower than initial dew-point;

- Compression to a greater pressure following by cooling and phase separation. As the pressure increases, the saturated water vapor decreases at a certain temperature.

5.2.1.1 Compression and Cooling
For a production system/gas gathering and process management, separation is very common and in the case of natural gases, an additional technique such as drying is incorporated.

In some cases, a compression and cooling strategy might be efficient in gas lift operations at the field scale (theoretical and practical experiences suggest that the operation is better if the free water is prevented [12]).

5.2.1.2 Cooling Below Initial Dew-Point
Low temperature usually corresponds to low water vapor saturation of gas. This technique generally requires some means of hydrate prevention and is employed as low-temperature separation (LTS). For hydrate prevention and subsequent gas dehydration, ethylene glycol is usually used for the lower-temperature LTS processes. This strategy is usually linked with direct glycol injection on the front end of refrigeration systems or/and lean oil absorption processes. New expansion technologies such as TWISTER have been also combined with direct injection to achieve dehydration [12].

5.2.1.3 Absorption of Water With a Liquid Desiccant
This usually uses one of the glycols, with contacting in an absorber column at ambient temperature. It is also applied in combination with cooling, at temperatures lower than the ambient temperature. It is the most widely employed in production operations and in many refinery and chemical plant operations [12].

5.2.1.4 Adsorption of the Water With a Solid Desiccant
Molecular sieves have found acceptance in the gas processing industry for cryogenic plant feed conditioning applications and some sour gas applications with special acid-resistant binder formulations. Dehydration of natural gas to the usual pipeline requirement of 7 lb water/MMSCF is normally least costly, utilizing a liquid desiccant such as glycol rather than using solid desiccants. Activated alumina and silica gel have been also successfully used for many years in production and processing applications that require lower dew point than is achieved by conventional glycol. It is possible to simultaneously remove hydrocarbons and water in so-called short cycle units by silica gel [12].

5.2.1.5 Deliquescent Systems

A deliquescent system is very viable for smaller volumes, which can be an isolated production system or/and fuel gas. In general, deliquescent desiccants are constructed using different mixtures of alkali earth metal halide salts such as calcium chloride, which are naturally hygroscopic. As natural gas flows through the beds of desiccant tablets in a pressured vessel, water vapor is removed. This is achieved because the moisture is attracted to salts in the deliquescent tablets, and coats them with hygroscopic brine. This phenomenon continues as the attracted moisture forms a droplet, which then flows down the desiccant bed into a liquid sump. Upon attracting and absorbing water, the desiccants dissolve because of their deliquescent behavior and, due to this fact, more desiccant is added to the vessel when needed [12].

Any of these techniques might be utilized to lower the water content of gas to a specific level. In general, the most appropriate dehydration method is determined considering various factors such as initial water content, operational nature, water content specification, process characteristics, economic considerations, or/and a combination of these. In most cases, the commonly used dehydration techniques are adsorption with a solid desiccant (e.g., molecular sieve), silica gel, and liquid desiccant such as glycol in recent times for natural gas processing beyond compression and cooling [12].

5.2.2 Flow Diagram of Physical Absorption Process: Description

TEG is the most commonly used glycol. Water dew point depressions range from 20 to 70°C, while inlet gas pressures and temperatures vary from 5 to 170 bar and from 15 to 60°C, respectively [13].

The entering wet (rich) gas, free of liquid water and liquid hydrocarbons, enters the bottom of the absorber (contactor) and flows countercurrent to the glycol (see Fig. 5.6). Glycol—gas contact occurs on trays and packing. Bubble cap trays have been used historically but structured packing is more common today. The dried (lean) gas leaves the top of the absorber. The lean glycol enters on the top tray or packing and flows downward, absorbing water as it goes. Glycol leaves rich in water, so it is also called "rich glycol."

The rich glycol leaves the bottom of the absorber and flows first to a reflux condenser at the top of the regenerator still column, then to a glycol—glycol heat exchanger. The rich glycol then enters a flash drum where most of the volatile components (entrained and soluble) are vaporized. Flash tank pressure is typically 300—700 kPa (3—7 bar) and temperatures are between

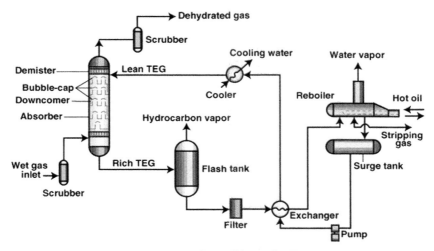

Figure 5.6 Typical flow scheme of a TEG unit.

60 and 90°C. Leaving the flash tank, the rich glycol flows through the glycol filters and the rich—lean exchanger where it exchanges heat with the hot lean glycol. The rich glycol then enters the regenerator still column where the water is removed by distillation. This is where the glycol concentration is increased to the lean glycol requirement [14].

The regeneration unit is designed to operate at prevailing atmospheric pressure, the rebuilder to a temperature very close to 204°C for TEG. This is the temperature at which measurable decomposition for TEG begins to occur. Steam generated in the rebuilder strips water from the liquid glycol as it rises up the still column packed bed.

The hot reconcentrated glycol flows out of the rebuilder into the accumulator, and then it is cooled by heat exchanger using rich glycol. Finally, the lean glycol is pumped back into the top of the absorber through a cooler to control its inlet temperature in the absorber.

In adsorption, water molecules from the gas are held on the surface of a solid by surface forces. Activated alumina, silica gel, and molecular sieves are materials which meet adsorption characteristics and they are common in commercial use. Molecular sieves possess the highest water capacity and produce the lowest water dew point. These characteristics made them the most common desiccant used in the industry nowadays.

Molecular sieve dehydrators are commonly used ahead of NGL recovery plants where extremely dry gas is required (see Fig. 5.7). Cryogenic NGL plants designed to recover ethane produce very cold temperatures (around

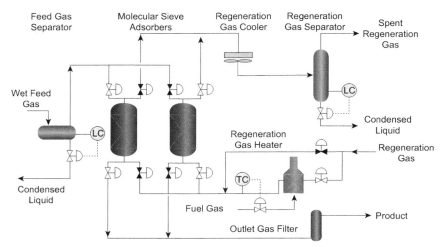

Figure 5.7 Typical flow scheme of a molecular sieve unit.

−80°C) and require very dry feed gas to prevent formation of hydrates. Comparing these with the other gas dehydration processes, the molecular sieve is the most expensive [14].

Multiple desiccant beds are used in cyclic operation to dry the gas on a continuous basis. The number and arrangement of the desiccant beds may vary from two towers, adsorbing alternatively, up to many towers. Three separate functions or cycles must alternatively be performed in each tower. These are an adsorbing cycle, a heating or a regeneration cycle, and a cooling cycle (as per the conditions). The important constituents of any solid desiccant dehydration system are:

- Inlet gas separator;
- Two or more adsorption towers (contactors) filled with a solid desiccant;
- Dry gas filters;
- A high-temperature heater to provide hot regeneration gas to reactivate the desiccant in the towers;
- Gas cooler for regeneration to condense water from the hot regeneration gas;
- Gas separator for regeneration to remove the condensed water from the regeneration gas;
- A regeneration gas blower or compressor if the regeneration gas is recycled;
- Piping, manifolds, switching valves, and controls to direct and control the flow of gases according to the process requirements.

6. GAS SWEETENING
6.1 Process Description

The gas sweetening process involves the removal of acid gases, namely, hydrogen sulfide (H_2S) and carbon dioxide (CO_2), from the gas mixture [15]. Not only are these gases toxic but they also lower the commercial value of natural gas, which makes removing them desirable. Gas sweetening can be accomplished through either the process of adsorption or absorption [15]. The adsorption process uses stacked beds of salt that are able to take up both acidic gases and water at the same time [15]. With adsorption processes, the contaminants are taken by the surface of salt medium [15].

In the absorption process, the contaminants are taken up inside the medium itself. The group of chemical species most commonly used in gas sweetening absorption processes is alkanolamines [15]. The two alkanolamines widely used are monoethanolamine (MEA) and diethanolamine (DEA) [15]. These aqueous alkanolamines react with H_2S and CO_2 and take them up into the aqueous phase [15]. The following reaction shows how H_2S reacts with MEA to become a part of the alkanolamine's aqueous phase [15].

$$RNH_2 + H_2S \leftrightarrow RNH^{3+} + SH^- \tag{5.1}$$

6.2 Gas Sweetening: Traditional and New Technologies

The acid gas content varies greatly with each natural gas stream. It is important to understand the acid gas components and their content in the inlet natural gas stream and required specifications in the treated gas to allow proper design and selection of the technology.

Shale gas found in the US contains moderate levels of CO_2. It ranges from less than 1 mol% to over 3 mol%. The H_2S concentration levels from most US shale gas have been generally low, ranging from less than 100 parts per million by volume (ppmv) to as much as 750 ppmv. Although the H_2S content is low, it is still high enough to require treatment to meet pipeline gas specification of less than 4 ppmv [16].

The gas sweetening process is required to dispose of the acid gas for different reasons which are [16]:
- To remove H_2S content of the natural gas stream to make it safe for refining;

- To satisfy a shale's gas specification: H_2S content of the shale gas must be below 4 ppmv (it is around 5.7 mg of H_2S/Sm^3 of gas or 0.25 grains of $H_2S/100$ SCF of gas). CO_2 content must be adjusted to allow the shale gas to fit with the required range of gross calorific value;
- To make the downstream processing reliable. This is the case for a cryogenic process (CO_2 can freeze at $-70°C$).

Traditional gas sweetening plants use amine units that contain alkanolamine solvents as the absorbers. Fig. 5.8 depicts a basic schematic of a gas sweetening plant.

The sour gas introduced into the bottom absorber column is forced to flow upstream. The lean alkanolamine solvent enters the top of the column and is directed to flow to the bottom countercurrent to the sour gas [15]. The sweet gas is released at the top of the tower and the rich alkanolamine solvent exits in the bottom of the column where it is transferred through a heat exchanger to a gas stripper column [15].

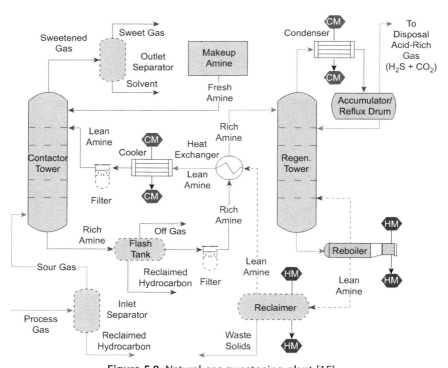

Figure 5.8 Natural gas sweetening plant [15].

In the gas stripper, heat is applied and pressure is reduced, resulting in the evolution of the acid gases into the gas phase, thereby regenerating the lean alkanolamine solvent [15]. The lean solvent is transferred back to the absorber column. The absorber column contains a large stack of trays onto which most of the absorption takes place [15].

The adsorption process works in a similar manner except it uses solid salt beds that physically take up the acid gases to their surface as oppose to inside the solid bed.

An application of adsorption in gas sweetening is through the use of molecular sieves [15a]. Molecular sieves use a porous crystalline alumina silicate bed where sour gases as well as water can stick [15a]. Alumina silicate beds can be manufactured to be specific for just one type of molecule [15a].

6.2.1 Flow Diagram of Adsorption Processes for Gas Sweetening: Description

For gas sweetening process, several chemical solvents are readily available and most are based on alkanolamine products which are those that contain hydroxyl (-OH) and amino functional groups.

These are used by making their aqueous solutions. The process of chemical absorption is on the basis of the contact between the targeted feed gas and an aqueous solution that includes one of the solvents, namely MEA, DGA, DEA, MDEA, etc. [16].

A weak acid is fed in, which reacts with the alkanolamine (alkaline product) to give bisulfide (with H_2S) and bicarbonate (with CO_2). This chemical reaction (chemical absorption) takes place in a fractionation column (absorber or contactor), which is equipped with trays and sometimes packing. In the column, gas enters at the bottom tray or at the bottom part of the packing and the aqueous solution comes from the top tray (or from the top section of the packing). There is a heat of reaction between the solvent and the acid gas during this absorption, which is exothermic. The treated gas exits the unit at a higher temperature than the feed gas temperature, implying that the water content will be higher in the treated gas, compared to the feed gas. As a consequence, dehydration of the gas is necessary, as well. This unit is always installed downstream of the sweetening unit. The alkanolamine salt is then retransformed into alkaline solution in a regeneration section and the cycle is repeated again [16].

The MEA/DGA solution is utilized for significant removal of CO_2 (when the feed gas is free from H_2S), H_2S (when the feed gas is free from

CO_2), or both H_2S and CO_2 when both components exist in the feed gas. Hence, this is not an apposite process for selective elimination of H_2S when both H_2S and CO_2 are available in the gas [17]. DEA is nonselective and will remove both H_2S and CO_2. Both impurities can be eliminated at any low level, regardless of the initial concentration in the feed gas stream. Thus, this is not proper for selective removal of H_2S when the gas includes both H_2S and CO_2. The processing scheme of DEA is similar to the MEA processing scheme with the exception of the reclaimer which is not required. If the mercaptans are present in the feed gas, DEA solution will be removing from 10% to 60% of it, based on their boiling points. MDEA is a tertiary amine. It is a novice to the group of ethanolamines utilized for sweetening of natural gas so that it has recently received lots of attention as it can selectively react with H_2S in the presence of CO_2. When MDEA is utilized as the sole pure solvent in the aqueous solution, the selection is based on this feature. MDEA is also useful for the bulk removal of CO_2 as its reaction with CO_2 requires a low amount of reaction heat which saves energy in the regeneration section. For significant CO_2 removal, activated agents must be added to the MDEA to enhance the extraction. For most shale gas processing, selective H_2S removal is preferred to meet the pipeline specification and also to reduce CO_2 to a desired level. Fig. 5.9 shows a typical process flow diagram (PFD) of an amine sweetening unit [16].

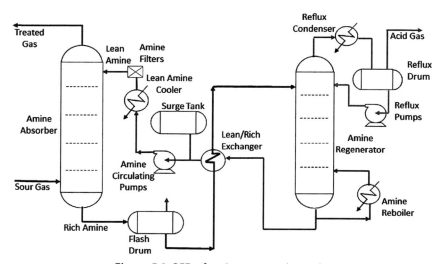

Figure 5.9 PFD of amine sweetening unit.

Natural gas to be treated (sour gas) is introduced at the bottom of a high-pressure absorber contacted countercurrently with regenerated lean amine solvent. Treated gas, now free of the acid gases, exits at the top of the column. The rich amine solvent containing acid gas is sent from the bottom of the absorber column to a lower-pressure flash drum to separate some dissolved gases from the rich amine. The rich amine separated is preheated in a lean/rich amine exchanger prior to feeding to the amine regenerator. The amine reboiler supplies the heat to an amine regenerator to dissociate the chemically bound acid gases from the amine solvent. The low-pressure stripped acid gases leave the amine regenerator (also referred to as a stripper) from the top where depending on the H_2S content and local environmental requirement is diverted to the sulfur recovery plant when a larger quantity of H_2S is present or sent to a thermal oxidizer. The lean regenerated amine solvent is pumped back to the absorber column via a lean/rich exchanger to recover heat [17].

6.2.1.1 Presence of Mercury in Feed Gas
Mercury forms an amalgam with aluminum which causes corrosion in heat exchangers. To prevent rapid deterioration of equipment and contamination of the rest of the gas processing equipment, mercury must be detected in the feed stream and trapped before gas processing in an adsorbent bed downstream of the dehydration unit [18].

6.2.1.2 NGL Extraction and Fractionation
The recovery of hydrocarbon liquids from the natural gas stream ranges from a simple dew point control to deep ethane extraction. The desired level of liquids recovery has a significant impact on the process selection, complexity, and cost of the processing facility.

There are different NGL recovery processes encountered in gas processing plants. The term NGL is a general term which applies to liquids recovered from natural and associated gases such that it refers to ethane and heavier products (mixture of ethane, propane, butanes, condensates, or natural gasoline). The term LPG (liquefied petroleum gas) describes hydrocarbon mixtures in which the main components are propane, iso- and normal butanes, propene, and butenes. Usually NGLs are produced by cooling down gases at saturated or hydrocarbon dew conditions or by pressure drop when gases are at saturated or hydrocarbon dew conditions and at retrograde zone conditions. The extraction of heavy components (propane + butane + condensate) from rich gas produces lean or poor

gas. In general, NGL can contain pentanes and heavier (condensate), propane and heavier, ethane and heavier [19]. Three basic objectives of NGL processing are:
1. To produce a transportable gas stream;
2. To produce a salable gas stream;
3. To maximize natural gas liquid production.

Production of a transportable gas stream implies minimal processing in the field to transport gas through a pipeline to a final processing plant without trouble. Three components may be undesirable in transported gas: water, H_2S, and condensate.

In many cases (particularly for long distances), water must be removed to a level that prevents hydrate formation, and CO_2 and H_2S corrosion. This is achieved when water vapor content of transported gases remains lower than the amount required for the formation of hydrates and free water in transport conditions. Usually this water-vapor content of transported gases is expressed in gas water dew point defined by a couple of pressure and temperature values. Hydrogen sulfide is highly toxic and may be removed if present at high concentration to satisfy local safety regulations [20].

Condensate (heavy components) may or may not be recovered prior to pipeline transport. If the dew point of the transported gas is lower at any pressure value than the lowest expected temperature in the pipeline, no processing to remove NGL is required. On the other hand, if this hydrocarbon dew temperature value is higher at any pressure value than the lowest expected temperature in the pipeline, there are two options:
- No NGL or condensate removal is carried out at field production facilities and gas transport is uncontrolled by two-phase flow in pipelines;
- A condensate removal is carried out at field production facilities to control the two-phase flow of the transported gas in pipeline or to produce transported gas which remains single-phase flow in all parts of the pipeline.

The amount of heavy components contained in transported gas is also defined by the hydrocarbon dew point. Production of sales-quality gas requires all processing necessary to meet specifications of H_2S (15 mg/Nm^3), CO_2 (<2 or 2.5%mol) and water content. Note that for gas liquefaction CO_2 must be removed to an amount of maximum 50 ppm mol to prevent total plugging of cold heat exchangers by solidification of CO_2 at low temperature [21].

The necessity and importance of NGL recovery are fixed by the requirement to meet the following three specifications:
- Gross heating value (GHV) or high heating value (HHV);
- Wobbe index (WI);
- Hydrocarbon dew point.

Note that the importance of NGL recovery depends on the percentage of nitrogen present in shale gas. If the gas contains a high percentage of nitrogen, it may be necessary to extract fewer hydrocarbons heavier from the gas.

Three situations motivate maximum natural gas liquid production. First of all is to maximize condensate production; secondly, the propane content of rich gas, and third is extracting ethane. In the latter, the gas must be cooled to a level of temperature which allows recovering liquid ethane in the produced NGL. Ethane is recovered in a fractionation plant with demethanizer, de-ethanizer, depropanizer, and debutanizer columns (see Fig. 5.10).

The gas composition has a major impact on the economics of NGL recovery and process selection. In general, gas with a greater quantity of liquefiable hydrocarbons produces a greater quantity of products and hence greater revenues for the gas processing facility.

From the standpoint of the phase diagram, the condensate removal is obtained by cooling and/or refrigeration of the gas to move the representative point of the fluid to the left of the dew-point curve of the rich gas in the

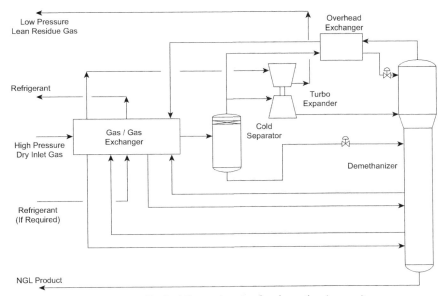

Figure 5.10 Typical flow scheme of a demethanizer unit.

two-phase area. The resulting condensate is then separated from the gas in one or several vapor—liquid separators. Frequently the last one in gas processing is called "cold separator" and the gas leaving this last separator is named "lean gas." Its hydrocarbon dew point is lower than the hydrocarbon dew point of the "rich gas." Frequently the unit in which hydrocarbon liquids are recovered by cooling and/or refrigeration is called a "low-temperature extraction" unit [22]. Three basic techniques are used today for condensate recovery:

1. External or mechanical refrigeration that is supplied by a vapor—compression cycle that typically uses propane as the refrigerant or working fluid;
2. Joule—Thomson expansion rich gas passes through the gas-to-gas exchanger, thence to an expansion or "choke" valve, frequently called the JT valve ("self-refrigeration" process). The JT process is most favored when the wellhead gas is produced at a very high pressure and can be expanded to export-line pressure with no recompression. If the gas must be recompressed, the JT process is penalized by the recompression horsepower requirement;
3. Expansion turbine, in this case the JT valve is replaced by an expansion turbine (so-called turbo expander). The plant using an expansion turbine may be sometimes called a cryogenic or expander plant ("self-refrigeration" process). A JT valve is usually installed in parallel with the turbo expander.

To prevent hydrate formation rich gas must be dehydrated or inhibited. An important number of variations and/or combination of the basic techniques (external refrigeration, JT expansion, and expansion turbine) can be considered in the gas processing plants.

6.2.1.3 Particular Arrangement for Gases with High Nitrogen Content

Virtually all natural gas contains some amount of nitrogen, which lowers the heating value of the gas but is not a particular problem. However, in some reservoirs gas has been discovered to contain larger amounts of nitrogen than can be tolerated due to contractual considerations on heating value. In these cases, the operator has three options:

1. Blend the gas with richer gas to maintain overall heating value;
2. Accept a reduced price or less secure market;
3. Remove the nitrogen to meet sales specifications.

Options 1 and 2 are reasonable approaches to the problem but are very location-specific. Option 3 is always a costly option. When a nitrogen rejection unit is selected as a process option for a gas stream, it is often combined with NGL recovery in an integrated plant design.

Many technologies can be used to remove nitrogen from feed gas but cryogenic technology is usual. Based on cryogenic distillation technology, the exact design of this nitrogen rejection unit is a strong function of the nitrogen content.

6.2.1.4 Condensate Stabilization

Separated liquids at inlet facilities of gas treatment plants are sent respectively to the condensate stabilization unit. The function of this unit is to remove the lightest components from the raw feed and to produce a liquid product, which after mixing with the light condensate from the NGL fractionation unit, will give a stabilized condensate having a Reid vapor pressure of 10 psia in summer and 12 psia in winter as a general specification. Raw condensate is first preheated in the exchanger with the stabilized condensate. The steam flows to the preflash drum that is a three-phase separator.

Hydrocarbon liquid is pumped to the condensate desalter. Demulsifying chemicals are injected in the pump suction line by means of a demulsifier injection package. Operating temperature in the desalter is maintained at $\sim 70°C$ to ensure an efficient separation of water from condensate. This is achieved by heating the fluid with stabilized condensate in the condensate desalter preheater. To prevent the flashing of light components, the desalter operating pressure is set above the condensate bubble point with sufficient margin. Fresh water is injected upstream of the desalter mixing valve. The desalter uses the electrostatic effect to achieve a very good phase separation. At the desalter outlet, the condensate salt content shall be less than 10 mg/L [23].

Then condensate enters the stabilizer. Raw condensate is treated in the condensate stabilizer operating at a pressure ~ 10 bar. Lighter components are removed as vapor overhead product with the condensed liquid serving as reflux. There is no overhead hydrocarbon liquid distillate. The stabilizer column bottom temperature is $\sim 190°C$ for summer case and $\sim 180°C$ for winter case. The water phase from the reflux drum is sent to the sour water stripper. Corrosion inhibitor is injected in the feed to prevent any acid corrosion at the top of the column [23].

6.2.1.5 Mercaptan Removal

Mercaptans occur in all fractions from methane and heavier, but cannot be removed by distillation because of which it requires chemical treatment for removal. The amount generally ranges from a few ppm to several thousand ppm.

Natural gas includes H_2S, CO_2, and other sulfur components such as mercaptans. The overall formula of mercaptans is RSH. Here R means the hydrocarbon chain. Mercaptans are the organic components of hydrocarbons with sulfur; they also have a bad smell and corrosive properties in pipeline transmission. Therefore, if mercaptans are in high concentration in natural gas, these must be removed to reduce the value to an acceptable limit. To separate the mercaptans from natural gas in the sweetening plant, normally the Merox process is used [24].

The Merox process is an efficient and economical catalytic process developed for the chemical treatment of LPG and condensate for removal of sulfur present as mercaptans. The process relies on a special catalyst to accelerate the oxidation of mercaptans to disulfides at or near economical product rundown temperature (see Fig. 5.11).

The extraction version of the Merox process removes caustic-soluble mercaptans from hydrocarbon feed stocks. For extraction of fuel gas and mercaptans from natural gas, the Merox process can remove virtually all the

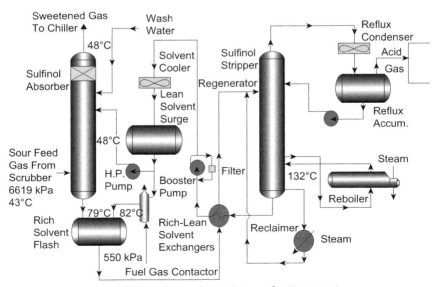

Figure 5.11 Typical flow scheme of a Merox unit.

mercaptans. The extracted mercaptans are catalytically oxidized to disulfides, which are sent for appropriate disposal. The caustic soda is then regenerated by bringing it into contact with air and catalyst to bring about oxidation of mercaptans to disulfides at or near ambient temperatures and pressures. Finally, the disulfides, which are insoluble in the caustic, are separated from the regenerated (lean) caustic. Hydrogen sulfide cannot normally be tolerated in Merox unit feed stocks in concentrations exceeding about 5 wt ppm [24].

This version of the Merox process utilizes gas—liquid contacting to extract low-molecular-weight mercaptans from the gas with a strong aqueous alkali solvent. The mercaptan-rich solvent, which also contains the dispersed Merox catalyst, is delivered to a regeneration section where air is injected to oxidize mercaptans. The product of this reaction gives disulfides. The disulfides are then separated from the solvent by coalescing, gravity settling, and decanting. Finally the regenerated lean solvent is recycled back to the extractor. The unit consists of two sections: one is the extraction section and the other is the caustic regeneration section.

6.2.1.6 Sulfur Recovery Process

If a large amount of sulfur contaminants is present, the facility requires a sulfur recovery unit to recover elemental sulfur from the off gas stream of this process. When the amount of sulfur is small, the off gases can be either incinerated or vented from the process to the atmosphere.

Normally a process package considers the Claus units for recovery of sulfur from acid gas produced by sweetening units. The sulfur recovery units are normally designed to achieve an overall recovery of 95% with purity of 99.8% wt. It mainly depends on H_2S content of the feed gas and variety of operating cases [25].

The sulfur recovery units (SRUs) are designed to recover commercial liquid sulfur from acid gas streams produced in the amine regenerators. The feed gas contains H_2S, CO_2, and some organic sulfur compounds (mainly mercaptans), together with a small amount of hydrocarbons and water vapor. The SRUs shall recover sulfur from this stream down to the required specification.

Sulfur recovery units will consist of:
- Air and acid gas feed preheaters;
- Thermal reaction furnace;
- Three Claus converters in series;
- One sulfur degassing section.

There is one incinerator. The incinerator is designed to incinerate the tail gas of SRU at $\sim 540°C$ prior to release to the atmosphere. Basically, sulfur recovery units are fed from an acid gas header receiving from a gas sweetening unit. The acid gas shall be routed to an acid gas knock-out drum. The liquid sulfur ($\sim 150°C$) from the sulfur recovery unit's storage and degassing pits will be sent to liquid sulfur storage [25].

Sulfur recovery by the Claus process involves conversion of a poisonous waste byproduct of natural gas treating operations, namely acid gas into salable sulfur and a relatively innocuous off gas suitable for discharge to the atmosphere. A net credit of the process is useful heat in the form of steam generation. In chemical terms, the objective is to oxidize the H_2S in the acid gas to sulfur and H_2O, while still burning any hydrocarbons present to sulfur-free products without soot formation.

At the combustion zone of the reaction furnace, one-third of total acid gases burns with air. After that, reaction of the remaining H_2S and SO_2 (produced at combustion zone) begins immediately in the combustion zone of the reaction furnace to form sulfur, but it requires further contact of the process gas with a Claus catalyst at controlled temperatures, in converters following in series, to carry the reaction toward completion [25].

The produced sulfur vapor is condensed and recovered as liquid sulfur after each step of the reaction (reaction furnace and subsequent catalytic converters).

Acid gas containing hydrocarbons and carbon dioxide may undergo side reactions to form carbonyl sulfide (COS) and carbon disulphide (CS_2).

Control of the H_2S to SO_2 ratio is essential to be as close as possible to the exact stoichiometric ratio of 2 throughout the unit for maximum conversion and thus minimum sulfur losses in the tail gases. The air flow rate for oxidation of one-third of H_2S must be set at the best ratio that is supplied by an air blower. Due to the high variations of acid gas composition the reaction furnace burner is fitted with fuel gas tips to burn continuously a small amount of fuel gas. The fuel gas burner acts as a pilot flame to maintain the acid gas flame in case of perturbation. The fuel gas flow will be adjusted from acid gas feed to achieve a reaction furnace temperature up to $\sim 925°C$ [25].

The sulfur gas is cooled in passing through the reaction furnace boiler where LP steam is generated. The condensed sulfur is removed through a bottom seal drain to the sulfur degassing pit. The tail gas from the last condenser will be routed directly to an incinerator or first to tail gas treating (TGT) units to recover as much sulfur as possible from tail gases and after that to the incinerator. TGT unit increases the sulfur recovery of a Claus unit from 95% to about 99.7%. This unit consists of reduction section where all sulfur components present in tail gas are hydrogenated to H_2S by means of catalyst, a quench section for process gas cooling and an amine solvent absorption section where the greatest part of the H_2S is selectively removed from the process gas stream. In the subsequent regeneration section the H_2S is stripped and the rich solution is regenerated. The H_2S released stream from the regeneration section is also returned to the inlet of the Claus unit and combines with the inlet of acid gas feed stream [25].

The sulfur unit is producing LP steam only in reaction furnace boiler and sulfur condensers.

Gas which is overburdened with sulfur such as that produced as a byproduct of gas sweetening is transferred to a sulfur recovery plant to remove the sulfur content. The Claus sulfur recovery process is the most widely used for sulfur recovery in many industries including the oil and gas industry [26]. The Claus recovery process involves the conversion of hydrogen sulfide into elemental sulfur as shown in the two consecutive reactions below [26].

$$H_2S + 3/2O_2 \rightarrow SO_2 + H_2O \tag{5.2}$$

$$2H_2S + SO_2 \rightarrow 3S + 2H_2O \tag{5.3}$$

The simplified setup of the Claus sulfur recovery plant is depicted in Fig. 5.12.

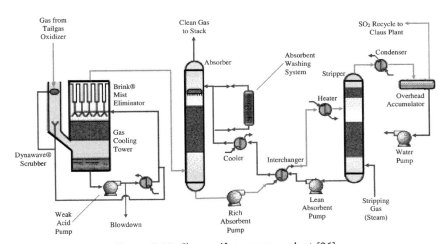

Figure 5.12 Claus sulfur recovery plant [26].

H_2S gas is introduced into a burner where it undergoes a combustion reaction to produce sulfur dioxide (SO_2) [26]. Only a small part of the H_2S gas is converted to SO_2 in the burner. The mixture is then transferred to a reactor where the H_2S reacts with SO_2 to produce elemental sulfur [26].

The exiting stream containing elemental sulfur, H_2S, and SO_2 will then move to a condensor. In the condensor, the elemental sulfur is condensed off the mixture and is removed [26].

The gas mixture leaving the condensor is reheated and moved to another reactor where the same reaction takes place once again and the condensor separates off the elemental sulfur [26]. The purpose of reheating the gas mixture is to prevent the condensation of elemental sulfur on the bed containing the catalyst as it will greatly compromise the ability of the reaction to go to completion in the reactor [26]. A typical circuit usually contains three reactors, each of which is responsible for a certain amount of conversion.

There are several advantages to sulfur recovery. First, it will decrease the amount of gaseous sulfur emitted to the atmosphere [26]. Also, the sulfur recovery unit produces steam of moderate pressure that can be used to power other parts of the gas processing plant [26]. The elemental sulfur recovered can be utilized for other purposes, such as for producing road asphalts and rubber polymer materials [26].

The disadvantage of sulfur recovery is that it poses serious health hazards to the employees running the facility. H_2S is extremely toxic and results in inflammation of the respiratory tract [26]. There is a possibility for the accumulation of H_2S in the void space at the top of sulfur storage tanks, which is very dangerous due to the flammability of the gas [26].

There is also a problem concerning the disposal of excess solid elemental sulfur that is produced. The sulfur must be securely disposed of in a location where it is not exposed to external weather conditions [26]. If exposed to the atmosphere, sulfur will react with oxygen and water to produce acidic runoff that can leach out heavy metals and enter water systems [26].

6.2.1.7 Sour Water Stripper Unit
The sour water streams from all units are routed to the sour water feed drums operating at low pressure. These drums are three-phase separators designed to remove oil from the inlet water streams. The vapor phase is vented from the feed drums to the low-pressure flare. The oil flows over an internal weir and is collected at one end of the drums. The water which

is collected at the other end of the feed drums is then pumped and sent to the sour water stripper.

The sour water stripper is normally equipped with internal random packing. In the stripping section, H_2S, CO_2, and the light hydrocarbons contained in the sour water are stripped by steam, generated in the reboiler. Caustic soda can be injected in the column if needed for the dissociation of ammonium compounds [27].

The stripper overhead vapors are partially condensed in the air condenser. The acid gas is separated in the reflux drum and sent to the low-pressure flare. Condensed water, free of oil, is recycled to the top of the stripper column by the pumps. The stripper bottom water is pumped and cooled in the air cooler. Then the stripped water is sent to the observation basin of the water treatment [28].

7. PROCESS DESIGN OF GAS PROCESSING PLANTS

Gas processing plants generally have similar setup of equipment but can have slight variations in the type of equipment used based on purification levels desired, the budget available and environmental restrictions. A typical plant with the corresponding equipment that can be used for each treatment stage is shown in Fig. 5.13.

Gas processing covers a broad range of interconnected operations to prepare natural gas for transportation to gas markets. Processing varies from simple separation and dehydration to compression, sweetening, and NGL recovery.

Accurate engineering design/calculation can be attained using the flow rate, composition, temperature, and pressure at wellhead, and conditions of the produced gas (components and/or impurities) that must be adjusted to meet pipeline specifications for delivery [22].

The number of utilities, storage, and off-site facilities should be considered for any gas treating plants. Some of the mandatory units are:

1. Electrical generation and distribution: This fulfills the electric power requirements of the plant during peak demand under the most severe atmospheric condition.
2. Steam and condensate: This unit supplies steam to all consumers throughout the facility. For this purpose normally high-pressure boilers (generating high-pressure and low-pressure steam) are installed.

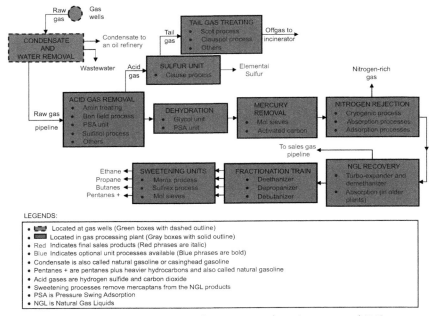

Figure 5.13 Gas processing plant setup and equipment used [29].

3. Fuel gas system: This supplies gaseous fuel to various users in the treatment plant, normally from export gas header and other possible sources depending on the location of the plant.
4. Instrument and service air system: The number of sets of air compressor packages, air drying packages, and instrument air receivers is integrated into a unit of instrument and service air system, supplying the instrument air and service air required by treatment plant.
5. Nitrogen generation system: This consists of a packaged facility for generating both gaseous and/or liquid nitrogen.
6. Water supply and treatment system including raw water, cooling water, potable water, fire water, plant water, etc., are required at the facility.
7. Waste effluent disposal including sanitary sewer package, neutralization package, oily water treatment, etc.
8. Flare and blow down system: The function of the flare system is to collect and burn all gases that are vented from the other units of the plant due to both continuous and emergency operations. It must also

cater for the possibility of depressurization of equipment during emergencies.
9. Drain system: This unit is designed to provide a hold-up volume for hydrocarbon drains from the equipment in all areas during maintenance. During normal operation, no liquid is sent to this system.
10. Burn pit system: The major waste liquid inventory to burn pit is the nonrecoverable liquid collected in the sump drums (or drain drum) located in process and utility area and liquid from flare knock out drums.
11. Product storage and export (condensate, propane, butane): numbers of storage facilities and related loading pumps shall be considered for treatment plants.
12. Sulfur storage and solidification unit (if any): The liquid sulfur received from the sulfur recovery unit flows to the liquid sulfur storage tank that is equipped with heating coils. The formation of sulfur granules can be achieved by the rotoform pastillator process.
13. Chemical storage: The chemicals storage area consists of independent storage systems ensuring the make-up requirements in the process and utilities units.

8. MODELING AND OPTIMIZATION OF EQUIPMENT OR/AND UNITS IN GAS PROCESSING PLANTS

Raw gases are fractionated into differently proportions which mainly include ethane, butane, propane, natural gasoline products, residue gas, etc. This separation is largely based upon the boiling point of the individual component gas. During the separation process, some of the components are not affected by normal variations in process conditions.

In the demethanizer column, separations occur mainly between methane and ethane. Carbon dioxide and propane may also appear in significant quantities in either the vapor or liquid product stream. Since butane and heavier components have a low true boiling point, 99.5% of these components always are in the demethanized liquid product stream. The same phenomenon/operating mechanism applies in the dethanizer, depropanizer, and debutanizer columns. Furthermore, some components have greater boiling points, compared to the others, so that they cannot exist in the system over separation operations. The assumptions made based on this basis approach are listed in Table 5.1.

Table 5.1 Gas Plant Fractionation Simplifying Assumptions

Column	Components in Vapor Product Only	Component in Liquid Product Only	Component Not Present in Liquid
Demethanizer	N_2	C4, C5, C6+	None
Deethanizer	C1	C5, C6+	N_2, CO_2
Depropanizer	C2	C6+	N_2, CO_2, C1
Debutanizer	C3	None	N_2, CO_2, C1, C2

It is also worthy of note that the purity of the product could be used as a constraint in approximation to the composition of light components in the vapor products. The plant works in such a way as to reach the magnitude of impurities in a product stream to the maximum level. The component material balance equations for an entire plant could be reduced considerably and combined into a linear form making use of these assumptions. This set of linear equations is very powerful because it relates each product stream composition and flowrate to the residue composition. Consequently, the solution reaches the composition of only a few components in the residue gas, including C_3, C_2, and CO_2.

Following this step is to develop a simple relationship between selected crucial plant control variables to overhead product of the demethanizer.

Certain variables including feed location, composition, temperature, and flowrate affect the performance and are used for sensitivity analysis of the demethanizer column.

These variables could generically be referred to as feed parameters and their fluctuations result from variations in upstream process variables. The inlet pressure, refrigeration load, and plant inlet fluctuations are the common process variables that affect the demethanizer unit. It is useful to relate the demethanizer performance to the important upstream variables, while it is attempted to remove or lower the difficulties associated with the complicated modeling methods.

It is important to develop a method that can be applied to offer the accuracy of a complex simulation but without disregarding its complexities by employing rigorous simulations of the gas plant to introduce a correlation that shows a relationship between the residue stream composition and the most dominant control variables. A minimum number of simulations in a given order are selected by employing a statistical response surface modeling design. For each unknown component, a correlation is developed in the residue stream as a function of all the key process variables. With the linear

9. TRANSPORT PHENOMENA IN GAS PROCESSING

Normally, the wellhead outlet is equipped with a choke valve to reduce the pressure of the reservoir fluids from wellhead flowing pressure. Then the reservoir fluids are collected in the production manifold. The condensed and free water can be knocked out in a three-phase separator and then water routed to a water treatment unit prior to reusing. In order to test individual well performance, each well can be routed to a test manifold.

Requirement of methanol injection immediately downstream of the Xmas tree should be studied to avoid hydrates formation. To avoid hydrate formation and corrosion of pipeline, a number of chemical injections should be considered. The pipeline regime is achieved possibly in multiphase mode since the fluid is a mixture of gas, hydrocarbon condensate, and water, which are condensed in the pipeline during the transportation. There are some specific requirements in the pipeline design and for operations to cope with hydrate prevention, minimum transfer flow rate, etc.

The main important consideration is flow regime in the pipeline for a flow rate within the normal operating range. During fluid transport to the gas processing plant the fluid undergoes pressure and temperature conditions for which hydrate may form once the saturation water is condensed. The fluids being transported contain considerable amounts of CO_2, H_2S, and condensed water. It is therefore corrosive and means of preventing internal corrosion are required. One slug catcher should be come into service at the entrance of the plant so as to separate gas, condensate, and water. The gas stream is sent to the gas treatment section for further processing. Liquids, hydrocarbons, and water are drawn off separately to prevent emulsion formation in the control valve, mixed downstream of the expansion, and routed to the condensate stabilization unit [30].

10. CORROSION IN GAS PROCESSING PLANTS

Corrosion is the destruction, corruption or changes in the properties and characteristics of materials due to their reaction with the environment. The corrosion process is caused by one of a combination of the following parameters: temperature, moisture percent and oxygen, Redox potential, soil pH, pressure, time, mineralogical composition.

Corrosion in the oil and gas industry has always been one of the biggest issues that not only causes significantly high maintenance and repair costs, but also could cause interruptions in production. Safety issues and environmental effects of any possible spills also could have significant tangible and intangible negative impacts as well.

Corrosion of facilities in operation could be caused by a combination of water and oxygen and/or bacteria in the environment. Corrosion can be divided into two main categories:

1. Internal corrosion, which refers to the interior side of the surface of the equipment. The main corrosion protection system of inside surface which is used in the oil and gas industry is inhibitors (i.e., epoxy paints).
2. External corrosion, which refers to the exterior surface of the equipment. External corrosion of the equipment depends on the natural factors and environmental conditions around the equipment. These factors cannot be completely removed, but should be minimized. Some of the common methods in the industry are using coatings and/or cathodic protection systems that are suitable for protecting the exterior of the equipment.

Corrosion in gas processing plants greatly reduces the efficiency of shale gas treatment. The plant is forced to shut down at unscheduled intervals for corrosion clean up. Corrosion can cause a decrease in the life time of the equipment or units used, which will increase the capital cost of the gas processing plant in the long run [31].

The particular area of the plant where corrosion is most likely to occur is at the amine unit. As described above, an alkanolamine solvent is used to strip away H_2S and CO_2. Due to their acidic nature, they can result in the corrosion of the equipment's internal surfaces [31]. There is a direct correlation between the corrosion rate and the types of acid gases as well as the relative ratios of each acid gas in the mixture. This relationship is indicated in Fig. 5.14.

The corrosion rates are higher when only one type of acid gas is present as opposed to the presence of more than one type in the mixture [31]. The sole presence of H_2S produces the highest corrosion rate of carbon steel [31]. When H_2S and CO_2 are present in a 1:3 ratio, respectively, the rate of corrosion is the lowest [31]. It is evident from these data that H_2S is the main culprit for accelerating corrosion rates in gas processing plants [31]. Corrosion rates are high in parts of the plant where considerable amounts of these acid gases are present.

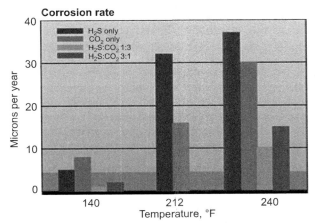

Figure 5.14 Corrosion rate using monoethanolamine (MEA) as the adsorbent with carbon steel as the surrounding wall material [31].

The initial amounts of acid gas loaded as well as the loading temperatures also impact the corrosion rate. The relationship between corrosion rates and CO_2 loading is shown in Fig. 5.15.

The higher the loading temperature, the greater the rate of corrosion of the material as presented in the data above [31].

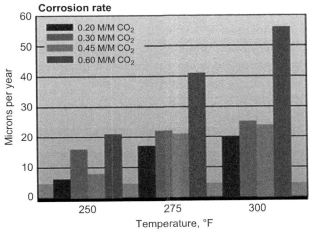

Figure 5.15 Corrosion rate of carbon steel with varying CO_2 loadings [31].

There are different types of corrosion that can occur in gas processing plants which may not be directly linked to acid gases. Some examples of these include general corrosion, galvanic corrosion, and erosion corrosion [31].

General corrosion simply refers to the general degradation of the equipment over time which is expected with continuous operation. General corrosion impacts the entire internal surface area of the material in direct contact with the corrosive chemicals or medium [31].

A second type of corrosion that may occur is galvanic corrosion. It involves two different types of metals which are electrically connected to one another through a conducting salt solution. A potential difference exists across the conducting electrolyte medium connecting the different metals. The metal which has a lower resistance (acts as the anode) will begin to dissolve into solution and will be deposited on the other metal which acts as the cathode [31].

Erosion corrosion results due to the movement of a fluid with corrosive properties at high velocities through the treatment vessel. There will be degradation of the metal surface as a result of the corrosive nature of the chemical species in the fluid as well as from the frictional forces due to relative motion [31].

11. TYPICAL COSTS FOR SHALE GAS PROCESSING

The cost of shale gas processing will depend on several factors which will vary from project to project. Some of these variables include [32]:
- The total recovery from the reservoir;
- Vertical and horizontal well drilling expenses;
- The cost for permission to set up production on the land;
- Cost for constructing facilities, pipelines, and shipment to market;
- Cost for operating the gas processing facilities;
- Taxes and royalties;
- The overhead expenses.

Each of these factors will differ by state. An example of the typical costs for production from different shale plays around the United States as well as the amount of production of shale gas on a daily basis is illustrated in Fig. 5.16.

The average typical cost of production ranges from $4–6 per mcf of gas [32]. However, there are some shale plays where production of mcf of gas exceeds $8, making production uneconomical [32]. Drilling a well with

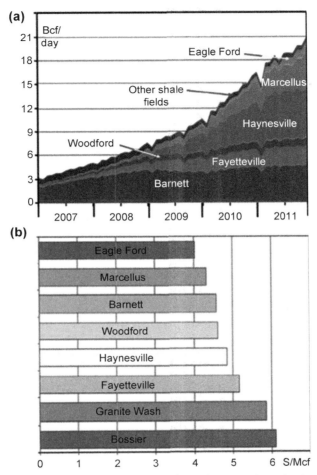

Figure 5.16 (a) Different shale plays in the United States and production levels from 2000 to 2011. (b) The cost for production at each gas play is given per million cubic feet (mcf) [32].

vertical depth of 10,500 ft and a horizontal length of 4000 ft costs around $8 million [32].

When analyzing the economics for a gas processing plant one must consider the volumes of gas to be treated, the energy expenditure by the plant, the flow rates of the different product streams, the market price for the natural gas, and terms and conditions listed in the contract with local governments and other companies [32].

Table 5.2 Shale Gas Project Cost Estimates

Drilling cost, MM$/well	4.0
Well tie-in and manifolds, MM$/well	0.15
Pipeline, MM$	250
Gas processing facility, MM$	600
OPEX	
Well work over cost	1% of CAPEX
Production facility	1% of CAPEX
Pipeline	1% of CAPEX
Processing plant	2.5% of CAPEX
All costs inflated by 2.5% per year	
100% working and net revenue interest.	
2013 condensate price of $90/bbl with 2.5% inflation per year	

*MM$*, Million US$.

The cost of gas development and technological challenges increase as we move from associated gas, nonassociated gas, tight gas, and coal bed methane shale gas hydrates. The cost of the nonassociated gas is further dependent upon the quality of gas, i.e., rich gas versus lean gas and whether this gas is cleaner or it has lot of impurities such as hydrogen sulfide (H_2S), carbon dioxide (CO_2), and nitrogen (N_2) (see Table 5.2) [33].

REFERENCES

[1] Linley D. Fracking under pressure: the environmental and social impacts and risks of shale gas development. Toronto: Sustainalytics; August 2011.
[2] Scott Institute & Carnegie Mellon University. Shale gas and the environment. Pittsburgh (PA): Wilson E. Scott Institute for Energy Innovation; March 2013.
[2a] US Energy. 2011. http://www.eia.gov/todayinenergy/detail.php?id=110.
[3] Allan D. Understanding shale gas. Mount Royal University; 2013. http://www.csur.com/images/CSUG_presentations/2013/Understanding_Shale_Gas_in_Canada_MRU_presentation.pdf.
[4] Energy Information Administration. Natural gas processing: the crucial link between natural gas production and its transportation to market. Natural Gas Processing: The Crucial Link Between Natural Gas Production and Its Transportation to Market. 2006. http://www.eia.gov/pub/oil_gas/natural_gas/feature_articles/2006/ngprocess/ngprocess.pdf.
[5] Avestar. Avestar – shale gas processing (SGP). 2013. http://www.alrc.doe.gov/avestar/simulators_shale_gas.html.
[6] Shah K. Transformation of energy, technologies and purification and end use of shale gas. Journal of Combustion Society of Japan 2013;55(171):13–20.
[7] Bland R. Glycol derivatives and blends thereof as gas hydrate inhibitors in water base drilling, drill-in, and completion fluids. Google Books; 1998. http://www.google.com/patents/WO1998040446A1?cl=en.
[8] Nemati Rouzbahani A, Bahmani M. Simulation, optimization, and sensitivity analysis of a natural gas dehydration unit. Journal of Natural Gas Science and Engineering 2014;21:159–69.

[9] EPA. Natural gas dehydration. Environmental Protection Agency; 2007. http://epa.gov/gasstar/documents/workshops/college-station-2007/8-dehydrations.pdf.
[10] Pall. Glycol dehydration. Membrane Technology 1991;12:14. http://www.pall.com/pdfs/Fuels-and-Chemicals/HCP-10.pdf.
[11] EPA. Zero emissions dehydrators. Natural Gas EPA Pollution Prevention; 2006. http://www.epa.gov/gasstar/documents/zeroemissionsdehy.pdf.
[12] huffmaster M. Gas dehydration fundamentals introduction. In: Laurancereid gas conditioning Conference. Hell Global Solutions; 2004.
[13] Elodie Chabanon, Bouchra Belaissaoui. Gas–liquid separation processes based on physical solvents: opportunities for membranes. Journal of Membrane Science 2014; 459:52–61.
[14] Scholes CA, Anderson CJ. Membrane gas separation – physical solvent absorption combined plant simulations for pre-combustion capture. Energy Procedia 2013;37: 1039–49.
[15] Beychok M. Amine gas treating. NuTec; 2012. http://www.eoearth.org/view/article/51cbf2257896bb431f6a805c [Molecular Sieve].
[16] Qiu K, Shang J. Studies of methyldiethanolamine process simulation and parameters optimization for high-sulfur gas sweetening. Natural Gas Science and Engineering 2014;21:379–85.
[17] Fouad WA, Berrouk AS. Using mixed tertiary amines for gas sweetening energy requirement reduction. Natural Gas Science and Engineering 2014;21: 26–39.
[18] Azman Shafawi LE. Preliminary evaluation of adsorbent-based mercury removal systems for gas condensate. Analytica Chimica Acta 2000;415(1–2):21–32.
[19] Johnson K. Natural gas processesors respond to rapid growth in shale gas production. August 10, 2014. Retrieved from: The American Oil & Gas Reporter: http://www.aogr.com/magazine/cover-story/natural-gas-processors-respond-to-rapid-growth-in-shale-gas-production.
[20] Dart Energy. The shale extraction process. August 10, 2014. Retrieved from: Dart Energy: http://www.dartgas.com/page/Community/The_Shale_Process/.
[21] Chevron Energy. How we operate. August 10, 2014. Retrieved from: Developing Shale Gas Wells, From Leasing to production: http://www.chevron.com/deliveringenergy/naturalgas/shalegas/howweoperate/.
[22] Nematollahi M, Esmaeilzadeh F. Trace back of depleted-saturated lean gas condensate reservoir original fluid: condensate stabilization technique. Journal of Petroleum Science and Engineering 2012;98–99:164–73.
[23] Masoumeh Mirzaeian AM. Mercaptan removal from natural gas using carbon nanotube supported cobalt phthalocyanine nanocatalyst. Natural Gas Science and Engineering 2014;18:439–45.
[24] Stephen Santo, Mahin Rameshni. The challenges of designing grass root sulphur recovery units with a wide range of H_2S concentration from natural gas. Natural Gas Science and Engineering 2014;18:137–48.
[25] Street R, Rameshni M. Sulfur recovery unit expansion case studies. Worley Parsons Resources and Energy; 2007. http://www.worleyparsons.com/CSG/Hydrocarbons/SpecialtyCapabilities/Documents/Sulfur_Recovery_Unit_Expansion_Case_Studies.pdf.
[26] Torres CM, Mamdouth Gadalla, Mateo-Sanz JM. An automated environmental and economic evaluation methodology for the optimization of a sour water stripping plant. Journal of Cleaner Production 2013;44:56–68.
[27] Daniel Sujo-Nava LA. Retrofit of sour water networks in oil refineries: a case study. Chemical Engineering and Processing: Process Intensification 2009;48(4):892–901.
[28] Wikipediaorg. Shale oil extraction. August 15, 2014. Retrieved from: Wikipedia.org: http://en.wikipedia.org/wiki/Shale_oil_extraction.

[29] Wang H, Duncan I. Understanding the nature of risks associated with onshore natural gas gathering pipelines. Journal of Loss Prevention in the Process Industries May 2014; 29:49—55.
[30] Dupart, Bacon, Edwards. Understanding corrosion in alkanolamine gas treatment plants. 1993. http://www.ineosllc.com/pdf/Hc%20Processing%20Apr-May%201993.pdf.
[31] Mearns E. What is the real cost of shale gas?. Energy Matters; 2013. http://euanmearns.com/what-is-the-real-cost-of-shale-gas/.
[32] U.S. Energy information (EIA). World shale gas resources: an initial assessment of 14 regions outside the united estate. U.S. Energy information (EIA); April 2011.

CHAPTER SIX

Shale Oil: Fundamentals, Definitions, and Applications

1. INTRODUCTION

Mankind was using oil shale because of its combustible property, which does not require complicated processing. Its utilization began in ancient times, when it was employed for decoration and construction purposes. Shale oil was used in medical and military sectors as well [1—4]. The oil shale was exploited by different countries until the 17th century. One of the interesting oil shales is the Swedish alum shale of Cambrian and Ordovician age because of its alum content and high concentrations of metals including uranium and vanadium. As early as 1637, the alum shales were roasted over wood fires to extract potassium aluminum sulfate, a salt used for tanning leather and also for fixing colors in fabrics. Late in the 1800s, the alum shales were retorted on a small scale for hydrocarbons. Production continued through World War II but ceased in 1966 because of the availability of cheaper supplies of petroleum crude oils. A part of uranium and vanadium was also extracted from alum shales. By the mid 19th century, the modern application of oil shale was initiated and started growing before World War I because of bulk production of automobiles. Due to the massive demand of this type of transportation, the consumption of the gasoline was increased and the supply was decreased. Hence, during World War I, the oil shale projects were developed in various countries. Because of the easy, convenient, and economic accessibility of conventional crude oils, the decline in production of the oil shale formations was observed after World War II. As of 2010, oil shale was commercially employed in Estonia, China, and Brazil, while several countries are thinking about starting and restarting commercial use of oil shale [1—4].

Naturally occurring shale oil is stored in shales. Shale is similar to other naturally occurring organic sediments. It has diverse composition with changeable properties according to the site of deposit and type of formation. Primarily shale is combination of a large amount of kerogen, which is a recipe of organic compounds. Its key composition is kerogen, quartz, clay, carbonate, and pyrate, whereas uranium, iron, vanadium, nickel, and molybdenum exist in shales as the secondary components. From shale rock, the shaly liquid oil and gas are extracted. Shale oil is the replacement for artificial crude oil; however, extraction from oil shales is expensive compared to that from usual crude oils. Compared to crude oils, a sample of shale oil also has a higher concentration of sulfur and nitrogen. Hence, these components should be removed in order to attain favorable oil through hydrotreating, hydrocracking, and delayed coking. The composition of crude oils is not the same throughout the world so that it mainly depends upon geographical structure and other factors such as temperature and depth. Its viability is a strong function of the cost of the conventional crude oil [1—5].

The naturally occurring, solid, insoluble organic matter, which occurs in source rocks, can yield oil upon heating. Kerogen is a portion of naturally occurring organic matter, but is nonextractable using organic solvents [5,6]. It is typically insoluble in most organic solvents because its components have a high molecular weight (more than 1000 Daltons; 1 Da = 1 atomic mass unit) [5,6].

Every kerogen particle is exceptional, since it has patchwork structures framed by the irregular mixtures of a lot of smaller molecular parts. The physical and chemical characteristics of kerogen are strongly impacted by a variety of biogenic particles that result in the formation of kerogen through transformation of those constituent molecules [1—4].

Components of kerogen are influenced by the thermal maturation processes (e.g., catagenesis and metagenesis) that change the initial kerogen. Heating kerogen below the surface leads to chemical reactions that consequently break off fragments of kerogen into gas or oil molecules. The leftover kerogens additionally experience paramount changes, which are reflected in their chemical and physical properties [7,8].

In the mid-2000s, a valuable plan for portraying kerogens was created by the French Petroleum Institute, which is still considered a standard. They distinguished three main kinds of kerogen (called Types I, II, and III) and have investigated the compound characteristics [2,3,6]. Type IV kerogens have also been identified after subsequent experiments. The first kind of kerogen (Type I) consists of mainly algal and amorphous (but presumably algal) kerogen so that it is highly likely to generate oil. The second category (Type II) includes mixed terrestrial and marine source materials that can produce waxy oil. The third group (Type III) is attributed to woody terrestrial source materials that typically result in gas production. The last type of kerogen (Type IV) has no hydrocarbon potential as the ratio of hydrogen/carbon is very low, while the kerogen has a high magnitude of oxygen/carbon ratio. Hence, even maturation process leads to producing just dry gas. Generally, the Type I and Type II kerogens in most oil shales are not yet mature enough to generate hydrocarbons. As these kerogens mature (normally via geologic burial and the increased heat associated with it), they turn into oil, and then with more heat, to gas. Methods that accelerate the maturation process attempt to control input heat, and eventually produce the desired types of hydrocarbons [6–9].

This chapter describes fundamentals, definitions, and applications of shale oil through a systematic manner.

2. TYPES OF CRUDE OIL AND OIL RESERVOIRS

Crude oil is not a single compound. It is a mixture of hydrocarbon molecules. Depending on the mixture of hydrocarbon molecules, crude oil varies in color, composition, and consistency. Different oil-producing areas yield significantly different varieties of crude oil. The words "light" and "heavy" describe a crude oil's density and its resistance to flow (viscosity) [10,11]. Those known as light are low in metals and sulfur content, light in color and consistency, and flow easily. Less expensive, low-grade crude oils, which are higher in metals and sulfur content, and must be heated to become fluid, are known as heavy. The term "sweet" is utilized to describe crude oil that is low in malodorous sulfur compounds such as hydrogen sulfide and mercaptans, and the term "sour" is used to describe crude oil containing high malodorous sulfur compounds. There are four main categories for petroleum oils as described next [11–13].

2.1 Category 1: Light, Volatile Oils

These oils are highly fluid, usually flammable, often clear, and spread rapidly on solid or water surfaces. In addition, they have a strong odor as well as high evaporation rate. Most refined products and a number of the highest quality light crudes can be placed in this group [11–13].

2.2 Category 2: Nonsticky Oils

These oils have a waxy or oily feel. This category includes less toxic oils, and they adhere more firmly to surfaces compared to category 1, although they can be removed from surfaces by vigorous flushing. Medium to heavy paraffin-based oils fall into this class [11–13].

2.3 Category 3: Heavy, Sticky Oils

Characteristically, the oils in this category are brown or black, sticky or tarry, and viscous. This class includes residual fuel oils and medium to heavy crudes. The density of this group of oils might be close to that of water. Hence, they often sink [11–13].

2.4 Category 4: Nonfluid Oils

These oils do not penetrate porous substrates. They are relatively nontoxic and usually black or dark brown in color. High paraffin oils, residual oils, weathered oils, and heavy crude oils are included in this class [11–13].

It should be noted that shale oil is petroleum that consists of light crude oil contained in petroleum-bearing formations of low permeability called shale [2,3]. Therefore, shale oil can fall into category 1.

Reservoir rock is defined as a permeable subsurface rock that contains petroleum. The rocks should be porous and permeable. Reservoir rocks are dominantly sedimentary (sandstones and carbonates); however, highly fractured igneous and metamorphic rocks have been known to produce hydrocarbons, albeit on a much smaller scale. The three sedimentary rock types most frequently encountered in oil fields are shales, sandstones, and carbonates [12,13].

3. SHALE OIL

The rate of silt and organic matter gathering would on occasion surpass the bowl floor subsidence and fill the lake bringing the water level to shallow depths and periodically drying out of the lake [1–4]. The new subsidence would permit the lake to change. After some time, this cycle of filling and subsidence brought about the collection and safeguarding of the thick grouping of algae-rich lake silt, which is now called oil shale. This oil shale contains kerogen, an organic chemical compound from which we can generate oil [6,9].

Shale oil was discovered in the early 1900s. Although several pilot plants were developed, no appreciable amounts of shale oil were actually produced or shipped until 1921, when 223 barrels of shale oil were produced [1–3,6]. Interest in shale oil production was extremely low from 1921 to 1944 due to the discoveries of much more profitable conventional oil sources found elsewhere [1–3,6]. However, there was renewed interest in oil shale due to fuel shortages caused by World War II. From 1944 to 1969, research activity was steady, and many different proposals were made for extracting shale oil, including using a nuclear device to fracture shale oil. Over this time period, the first large-scale mining and retorting operations for shale oil were developed. Research studies conducted about retorting and shale oil technologies were relatively rare from 1969 to 1973 [1−4]. The oil embargo of 1973 again renewed interest in shale oil, with research programs established in Colorado [1−4]. However, all research in the United States was halted in the 1980s due to the considerable project costs, and low global oil prices [1,14]. Contrary to the United States, Canada continued to develop the oil sands during the 1980s and is producing over one million barrels/day, mostly exported to the United States [14,15].

Shale oil is unconventional oil, which is produced through oil shale pyrolysis, hydrogenation, or/and thermal dissolution [1,9,16]. Generally, the term oil shale refers to any type of sedimentary rock that contains solid bituminous materials (called kerogen) that are released as petroleum-like liquids when the rock is heated in the chemical process of pyrolysis [9,16].

Formation of oil shale was from the same process that crude oils were generated millions of years ago, mainly by deposition of organic debris on ocean, lake, and sea beds. Indeed, they were formed in long periods of time under high temperature and pressure conditions, which were also responsible for formation of crude oil and natural gas; however the temperature and pressure were not that severe in oil shale cases. The oil shale is sometimes called "the rock that burns" as it contains enough oil to burn itself [1−3,9].

Shale oil can be obtained by mining and processing the shale rocks. However, extracting oil from oil shale is not as simple as extracting oil from underground oil reservoirs, since shale oil naturally occurs as solid particles that cannot be directly pumped to the ground. Extracting shale oil is much more expensive than conventional methods used to produce crude oil as oil shale has to be mined first and then heated to a high temperature (process called retorting). Heating oil shale melts it and resultant fluids are separated and collected. These days, experiments are being conducted to develop a process for in situ retorting, which includes heating oil shale, while it will occur underground, and then the consequential fluid is pumped to the ground [1,9,16]. Fig. 6.1 shows a sample of oil shale.

Figure 6.1 A sample of oil shale [1,3].

4. SHALE OIL COMPOSITION

The oil recovered from the processing of the kerogen-rich oil shale is called shale oil. In some prospects, shale oil is similar to conventional oil; however, there are a number of differences. For instance, the quality of shale oil tends to be lower as it contains higher concentrations of impurities (e.g., sulfur and nitrogen) in contrast with conventional oil (see Table 6.1). Also, shale oil produced via pyrolysis appears to be unsaturated and deficient in hydrogen [15,17]. Using standard refining practices such as hydrotreating, these impurities can be removed, and the hydrogen deficiency is eliminated. This process and other treatment methods to improve the quality of nonconventional oil resources are known as upgrading. The upgraded oil product can be sold as a synthetic crude oil or/and finished products, including gasoline and diesel [1,9,17]. A proper comparison between elemental composition of shale oil, crude oil, and coal is presented in

Table 6.1 Elemental Composition of Oil Shale, Crude Oil, and Coal [17]

Type/Location	% Organic C	%S	%N	%Ash
Oil shale, Piceance, Colorado	12.4	0.63	0.41	65.7
Oil shale, Dunnet, Scotland	12.3	0.73	0.46	77.8
Oil shale, Alaska	53.9	1.50	0.30	34.1
Coal, subbituminous C., Wyoming	61.7	3.40	1.30	14.9
Coal, bituminous C., Colorado	73.1	0.60	1.50	7.2
Coal, anthracite, Pennsylvania	76.8	0.80	1.80	16.7
Crude oil, Texas	85.0	0.40	0.10	0.5
Shale oil, Colorado	84.0	0.70	1.80	0.7

Table 6.2 Chemical Composition of a Sample of Shale Oil and Corresponding Artificial Crude Oil [17]

Component/Ratio	Shale Oil	Synthetic Crude Oil
C/H ratio	1.6	~1.25
Nitrogen	2%	<0.5%
Sulfur	10%	1—3%

Table 6.1. For more clarification, the significant difference between characteristics (C/H, N2%, and S %), a sample of shale oil and corresponding artificial crude oil is depicted in Table 6.2. Table 6.3 also tabulates the elemental composition of various oil shales in different regions.

Oil shale, peat, and tar sands have significant ordinary individuality. Shale oil is less combustible than natural gas. Therefore, its weight basis energy content is lower than natural gas, petroleum, and coal. Its organic content is 50% by weight of a sample and the inorganic content is 30% for very rich shales and 95% for very lean shales. Its physical properties show that shale rocks are brittle and rigid enough to maintain open fractures. It is mostly composed of solid organic materials of mean molecular weight of 3000 g/gmol [1—3,6,9].

It is clear that the oil shale composition varies place to place. Hence, shale oil's composition depends on its parent source material (shale) from which it was extracted or harnessed. In a comparative analysis, shale oil is heavier, more viscous, and has a greater composition of nitrogen and oxygen in contrast with natural or conventional petroleum.

Shale comprises of low-molecular-weight oxygen compounds which are principally carboxylic acids, nonacidic oxygen compounds (e.g., ketones), and phenolics [1—3,9]. Shales also contain strong basic nitrogen compounds

Table 6.3 Elemental Composition of Various Oil Shale [17]

Location of Sample	% Organic C	%S	%N	%Ash	Fischer Assay
Piceance, Creek, CO	12.4	0.63	0.41	65.7	28
Elko, NÉE	8.6	1.10	0.48	81.6	8.4
Lone, CA	62.9	2.10	0.42	23.0	52
Soldier Summit, UT	13.5	0.28	0.39	66.1	17
Rifle, CO	11.3	0.54	0.35	66.8	26.2
Cleveland, OH	11.3	0.84	0.40	72.3	7.9
Condor, Australia	15.9	0.22	0.39	64.1	32

Table 6.4 Composition of Organic Matter in Shale Oil [1,9,17]

Element	Range (%)	Average (%)
C	77.11–77.80	77.45
H	9.49–9.82	9.70
O	9.68–10.22	10.01
N	0.30–0.44	0.33
S	1.68–1.95	1.76
Cl	0.60–0.96	0.75

such as amine, acridine, quinolone, pyridine, and their alkyl substituted derivatives. Carbazole, indole, pyrrole, and their derivatives are considered as the weak basic nitrogen compounds. The nonbasic constituents are nitrile and amide homologues. Thiols, sulfides, and thiophenes comprise the sulfur compounds in the shale oils. Elemental sulfur is present in some crude shale oils but is absent in others [1–3,15].

Typical ranges of concentration of organic matter in shale oil are given in Table 6.4.

Millions of years ago, deposition of silt and organic debris on lake beds and sea bottoms resulted in formation of oil shale. Like the process that forms conventional oil, these materials have been transformed into oil shale over a very long period of time by heat and pressure. Although the process is similar to the conventional oil formation, the process was not complete due to lower intensity of pressure and heat; leading to a flammable rock (shale) rather than a liquid. Oil shale can be differentiated from coal since the organic matter in coal has a lower atomic H:C ratio and the OM:MM ratio of coal is usually greater than 4.75:5.0, where OM and MM stand for organic matter and mineral matter, respectively [1–3,15–17]. Table 6.5 is provided for further comparison of shale oil, coal, and crude oil, as well.

In the natural process, it will take millions of years of burial at temperatures between 100 and 150°C for shale rocks to generate oil. However, quick heating of the kerogen-rich rock at greater temperatures reduces this time to a few years or even minutes. Therefore, liquid hydrocarbons are generated in much less time because of this accelerated operation [1–5].

As an example, the typical composition of Green River Oil Shale is listed in Table 6.6.

Table 6.5 Properties of Shale, Coal, and Crude Oil [1,9,17]

Parameter		Crude Shale Oil	Coal Liquids		Arabian Light Crude Oil
			COED	H-Coal	
Specific gravity		0.92	1.13	0.92	0.85
API		22.2	−4	23.0	34.7
Boiling range, °C		60–540	—	30–525	5–575+
Composition	Nitrogen, wt%	1.8	1.1	0.1	0.08
	Sulfur, wt%	0.9	2.8	0.2	1.7
	Oxygen, wt%	0.8	8.5	0.6	—
C/H ratio		7.3	11.2	8.1	6.2
Pour point, °C		15.55	37.78	<−15	−26.11
Viscosity @ 37.8°C, cp		18.34	31.64	—	5.3
Hydrogen added	SCF/bbl product	0	0	6000	0
	wt%	0	0	10	—
Yield, Gallons/ton		25–35	30–48	60–90	—

COED, Char Oil Energy Development.

Table 6.6 Properties of Green River Shale Oil (Typical Composition) [1,9,17]

Parameter		Shale Oil
Kerogen content, wt%		15
Simple chemical formula (sulfur and nitrogen replaced by oxygen)		$C_{20}H_{32}O_2$
Mineral content, wt%		60–540
Kerogen composition	Nitrogen, wt%	2.4
	Sulfur, wt%	1.0
	Oxygen, wt%	5.8
	Carbon, wt%	80.5
	Hydrogen, wt%	10.3
Mineral composition	Carbonates, wt%	48.0
	Feldspares, wt%	21.0
	Quartz, wt%	15.0
	Clays, wt%	13.0
	Analcite and Pyrite, wt%	3.0

5. KEROGEN AND ITS COMPOSITION

Kerogen is described as a complex waxy mixture of hydrocarbon compounds that is the principal organic constituent of oil shale. Kerogen is the naturally occurring, solid, insoluble organic matter that occurs in source rocks and can yield oil by heating [1−3,6]. Kerogen is insoluble in water and also organic solvents such as benzene or/and alcohol. Therefore, kerogen cannot be extracted using organic solvents. Upon heating under pressure, however, the large paraffin molecules break down into recoverable gaseous and liquid substances resembling petroleum. This property makes oil shale a potentially important source of synthetic crude oil. The name was first used for the carbonaceous materials found in oil-bearing shales in Scotland [1−3,6,9].

In general, there are four categories for kerogen which implies their quality and likeliness to produce oil (see Fig. 6.2 known as Van Kerevelen diagram):

- Type I (Sapropelic): This type of kerogen is very uncommon, and is essentially derived from lacustrine algae growth. The well-known sample is the Green River Shale of Eocene age, from Wyoming, Utah, and Colorado. These oil shale reserves have created a lot of interest and have prompted numerous investigations about the Green River Shale

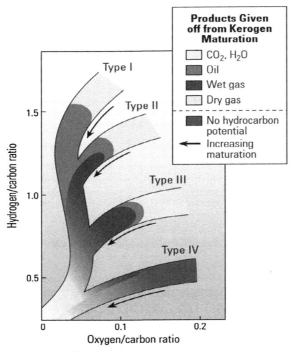

Figure 6.2 Different types of kerogen in terms of hydrogen/carbon and oxygen/carbon ratios [6].

kerogens so that it substantially increased reputation of Type I kerogens. Occurrence of Type I kerogens are restricted to anoxic lakes and to a couple of abnormal marine environments [1−3,6,9].

- Type II (Planktoni): This type of kerogen emerges from different sources, including leaf waxes, marine algae, dust and spores, and fossil resins. They likewise incorporate contributions from cell lipids of bacterial species. Various species of Type II kerogens are grouped together (in spite of their extremely divergent origins) so that they all have incredible abilities to produce liquid hydrocarbons. Most Type II kerogens are found in marine sediments under anaerobic conditions [1−3,6,9].
- Type III (Humic): This type of kerogen is composed of land-based organic matter that is deficient in fatty or waxy parts. Cellulose and lignin are the main components. This type of kerogen has much lower ability to generate hydrocarbons than do Type II kerogens. In case they have little incorporations of Type II material, there normally is a possibility of gas generation [1−3,6,9].
- Type IV (Residue): These kerogens are mainly constituted of restructured organic debris and much oxidized matter of various origins. In general, they do not have any capability to produce hydrocarbons. The type of kerogen is mainly correlated with the amount of hydrogen content present in it (H/C ratio). Type I kerogen contains much fewer rings or/and aromatic structures so they have the highest hydrogen-content (H/C $>$ 1.25). Type II kerogen also has high H/C ratios ($<$1.25). Type III kerogen includes much less hydrogen content (H/C $<$ 1) because of heavy aromatic structures existing in them. Type IV kerogen holds the lowest amount of hydrogen (H/C $<$ 0.5) as it has polycyclic aromatic content. In case of oxygen content, Type IV has the maximum amount of oxygen as the kerogens are highly oxidized, followed by Type III, Type II, and Type I [8,18].

It seems important to scientists and engineers working on oil and gas shale to characterize the shale fluids and their primary components through proper and accurate manners. One of the main characters of shale oil is the kerogen composition.

Composition of kerogens differs from place to place; there are no precise techniques available to find out detailed composition of kerogens, though some researchers made efforts to characterize various types of kerogens using simple but effective methods. For instance, in a research study conducted by Menzela et al. entitled "The Molecular Composition of Kerogen in Pliocene Mediterranean Sapropels and Associated

Homogeneous Calcareous Ooze" [7,19], the researchers employed compound identification on the basis of mass spectral and retention time data. Relative measures of individual components were obtained through integration of individual peak zones from summed mass chromatograms utilizing applicable characteristic mass fragment (m/z) values [7,19]. The measured peak zones were further multiplied by respective correction factors as these m/z values show very less ion count than actual. The correction factors were selected from the study performed by Hartgers et al. Part of the results are shown in Table 6.7 [7,19–21].

There is another study, "Investigations of the Hydrocarbon Structure of Kerogen from Oil Shale of the Green River Formation," by J.J. Schmidt Collerus and C.H. Prien [8]. According to their investigation, the Green River Formation constitutes about 16% of insoluble organic matter, which is known as kerogen [8,19]. This accounts for around 80% of total organic matter present; the remaining 20% of organic matter is the soluble bitumen [8,19]. Basically, two techniques in combination were used in this study: micropyrochromatography and mass spectrometry. The utilized apparatus is depicted in Fig. 6.3 [8].

Table 6.8 reports main fragmentation products of the kerogen. Figs. 6.4 and 6.5 also illustrate the generalized structure of kerogen and schematic structure of kerogen matrix, respectively.

In general, kerogen molecules are very large and complicated. Kerogen consists essentially of paraffin hydrocarbons, though the solid mixture also includes nitrogen and sulfur. Algae and woody plant materials are typical organic constituents of kerogen. Compared to bitumen or/and soluble organic matters, kerogens have a high molecular weight. During petroleum generation, bitumen is formed from kerogen. When the kerogen is heated to temperatures greater than 425–500°C (800–930°F) in an anaerobic atmosphere, the large kerogen molecules are normally broken into smaller pieces, similar to conventional oil and gas in properties and content. This heating and decomposition process is called pyrolysis [1,6,19].

6. TYPES AND SOURCE OF SHALE OIL

Shale oil is also known by various names such as cannel coal, boghead coal, alum shale, stellarite, albertite, kerosene shale, bituminite, gas coal, algal coal, wolongite, schistes bitumineux, torbanite, and kukersite. The classification of shale oil is mainly on the basis of the environment deposits, organic matter, and precursor organisms from which the organic matter is derived

Table 6.7 Characteristic m/z Values and Correction Factors Used to Compute Relative Abundances and Internal Distribution in Pyrolysates [7,19–21]

Compound Class	Characteristic Mass Fragment (m/z)	Correction Factor
n-Alkanes	55 + 57	2.9
n-Alkenes	55 + 57	4.9
Unsaturated isoprenoids	55 + 57	4.9
Saturated isoprenoid	55 + 57	2.9
Methyl ketones	58	3.6
Alkyl benzenes	78 + 91 + 92 + 105 + 106 + 119 + 120 + 130 + 133 + 134	1.6
Alkyl thiophenes	84 + 97 + 98 + 111 + 112 + 125 + 126 + 139 + 140	2.5
Alkyl phenols	94 + 107 + 108 + 121 + 122	2.2
Alkyl pyrroles	80 + 81 + 94 + 95 + 108 + 109 + 122 + 123 + 136 + 137	1.5
Oxygenated aromatics	78 + 91 + 92 + 105 + 106 + 119 + 120 + 133 + 134	2.6
Alkylindenes	115 + 116 + 129 + 130 + 143 + 144	1.8
Indole	117 + 132	7.0
Phytadiene/phytene	55 + 57 + 68 + 70 + 82 + 95 + 97 + 123	2.4
Prist-1-ene/prist-2-ene	55 + 57	4.9
3-Ethyl-4-methyl-1H-pyrrole-2,5-dione	53 + 67 + 96 + 110 + 124 + 139	2.9
6,10,14-Trimethyle-2-pentadecan-2-one	58	7.5
C20 isoprenoid thiophenes	98 + 111 + 125	2.6

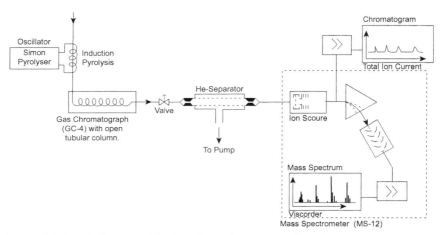

Figure 6.3 A technique combination of gas chromatograph and mass spectrometer to determine concentration of organic matter [8].

[1—3,9]. In 1987, Hutton classified oil shale into three broad groups such as terrestrial, lacustrine and marine. The terrestrial oil shale is composed of lipid-rich organic matter; lacustrine oil shale includes organic matter derived from algae; and marine shale oil is made of organic matter derived from marine animals [1—4,9,19].

In 1987, Hutton also specified six particular shale oils such as cannel coal, lamosite, marinite, torbanite, tasmanite, and kukersite [22]. However, little data is available about oil shale reserves, and the accurate evaluation of shale oil reserves is very difficult due to a variety of analytical techniques and apparatus employed. The unit/measure mostly used for the deposit is in US or Imperial gallons of shale oil per short ton of rock, liters of shale oil and kilocalories per kilogram of oil shale of gigajoules per unit weight of oil shale.

7. OCCURRENCE AND HISTORY OF SHALE OIL

Oil shale is regarded as an unconventional or alternate fuel source of hydrocarbons, which has not been fully tapped yet [1—4]. As oil shale burns without any dispensation or processing, it has been used by mankind since prehistoric times. It was also utilized as decorative, health, military, and construction purposes [1—4].

The thermal decomposition of kerogen yields oil production. Kerogen is not freely extractable as it is closely bound within the

Table 6.8 Main Fragmentation Products of Kerogen Concentrate [8,19]

No.	Name	Formula	Identified in Fraction
1	Aliphatic Hydrocarbons	n-C_{10} to n-C_{34} b-C_{10} to b-C_{36}	85 - 7 122 - 1
2	Alicyclic Hydrocarbons		
	Cyclohexanes	$C_{10-13}\,H_{21-27}$	123 - 1
	Decalins	$C_{5-8}\,H_{11-17}$	123 - 1
3	Hydroaromatic Hydrocarbons		
	Dialkyltetralins	$C_{2-5}\,H_{5-11}$	122 - 1 123 - 1
		$C_{8-12}\,H_{17-25}$	122 - 1
	Hexahydro-phenanthrenes	$C_{1-3}\,H_{3-7}$ + 6H	123 - 1
4	Dialkylbenzenes	$C_{8-13}\,H_{17-27}$	123 - 1
5	Dialkylnaphthalenes	$C_{3-4}\,H_{7-9}$	123 - 1 123 - 4
6	Alkylphenanthrenes	$C_{1-3}\,H_{3-7}$	121 - 4 123 - 2

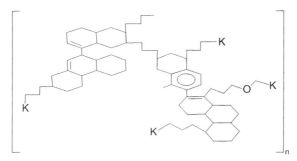

Figure 6.4 Generalized structure of kerogen of the Green River Shale formation [8,19].

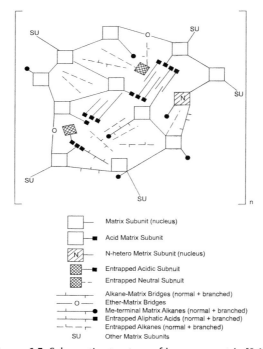

Figure 6.5 Schematic structure of kerogen matrix [8,19].

mineral matrix of shale [1,9]. Oil shale typically exists in shallower geological zones (less than 3000 ft), while warmer and deeper zones are required for the formation of conventional oils [1,4,9]. It is believed that the worldwide shale oil resources are capable of producing 4.8 trillion barrels. Hence, it seems logical to assume that the halfway point is more than three times the proven oil reserves in Saudi Arabia [1,4,9]. Considering current petroleum product demand (approximately

20 million barrels per day (Mbbl/day) for the United States alone), the oil shale exploitation can easily last for over 400 years [1,4,9].

The oil shale was exploited by different countries by the 17th century. One of the appealing oil shales is the Swedish alum shale since it contains high concentration of metals plus uranium and vanadium [1−3,6]. In early 1637, the alum shales were heated on wood fires to achieve potassium aluminum sulfate, which is a saline used in the leather tanning process and for fixing colors in clothes. In the 1800s, alum shales were used on a small scale as an alternate for hydrocarbons. The production of alum shale remained in stream line in World War II; however it was banned in 1966 because of the presence of cheaper supplies of petroleum crude oil [1−4,6]. A small quantity of uranium and vanadium were also obtained from shales (alum). By the mid 19th century, the modern use of oil shale was initiated and started expanding before World War I because of massive production of automobiles. Due to the huge increase in this transportation application, the consumption of gasoline increased and the supply decreased. Hence, the oil shale projects were initiated in different countries during World War I [1−4].

About a century ago, oil shale processing originated from Europe, and was produced by an industry in Scotland between the years 1850 and 1930 [9,23]. Toward the middle of the 20th century, shale gas was being produced by other countries since the oil price at that time justified the investment [23]. It should be noted that the supply of conventional petroleum was also limited [1,23]. Production history across various countries is depicted in Fig. 6.6 [23]. Unavailability of large oil fields in the United States in combination with the spike in oil prices allowed a shale oil boom after World War I [23]. Prospectors who staked claims in the west back in the day presently own the bulk of the privately owned oil shale in Colorado,

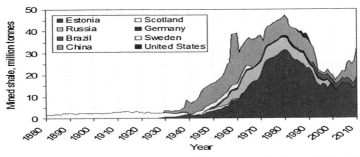

Figure 6.6 History of oil shale mining across different countries from 1880 to 2010 [23].

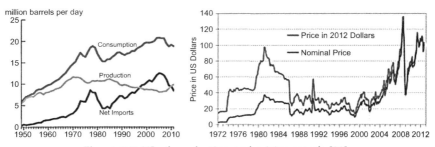

Figure 6.7 US oil production and pricing trends [23].

Utah, and Wyoming, up to 30% of the total resources [1,3–6,23]. The discovery of large oil fields through the 1920s dented the oil price, causing it to drop [6,23]. This halted the first boom as shale oil was unable to economically compete with conventional oils [9,23]. As oil production in the United States peaked in 1970, oil imports became almost double, as illustrated in Fig. 6.7 [23]. Since the start of Arab–Israel war, these oil supplies were cut off by the Organization of Petroleum Exporting Countries (OPEC), which led to an increase in oil prices (see Fig. 6.7) [23]. This prompted the US government to enforce a major push toward development of unconventional oil supplies, and become independent when it comes to energy [23]. In 1974, the federal government came forward with the prototype Oil Shale Leasing Program. This attracted companies for commercialization of oil shale reserves [23].

Over 20 modern technologies for oil shale processing had been conceived and tested by 1990. Most of the research focused on the practical applications of the technology, with a little research into the environmental effects [6,23]. Some of the studies encountered design or technology problems associated with scale-up, but others have been commercialized from the experimental stage. Around 2005, the US research effort for shale gas was rebooted due to the increasing importance of shale gas reservoirs evidenced by the projections for the US natural supply by source, with unconventional reservoirs rising dramatically (see Fig. 6.7) [24]. Furthermore the research today focuses on the practical applications as well as the environmental prospects of each technology.

Today, horizontal drilling and hydraulic fracturing are the two main technologies that have allowed tighter gas reservoirs to be economically viable. Another area of significant research is in the drilling fluids used, which has also helped to reduce drilling costs and improve environmental friendliness [24].

For instance, two important (and well-known) oil shale formations are briefly described next.

7.1 Central Queensland Region

Around 40 million years ago, the central Queensland region experienced the development of an arrangement of expanded, fault-bounded lake basins. At this point, the ocean levels were around 150 meters lower than present with the Gladstone area more raised, and around 200 kilometers southwest of the coastline as it was then. In the area west of Gladstone, around 500 meters of oil shale amassed in the lake bowl known as The Narrows Graben [1–4]. This happened over a 10-million-year period, from the middle of the late Eocene to early Oligocene (around 30–40 million years ago). Fig. 6.8 delineates a typical lake basin from that time. There was rise in water level of the lake, and then a fall because of a mix of (1) steady subsidence of the lake floor in light of tectonic forces and (2) variation in climate [1–4]. The lake water chemistry fluctuated because of the changing depth of water and natural activity that brought about occasional algal blossoms alongside a mixed bag of plants and aquatic creatures (crocodiles, turtles, and fish) [1,4,24]. Fig. 6.8 demonstrates the resultant oil shale as a succession of layer groupings as well. There are around 100 climatically controlled layers, from 1 to 7 m thick (generally, the thickness is around 3 m). It is assessed that each layer took around 75,000 years to gather on the lake floor [24,25].

Figure 6.8 Schematic of Queensland Energy Resources [24].

7.2 The Green River Formation

One specific underground structure, known as the Green River Formation, contains a huge quantity of oil shale. The Green River Formation was made around 48 million years ago amid the Eocene era [1–4]. The district did include an arrangement of intermountain lakes where, over 6 million years ago, fine sediment and organic matter were deposited. Considering very long time periods, the organic matter was converted into the oil shale reservoirs seen today. There are different formations in the strata, deposited in the same way, that also hold oil shale assets. These reservoirs likewise contain possibly reasonable oil shale reserves [6–9]. In March 2009, the United States Geologic Survey completed a reassessment of the present oil shale deposits in this area and expanded their evaluation from around 1 trillion barrels of shale oil to 1.525 trillion barrels of shale oil. The Department of Energy forecasts that around 1.8 trillion barrels of shale oil possibly underlay Colorado, Utah, and Wyoming in reserves more than 15 barrels/ton [24–26]. Fig. 6.9 shows the oil shale deposits in the three US regions [26].

Oil Shale Deposits in the Three-State Area

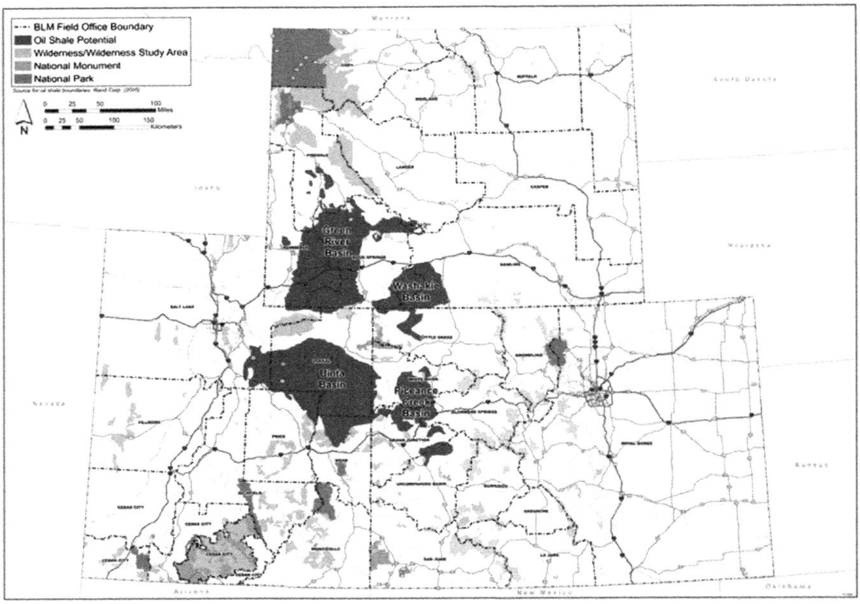

Figure 6.9 Oil shale deposits in the three different US regions [26].

8. DEFINITIONS OF MAIN FACTORS AND PARAMETERS IN SHALE OIL FRAMEWORK

8.1 Depth

The vertical distance between surface and shale reservoir is called depth. The depth of shale formations vary from place to place. For example, North Dakota's shale oil formation is two miles deep [1–4,9].

8.2 American Petroleum Institute Gravity

The lightness or heaviness of the shale oil is normally expressed by American Petroleum Institute (API) gravity, which is related to the specific gravity (or density). API is defined by the following equation [27,28]:

$$\text{API} = \frac{141.5}{\text{SG}} - 131.5 \qquad (6.1)$$

where SG represents the specific gravity with respect to water density at standard conditions.

8.3 Dielectric Constant

The dielectric constant of oil shales exhibits a functional dependency on temperature and frequency. Anomalously high dielectric constants are observed for oil shales at low temperatures and the high values are attributed to electrode polarization effects.

8.4 Self-ignition Temperature of Oil Shale

The self-ignition temperature is the temperature at which oil shale samples spontaneously ignite [1–4,9].

8.5 Porosity (Void Fraction)

This parameter is a measure of the void spaces in a material such as a reservoir rock. Porosity is a fraction of the volume of void space over the total volume, and is expressed as a fractional number between 0 and 1 or as a percentage between 0 and 100. Porosity is about 2–4% for shale formations [1,4,6].

8.6 Permeability

This property is associated with the presence of natural cracks/fractures in the rock, which enable the flow of reservoir fluids between pore spaces.

Permeability controls the flow of natural gas or oil into the borehole and their production. Compared to the conventional reservoirs, oil shale has very low permeability, in the range of 10^{-4} to 10^{-6} microDarcy [2,3,29].

9. OIL SHALE RESERVOIRS

Similar to shale gas, the development of unconventional oil shale is starting to grow across the world, particularly in North America. During the oil crisis of the 1970s, oil shale development came to prominence and was supported by the government as a way to mitigate the risk of foreign imported oil [1,6,9]. However, once oil prices crashed, most of these projects were deemed uneconomical due to the higher cost of extraction, production, and processing operations [1–4,9].

Oil shale is found in many places worldwide; however the Green River Formation in the United States is by far the largest deposit in the world. This formation covers portions of Colorado, Utah, and Wyoming. The oil resource within the Green River Formation is estimated in the range of 1.2–1.8 trillion barrels. While not all resources here are recoverable, even a moderate estimate of 800 billion barrels of recoverable oil from oil shale in the Green River Formation is three times greater than the proven oil reserves of Saudi Arabia [6–9]. If oil shale could be used to meet a quarter of the present US demand for petroleum products of about 20 Mbbl/day, the estimated 800 billion barrels of recoverable oil from the Green River Formation would be enough to last for more than 400 years [1–6,9].

More than 70% of the total oil shale acreage in the Green River Formation, including the richest and thickest oil shale deposits, are under federally owned and managed lands. Thus, the federal government directly controls access to the most commercially attractive portions of the oil shale resource base [6,9].

The world's top 12 countries with the most plentiful shale oil resources (see Fig. 6.10) known are as follows, in billion bbl [29,30]:

1. United States: 2085
2. Russia: 247
3. Congo: 100
4. Brazil: 82
5. Italy: 73
6. Morocco: 53
7. Jordan: 34
8. Australia: 31 (recoverable: 4)
9. Estonia: 16 (recoverable: 1)
10. China: 16 (proven: 1.8)
11. Canada: 15
12. France: 7

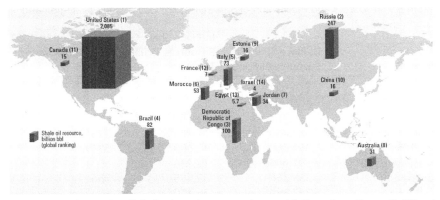

Figure 6.10 The main oil shale deposits across the world. *Data from Knaus E, Killen J, Biglarbigi K, Crawford P. An overview of oil shale resources. In: Ogunsola OI, Hartstein AM, Ogunsola O, editors. Oil shale: a solution to the liquid fuel dilemma. ACS symposium series, vol. 1032. Washington, DC: American Chemical Society; 2010. p. 3–20.*

It is not a simple task to estimate oil shale reserves because of its complexity and dependence on several factors [6–9,24]:
- The amount of kerogen present in oil shale deposited varies considerably [6,9,24].
- Some nations report as reserves the total kerogen content in place regardless of the economic constraints. Typically "reserves" refers only to those resources that are economically feasible and technically exploitable [6,9,24].
- Due to the developing situation of the shale oil extraction, only the recoverable amount of kerogen is estimated. The kerogen contents of oil shale reserves vary, depending upon geographical location. Almost 600 known oil shale deposits are spread diversely throughout the world, and are found in different continents except Antarctica, where the exploration of oil shale reserves has not been yet evaluated [6,9,24].

The estimated amount of shale oil is clearly shown in Table 6.9. It should be noted that the shale oil refers to the synthetic oil that can be obtained by heating the solid organic matter (e.g., kerogen at a high temperature like pyrolysis). Upon heating, the condensable vapors (shale oil), noncondensable vapors (shale gas), and a solid residue (spent shale) are obtained. The units of oil in-place and production presented in the table are barrels and metric tons. Further information for shale resources in each continent/region is separately summarized in the following subsections.

9.1 Asia

In Asia, the major oil shale deposits are located in China and Pakistan, with total estimated reserves of about 52 billion metric tons, out of which

Table 6.9 Resources and Production of Shale Oil at the End of 2008 by Region and Countries [31,32]

Region	In-Place Shale Oil Resources (Mbbl)	In-Place Shale Oil Resources (Million Tons)	Production in 2008 (Thousand Metric Tons)
Africa	159,243	23,317	—
Congo	100,000	14,310	—
Morocco	53,381	8167	—
Asia	384,430	51,872	375
China	354,000	47,600	375
Pakistan	91,000	12,236	—
Europe	358,156	52,845	355
Russia	247,883	35,470	—
Italy	73,000	10,446	—
Estonia	16,286	2494	355
Middle East	38,172	5792	—
Jordan	34,172	5242	—
North America	3,722,066	539,123	—
USA	3,706,228	536,931	—
Canada	15,241	2192	—
Oceania	31,748	4534	—
Australia	31,729	4531	—
South America	82,421	11,794	157
Brazil	82,000	11,734	159
Total	4,786,131	689,227	930

4.4 billion metric tons are technically exploitable and economically feasible. In 2002, China produced almost 90,000 metric tons of shale oil. Thailand has a reserve of 18.7 billion metric tons; the major deposit in Kazakhstan is located at Kenderlyk field, with 4 billion metric tons, and Turkey has the reserves of almost 2.2 billion metric tons. There are also some small oil shale reserves found in India, Pakistan, Armenia, Mongolia, Turkmenistan, and Myanmar [31,32].

9.2 Africa

As clearly shown in Table 6.9, almost 14.31 billion metric tons of shale oil deposits are located in the Democratic Republic of Congo and 8.16 billion metric tons are present in Morocco. The deposits in Congo still need more consideration to be exploited. In Morocco, the largest deposits are found in Tarafaya and Timahdite, and although the deposits are explored in these two areas, the commercial scale exploration has not been started yet. The oil shale reserves also exist in the Safaga-Al-Qusayr and Abu Tartour areas of Egypt, South Africa, and Nigeria [6,24,31].

9.3 Middle East

In the Middle East, the largest oil shale deposits are found in Jordan having 5.242 billion metric tons of shale oil. Israel has deposits of 550 million metric tons of shale oil. Although the sulfur content of Jordanian oil shale is high, its oil is considered high-quality oil compared to that of western US oil shale. The best deposits are found in El Lajjun, Sultani, and the Juref Ed Darawishare, which is located in west-central Jordan. The deposit in Yarmouk extends into Syria, located in the northern border. Most of Israel's deposits are placed in the Negev desert near the Dead Sea. Its oil is relatively low in oil yield as well as heating value [6,24,31].

9.4 Europe

Russia has the biggest oil shale reserves in Europe, approximately equal to 52.845 billion metric tons. Its major deposits are located in the Volga-Petchyorsk province and in the Baltic Oil Shale Basin. Other oil shale deposits are found in Italy having 10.45 billion metric tons, Estonia having 2.49 billion metric tons, France and Belarus having 1 billion metric tons each, Ukraine having 600 million metric tons, Sweden having 875 million metric tons, and the United Kingdom having 500 million tons of shale oil. Some reserves are also found in Germany, Bulgaria, Poland, Spain, Romania, Albania, Austria, Serbia, Luxembourg, and Hungary [6,24,31].

9.5 Oceania

The oil shale deposits in Australia are estimated to be about 58 billion metric tons or 4.531 billion metric tons of shale oil, of which about 24 billion barrels (3.8 billion cubic meters) are recoverable. Mostly deposits are in both eastern and southern states with the biggest potential in the eastern Queensland deposits. Some oil shale deposits have been also found in New Zealand [6,24,32].

9.6 South America

Brazil has the world's second-largest known oil shale resources and is currently the world's second largest shale oil producer, after Estonia. Oil shale resources occur in São Mateus do Sul, Paraná, and Vale do Paraiba. Brazil has developed the world's largest surface oil shale Pyrolysis retort at

Table 6.10 Major Shale Deposits in the United States in Terms of Richness [32]

Deposits	Richness (Gals/t)		
Location	5—10	10—25	25—100
Colorado, Wyoming, and Utah (Green River)	4000	2800	1200
Central and eastern states	2000	1000	N/A
Alaska	Large	200	250
Total	6000+	4000	2000+

Petrosix, with an 11 m (36 ft) diameter vertical shaft. The production of Brazil was almost 2000,000 metric tons in 1999. Some other small deposits of oil shale are also found in Argentina, Chile, Paraguay, Peru, Uruguay, and Venezuela [6,24,31].

9.7 North America

The United States, with oil shale deposits of almost 536 billion metric tons, is not only the largest in the North American region, but also in the world. There are two major deposits: the eastern US deposits, in Devonian-Mississippian shales, which cover 250,000 square miles (650,000 km^2); and the western US deposits of the Green River Formation in Colorado, Wyoming, and Utah, which are among the richest oil shale deposits in the world (see Table 6.10). In Canada, among 19 deposits that have been identified, the best-examined deposits are in Nova Scotia and New Brunswick [6,24,31].

10. PRODUCTION HISTORY OF SHALE OIL RESERVOIRS

The only tight oil play field to provide sufficient production history is Elm Coulee field. This field was identified in eastern Montana in the year 2000 and produces from the Bakken region. It peaked in 2006, producing 53,000 bbl/day from 350 wells or approximately 150 bbl/day/well. Today, this production rate has declined to 24,000 bbl/day, and cumulative production to 113 million bbl of light sweet crudes [31,32].

A brief review and analysis of the Bakken field reveal that more than 600 wells have been drilled, and of this over 1000 laterals so that some of them are as long as 10,000 ft. The thickness of shales in the region ranges from 10 to 45 ft with porosities of 3—9%, and average permeability of 40 µD. The field is to some extent overpressured, with a gradient of 0.52 psi/ft.

Table 6.11 Initial Production Rates for Wells in Tight Oil Plays [31,32]

Name of Shale	Initial Well Production Rate (bbl/day)	Early Well Decline Rates (%)
Barnett	2.0	70
Bakken	2000	65–80
Eagle Ford	1340–2000	70–80
Monterey	623	80
Niobrara	400–700	80–90
Avalon and Bone Spring	534	60
Elm Coulee (Bakken)	425 (multilateral)	65

Multilateral wells show a higher range of initial oil rates (200–1900 bbl/day) compared to vertical wells, which show 100 bbl/day less rate.

The initial well rates and first-year decline rates for Elm Coulee (and the six major shale plays) are depicted in Table 6.11. Note that the decline rate is typically greater than 80% for vertical wells.

All the plays started after 2008 with the exception of Bakken, which had an early start in 2000. This information is incomplete because (1) the data on the number of wells drilled and that are still actively producing is smaller and uneven among various operators, and (2) due to the intense economic activity connected with these plays, information is slow in fetching the public [1–4,6]. Therefore, data for the different parameters is given in the best possible range in contrast to the average values. It is not very easy to obtain accurate and reliable data for shale plays as it is very location sensitive, therefore many wells are drilled to give consistent information.

The general trend shows that well decline rates are characteristically high for tight oil plays, ranging from 65% to 90%, while initial well rates vary from 400 to 2000 bbl/day for the Bakken and Eagle Ford [1–4,9].

The wells from the Niobrara and Monterey regions are still under assessment in certain areas as seen by their high well density. For instance, there are 8 and 12 wells/sq mile, respectively, versus 2 for the Bakken region. As a rough estimate, it can be assumed initial well decline rates for conventional oil field wells are in the range of 5–10% [1–4,32].

As discussed before, shale and tight oil plays are continuous accumulations over widespread areas in contrast to conventional distinct oil fields that are geologically disconnected. Thus, oil fields like Elm Coulee can be better described as sweet spots that have better-quality matrix reservoir properties (e.g., permeability greater than 0.15 mD and extra natural fractures) than the adjacent areas in the play [1–4,9]. The Elm Coulee sweet spot has a recovery factor of 5.6% unambiguously, which refers to the in-place volume of oil. However, when this recovery factor is determined with reference to the complete area adjoining the sweet spot, which includes the productive and nonproductive areas, then the recovery factor declines significantly [1–4,9].

The location of sweet spots within a play is very random. For example, a specific area of the play may have good hydrocarbon potential but extensive achievement in locating the sweet spots is the final determinant [1,6–9].

11. ESTIMATES OF RECOVERABLE SHALE OIL RESOURCES

The correlation between potential production and recoverable oil for five predictable oil fields throughout the world can be shown in Fig. 6.11 [33].

When the values for Elm Coulee field are included they fall perfectly on the trend. The resultant equation is as follows [33]:

$$q_{peak} = 0.26\ K^{0.7088} \qquad (6.2)$$

in which

q_{peak} = potential of the shale play, million bbl/day
K = size of play in billions of barrels

It should be noted that the correlation coefficient (R^2) of this relationship is equal to 0.993.

This correlation is promising since it confirms that tight oil reservoirs do follow the dynamics of the Darcy flow regime as do conventional oil reservoirs.

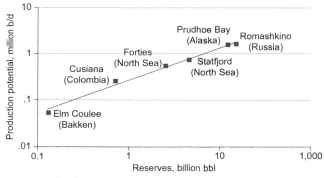

Figure 6.11 Production potential and reserves of giant oil fields [33].

The physical behavior of tight gas sand is the same as that of conventional gas reservoirs. On the other hand, their recovery factors are low, almost up to 6—10% [32,33].

The production potential of major plays can easily be estimated from the preceding algorithm, and this potential is based upon the current estimates of recoverable oil. The Bakken has an estimate of almost 815,000 bbl/day; Eagle Ford, 565,000 bbl/day; and Avalon Spring, 360,000 bbl/day. Niobrara has this value up to 1.03 Mbbl/day, and Monterey has 1.68 Mbbl/day. The huge challenges that lie ahead in these two plays can be resolved by these huge potentials [33].

12. IMPORTANCE OF OIL SHALE

Oil shale can be an important energy resource for many nations. The importance of oil shale is summarized as follows [1—6,31—33]:

- Oil shale is an enormous untapped resource for some countries (e.g., China, United States, and Canada) that can help them become less dependent on foreign sources of hydrocarbons and reduce their costs over the long term [32,33].
- There are looming oil shortages, higher gasoline prices, and political instability resulting from importing so much petroleum [1—4,6,31—33].
- Both the military and public will benefit through the stabilization of gasoline prices, the reduction in the trade deficit, the creation of jobs, the tax and royalty income for local communities, and a more secure future for children and grandchildren [6,31—33].

After required processes (such as processing/refinery and upgrading), shale oils can be employed as transport fuels, raw materials for chemical intermediates, pure chemicals and industrial resins, even kerosene, jet fuel, diesel fuel, and gasoline [1—4].

13. SHORT DESCRIPTIONS OF MAIN COMPANIES INVOLVED IN OIL SHALE DEVELOPMENT

Table 6.12 presents a list of oil shale companies in the United States that are focusing on project and technology development. The main technologies generally include in situ, surface, and upgrading processes.

There are over 34 companies that are actively engaged in shale research and technology development across the world. A list of companies is shown in Table 6.13 [34,35]. Additionally, high-level summary information of

Table 6.12 A Number of Companies Dealing With Various Stages of Production and Processing Operations of Oil Shale in the United States [34,35]

Company	Technology Type	Company	Technology Type
Anadarko Petroleum Corporation	N/A	Great Western energy Corporation	N/A
Chattanooga Corporation	Surface	Western Energy Partners	Surface
Chevron USA	In-situ	Syntec, Inc.	Surface
E.G.L Resources	In-situ	Shell Frontier Oil and Gas, Inc.	In-situ
Electro-Petroleum	In-situ	Raytheon Corporation	In-situ
Brent Freyer, Sc.D	Surface/In-situ	Red Leaf Resources	Surface
Exxon Mobile Corporation	In-situ	Phoenix -Wyoming, Inc.	In-situ
Earth-Search Sciences/Petro-Prob, Inc.	In-situ	Oil Shale Exploration Corporation (OSEC)	Surface
J.W. Bunger and Associates, Inc.	Surface	Natural Soda, Inc.	N/A
Independent Energy Partners	In-situ	Mountain West Energy Company	In-situ
Imperial Petroleum Recovery Corp.	Upgrading	Millennium Synthetic Fuels, Inc.	Surface
Global Resource Corporation	In-situ	James A. Maquire, Inc.	In-situ

these companies is tabulated [34,35]. This summary covers the company's role in development, scope and features of process technology involved, resource locations, technology status, and so on [34,35].

Descriptions of a few high-profile companies are as follows:

Exxon Mobil Corporation (ExxonMobil) — Headquartered in Texas, US, Exxon has operations located all across the world [34,35]. It is a prominent energy company that has been active in all aspects of oil and gas development [34,35]. It has been involved in the shale industry since the 1960s, as a resource owner, technology developer, and project developer [34,35]. Their ongoing research includes field testing and development of in situ technologies and advanced mining and surface retorting processes [34,35]. Their prime objective has been to achieve reduction in costs and environmental impact, thereby enhancing commercial feasibility [34,35].

Table 6.13 A Summary of Companies Involved in R&D of Unconventional Oil and Gas [34,35]

Company	Resource	Project Developer	Technology Developer	Technology Type	BLM	State	Private
Ambre Energy	S/C	*		Surface			*
American Shale Oil	S	*	*	In-situ	*		
Anadarko Petroleum Corp	S						*
Chattanooga Corp	S/T		*	Surface			
Chevron	S	*	*	In-situ	*		
Combustion Resources	S		*	Surface			
Composite Technology Development, Inc.	S/H		*	In-situ			
Electro-Petroleum	S/H		*	In-situ			
Enefit	S	*	*	Surface	*	*	
Enshale	S	*	*	Surface		*	
ExxonMobile	S	*	*	In-situ	**		
Brent Fryer	S		*	Surface/In-situ			*
General Synfuels International	S	*	*	In-situ			
Heliosat, Inc.	S		*	In-situ			
Imperial Petroleum Recovery Corp	H		*	Upgrading			
Independent Energy Partners	S	*	*	In-situ			*
James Q. Maquire, Inc.	S		*	In-situ			

(*Continued*)

Table 6.13 A Summary of Companies Involved in R&D of Unconventional Oil and Gas [34,35]—cont'd

Company	Resource	Project Developer	Technology Developer	Technology Type	Resource Holder BLM	Resource Holder State	Resource Holder Private
J.W. Bunger and Associates, Inc.	S/T		*	Surface			
MCW Energy	T	*	*	Surface		*	
Mountain West Energy	S	*	*	In-situ		*	
Natural Soda Inc.	S	*	*	In-situ	**	*	*
Phoenix - Wyoming, Inc.	S/T/H	*	*	In-situ			
PyroPhase	S/T	*	*	In-situ			
Quasar	T		*	In-situ		*	
RedLeaf	S	*	*	Surface			
Sasor	S/T		*	In-situ			
Schlumberger	S		*	In-situ			
Shale Tech International	S	*	*	Surface			*
Shell Frontier Oil & Gas	S	*	*	In-situ	*		*
Standard American Oil Co.	S/T		*	Surface			
Temple Mountain Energy Inc.	T	*	*	Surface		*	
U.S. Oil Sands	T	*	*	Surface		*	
Western Energy Partners	S	*	*	Surface			
Great Western Energy	S					*	

*Oil Shale = S, shale oil; T, tar sands; H, heavy oil; C, coal.
** means 2nd Round BLM R D Lease Application Pending.

Chevron U.S.A. Inc. — Headquartered in San Ramon, California, Chevron has operations in 180 countries worldwide [34,35]. One of the largest energy companies in the world, and has been active in all aspects of oil and gas development as well [34,35]. It is also a leader in shale development, and is known in the business as resource owner, technology developer, and project developer [34,35]. Their ongoing research includes developing an in situ process [34,35]. Depending on technical and economic feasibility, the project is expected to expand to commercial-scale production [34,35].

Schlumberger Inc. — Schlumberger is regarded as the world's leading oilfield services company, with operations in 80 countries worldwide [34,35]. They have a huge portfolio of products and services, which includes seismic acquisition and processing; drilling; formation, evaluation, and testing; simulation; cementing; completions; and so on [34,35]. With respect to shale-related services, Schlumberger provides reservoir characterization and monitoring services, which allows resource holders to plan and optimize their production processes [34,35]. Drilling, simulation, and completion services are also offered [34,35].

Shell Oil Company — Shell is regarded as one of the leading energy companies in the United States, and has business operations in oil and natural gas production, natural gas marketing, gasoline marketing, petrochemical manufacturing, wind, and biofuels [34,35]. With respect to shale development, over the course of the past few decades, Shell has been involved in extensive research related to the innovative In situ (in-ground) Conversion Process, to be able to recover shale oil or gas much more economically, with the least environmental impact [34,35].

14. ENERGY IMPLICATION OF SHALE OIL

There are number of documents in which the shale oil revolution has been discussed; however, according to Christof Ruhl, group chief economist, the real implications of shale oil have yet to sink in. A lot of work is needed for assessment of shale oil and its macroeconomic impact. He also mentioned that in case of shale development, the whole global scenario will change. The question whether the climate is better served by developing more shale gas or by the build-up of renewable alone also comes up [1–4].

In addition, why Canada and the United States are the only countries in which the shale oil revolution is more significant, compared to rest of the world having equally sufficient resources has been discussed. The simple answer was that it is all about competition [36—40].

A lot of innovation is involved to obtain these resources, and to bring them to markets in countries that do not allow for competition at home takes time. The large-scale production may be new for them but the principle is same as in the history for that industry.

Meanwhile, production in North America keeps rising and shale is changing the broader landscape. Thus, no one doubts anymore that the country will become a natural gas exporter in the not-so-distant future. Total imports of oil have been half now [36—41].

Small surprise that the shale revolution, or unconventional oil and gas in North America more largely, is slowly becoming documented as significantly ahead of this region, beyond energy markets, and for years to come. Yet, the evaluation of worldwide tactical implications—for energy safety, geopolitics, the universal economy, or for the atmosphere—is still sprawling real development; and maybe obviously so, given the insecurity around future production expansion.

To think about these likely strategic, global implications, we need a sensible point of reference.

Shale resources are different than those of conventional oil in two ways. First, they are widely distributed across regions. Assessments are still in their immaturity, but all availability shows precisely recoverable oil and gas shale resources spread across Asia, Australia, Africa, South America, Europe, and Eurasia. There is limited knowledge about economic recoverability of these resources [41—44].

The impacts of rising shale oil (and gas) production on markets and prices were the first concern upon its revolution. In a basic market trend, prices should fall and the only issue for an economist would be to calculate how fast and by how much. However, the oil market is not exactly normal as it is largely monitored by a producer cartel whose primary objective is to manage prices and production. In the oil markets, the question of the consequences of higher (shale) oil supplies seamlessly translates into the question of how OPEC is likely to react [41—44]. It seems totally logical to assume that OPEC members are able to decide on production cuts to rising non-OPEC supplies [36—39]. It is clear that the spare capacity normally results from production cuts. The build-up of spare capacity in OPEC countries needed to neutralize the extra supplies (e.g., oil sand production and biofuels) is considerable. Having the conservative production profile as a base, the additional capacity to accommodate the new supplies will be

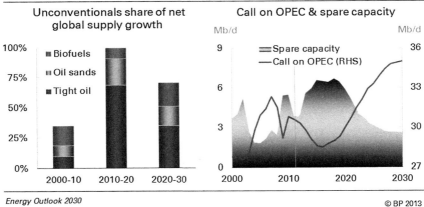

Figure 6.12 The call on OPEC to present share of unconventional oil in global supply energy [46].

over 6 Mbbl/day throughout this decade such that it is the highest since the late 1980s, as depicted on the right side of Fig. 6.12 [43—46].

This will be no easy task for OPEC. The cohesion of the organization introduces an important uncertainty, particularly in the 2010s. It is clear that Saudi Arabia, Russia, and the US will soon produce a third of global petroleum production. The only country among the three which is a member of OPEC is Saudi Arabia, which is likely to pay the cost of maintaining a great extent of additional production, compared to their planned capacity. The above political issue strongly affects the important decisions related to the development of shale resources. For instance, it governs whether, where, and when shale oil and gas resources will be available for considerable utilization [39—41].

The cost of tight oil production is very expensive. It is also exceptionally scalable, which is different from the conventional supplies in the North Sea or/and Alaska that were added in the 1980s. Following a basic market trend, supplies will change inversely proportional to the price change [41—46]. The revolution of shale gas and oil is expected to put pressure on prices by stressing the market configuration. In the case of gas, the oil indexation of gas prices is controlling the markets, while OPEC is the lynch pin of these markets in the case of oil [39—41]. It can be concluded that OPEC policies and projected trend of tight gas and oil production rates clearly imply the importance of shales in providing our future fuel demands.

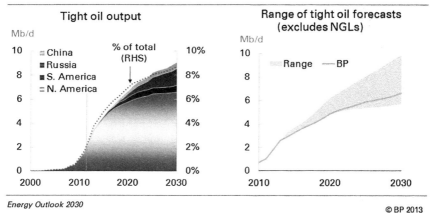

Figure 6.13 Global tight oil production rate within time period of 2000–2030 [46].

Both points are best illustrated in Figs. 6.12 and 6.13. Production profile that is taken from BP Energy is shown in the right of Fig. 6.13, which was published in 2013. Geologically, communications requirements, drilling movement, and on top of it, the speed with which shale resources are likely to be accessed in various locations, it clearly illustrates that shale oil will rise about 9% in global oil production by 2030. The rate of production growth will remain slow after 2020 [46].

Three characteristics are noteworthy: (1) production growth continues to be driven by North America; (2) growth will remain positive but slow down after about 2020 from the very high rates today; and (3) other countries will enter the game—notably Russia, China, Argentina, and Columbia—but their contribution will not grow fast enough to reverse the general pattern of slower growth after 2020 [41–46].

Next, the comparison of freshly minted forecast to others can be shown by the shaded areas on the right-hand side of Fig. 6.13. It turns out that BP's forecast is conservative indeed. It lies at the very low end of the range of all other forecasts [46].

REFERENCES

[1] Russell PL. History of western oil shale. East Brunswick, New Jersey: The Center for Professional Advancement; 1980.
[2] http://en.wikipedia.org/wiki/History_of_the_oil_shale_industry.
[3] Laherrère J. Review on oil shale data (PDF) Oil Shale A Scientific — Technical Journal (Hubbert Peak) 2005.

[4] Russell PL. Oil shales of the world, their origin, occurrence and exploitation. 1st ed. Pergamon Press; 1990, ISBN 0-08-037240-6. p. 162–224.
[5] Oil shale & tar sands programmatic. EIS; 2012.
[6] http://www.slb.com/~/media/Files/resources/oilfield_review/ors10/win10/coaxing.pdf [Schlumberger].
[7] Menzela D, van Bergena PF, Veldb H, Brinkhuisc H, Sinninghe Damstéa JS. The molecular composition of kerogen in Pliocene Mediterranean sapropels and associated homogeneous calcareous ooze. 2005.
[8] Schmidt Collerus JJ, Prien CH. Investigations of the hydrocarbon structure of kerogen from oil shale of the green river formation. 1974.
[9] Speight JG. Shale gas production processes. Gulf Professional Publishing 2013. http://dx.doi.org/10.1016/B978-0-12-401721-4.00001-1.
[10] Alberta Geological Survey, www.ags.gov.ab.ca.
[11] Barnes P. Oil and gas industry activity, trends and challenges, Canadian association of petroleum producers. 2007. Presentation to CAMPUT, 10-Sep-07.
[12] Canadian Association of Petroleum Producers, CAPP, www.capp.ca.
[13] Calhoun JC. Fundamentals of reservoir engineering. Norman: University of Oklahoma Press; 1960.
[14] Killen PM. New challenges and directions in oil shale development technologies. In: Olayinka AM, Ogunsola I, editors. Oil shale: a solution to the liquid fuel dilemma. Washington, DC: Oxford University Press; 2010. p. 21–60.
[15] Andrews A. Developments in oil shale. In: Oil shale developments. Washington, DC: Oxford University Press; 2009. p. 40–70.
[16] SHAH K. Transformation of energy, technologies in purification and end use of shale gas. Journal of the Combustion Society of Japan 2013;55(171):13–20.
[17] Branch MC. Progress in Energy and Combustion Science 1979;5:193. Pergamon Press. Great Britain with permission.
[18] Altun NE, Hiçyilmaz C, Hwang J-Y, Suat Bağci A, Kök MV. Oil shales in the world and Turkey; reserves, current situation and future prospects: a review. 2006.
[19] Lee S. Oil shale technology. CRC Press; 1990, ISBN 978-0-8493-4615-6.
[20] Hartgers WA, Sinninghe Damsté JS, Requejo AG, Allan J, Hayes JM, Ling Y, Xie T-M, Primack J, e Leeuw JW. A molecular and carbon isotopic study toward the origin and diagenetic fate of diaromatic carotenoids. In: Telnæs N, van Graas G, Øygard K, editors. Advances in organic Geochemistry 1993. Org. Geochem., vol. 22. GB: Elsevier Science Ltd; 1994. p. 703–25.
[21] Hoefs MJL, Versteegh GJM, Rijpstra WIC, de Leeuw JW, Sinninghe Damsté JS. Post-depositional oxic degradation of alkenones: implications for the measurement of palaeo sea surface temperatures. Paleoceanography 1998;13:42–9.
[22] Hutton AC. Petrographic classification of oil shales. International Journal of Coal Geology 1987;8(3):203–31. http://dx.doi.org/10.1016/0166-5162(87)90032-2. Amsterdam: Elsevier, ISSN: 0166–5162.
[23] Mackley AL, Boe DL, Burnham AK, Day RL, Vawter RG, Oil Shale History Revisited, American Shale Oil LLC, National Oil Shale Association, http://oilshaleassoc.org/wp-content/uploads/2013/06/OIL-SHALE-HISTORY-REVISITED-Rev1.pdf.
[24] Department of Energy. Annual energy outlook. 2009.
[25] http://www.qer.com.au/understanding/oil-shale-z/oil-shale-formation.
[26] http://www.eccos.us/oil-shale-in-co-ut-wy.
[27] Craig Jr FF. The reservoir engineering aspects of waterflooding. New York: American Institute of Mining, Matallurgical, and Petroleum Engineers; 1971.
[28] Dullien FAL. Porous media: fluid transport and pore structure. 2nd ed. San Diego: Academic Press; 1992.

[29] Rick Lewis DI. New evaluation techniques for gas shale reservoirs. Reservoir Symposium 2004;2004:1−11.
[30] Knaus E, Killen J, Biglarbigi K, Crawford P. An overview of oil shale resources. In: Ogunsola OI, Hartstein AM, Ogunsola O, editors. Oil shale: a solution to the liquid fuel dilemma. ACS symposium series, vol. 1032. Washington, DC: American Chemical Society; 2010. p. 3−20.
[31] Dyni JR. Oil shale. In: Clarke AW, Trinnaman JA, editors. 2010 survey of energy resources (PDF). 22 ed. World Energy Council; 2010, ISBN 978-0-946121-02-1. Archived (PDF) from the original on 2014-11-08.
[32] IEA. World energy outlook 2010. Paris: OECD; 2010, ISBN 978-92-64-08624-1.
[33] Sandrea R. Evaluating production potential of mature US oil, gas shale plays. Journal of Oil and Gas 2012.
[34] Biglarbigi K. Secure fuels from domestic resources, INTEK Inc. for the US Department of Energy, Office of Petroleum Reserves, Naval Petroleum and Oil Shale Reserves. September 2011. http://energy.gov/sites/prod/files/2013/04/f0/SecureFuelsReport2011.pdf, http://energy.gov.
[35] Oil Shale and Tar Sands Programmatic Environmental Impact Statement (PEIS) Information Center, http://ostseis.anl.gov/guide/oilshale/.
[36] Rice University, News and Media Relations. Shale Gas and US National Security July 21, 2011.
[37] The economic and employment contributions of shale gas in the United States. IHS Global Insight December 2011.
[38] Canada's energy future: energy supply and demand projections to 2035. The National Energy Board November 2011.
[39] IEA. Golden rules for a golden age of gas: world energy outlook − special reports on unconventional gas. 2012.
[40] Canada S. Report on energy supply and demand in Canada − 2008. Ottawa (ON): Statistics Canada; 2010. p. 29.
[41] Geochemistry W. Review of data from the Elm worth Energy Corp. Kennetcook #1 and #2 Wells Windsor Basin 2008. Canada. p. 19.
[42] US Office of Technology Assessment. An assessment of oil shale technologies. 1980.
[43] Dawson FM. Cross Canada check up unconventional gas emerging opportunities and status of activity. Paper presented at the CSUG Technical Luncheon, Calgary, AB. 2010.
[44] Gillan C, Boone S, LeBlanc M, Picard R, Fox T. Applying computer based precision drill pipe rotation and oscillation to automate slide drilling steering control. In: Canadian unconventional resources conference. Alberta, Canada: Society of Petroleum Engineers; 2011.
[45] Understanding Shale gas in Canada, Canadian society for unconventional gas (CSUG) Brochure.
[46] IEA. World energy outlook 2013. 2013.

CHAPTER SEVEN

Properties of Shale Oil

1. INTRODUCTION

Shale oil is unconventional oil, which is produced from oil shale pyrolysis, hydrogenation, or thermal dissolution [1]. Generally, the term oil shale is given to any type of sedimentary rock that contains solid bituminous materials (called kerogen) that are released from petroleum-like liquids when the rock is heated in the chemical process of pyrolysis [2].

Formation of oil shale was from the same process as that in which the crude oils were generated millions of years ago—mainly by deposition of organic debris on ocean, lake, and sea beds. The shale oils were formed in long periods of time under high-temperature and high-pressure conditions, which were also responsible for the formation of crude oil and natural gas. However, the temperature and pressure are not as high as in oil shale cases. The oil shale is sometimes called "the rock that burns" as it contains enough oil to burn itself [1,2].

Shale oil is obtained by mining and processing of the shale rocks. However, extracting oil from oil shale is not as simple as extracting oil from underground oil reservoirs as shale oil naturally occurs as solid particles, which cannot be directly pumped to the surface. Extracting shale oil is much more expensive than conventional methods which are used to obtain crude oil as oil shale has to be mined first and then it has to be heated to a high temperature (a process called retorting). Heating oil shale melts it and the resultant fluid is separated and collected. Nowadays, various experiments are being conducted to develop a process for in situ retorting, which includes heating oil shale (while it is underground) and then pumping the resultant fluid to the ground [2]. Fig. 7.1 shows oil shale samples.

In general, shale is a sedimentary rock which comprises of mud (clay and silt) and commonly by combining clay, silica (e.g., quartz), carbonate (calcite or dolomite), and organic material. In general, shale is a complex of a large amount of kerogen, which is a mixture of organic compounds. As the main composition, it has kerogen, quartz, clay, carbonate, and pyrate. Uranium, iron, vanadium, nickel, and molybdenum are the secondary components. From this rock, the liquid oil and shale gas are extracted. The synthetic crude oil is substituted by the shale oil but its extraction is expensive in contrast to

Figure 7.1 Oil shale samples [1,2].

that of conventional crude oil. The composition of crude oil is not the same throughout the world, since it depends upon geographical formations and some other factors. Its feasibility depends upon the cost of the conventional crude oil. Its price is greater than that of conventional crude oil. Hence, development and production of oil shales are strongly dependent on oil price.

Shale oil can be described through various ways in chemical and petroleum engineering. There is an enormous disparity in organic contents and mineral materials, it is not easy to simplify the findings and common sources for contrast. However, the in situ processes provide various qualitative and quantitative predictions of process variables, oil yields, and effluent product properties.

A sedimentary rock is the main source of shale oil, having porosity and permeability characteristics. The parts of shale formations which have high porosity and permeability are responsible for high productivity from shale oil and gas wells. Porosity is the property due to the tiny spaces in the rock that hold the oil or gas, while permeability is a flowing characteristic of the oil and gas through the rock. Mathematically, porosity is the open space in a rock divided by the total rock volume (solid + space or holes), while the permeability of a rock is a measure of the resistance to the flow of a fluid through a rock.

The depth of shale formations varies from place to place. For example, the North Dakota shale oil formation is 2 miles deep [3].

2. SHALE OIL UTILIZATION

In brief, the uses of shale oil are listed below:
- Before World War II, most shale oil was upgraded for use as transport fuels.
- Afterward, it was used as a raw material for chemical intermediates, pure chemicals, and industrial resins.

- It can be utilized as a railroad wood preservative.
- Shale oil with high-boiling point compounds is suited for the production of the middle distillates such as kerosene, jet fuel, and diesel fuel.
- Additional cracking can create the lighter hydrocarbons used in gasoline.

3. OIL SHALE FORMATIONS

This section describes a few known shale formations, in brief.

3.1 Central Queensland Region

Around 40 million years ago, the central Queensland region experienced the development of expanded and fault-bounded lake basins. At that time, the ocean levels were around 150 meters lower than present, with the Gladstone area more raised and around 200 kilometers southwest of the coastline as it was then.

Fig. 7.2 delineates a typical lake basin from that time. There was a rise in water level of the lake and then a fall because of a mix of (1) steady subsidence of the lake floor in light of tectonic forces and (2) variation in climate. The lake water chemistry fluctuated because of a change in the depth of the water and natural activity which brought about occasional algal

Figure 7.2 Schematic of Queensland formation structure (Queensland Energy Resources).

blossoms alongside a mixed bag of plants and aquatic creatures (crocodiles, turtles, and fish). Dead green growth poured down onto the lake floor, gathering with residue, mud, and dead crocodile bodies.

The rate of silt and organic matter gathering would occasionally surpass the bowl floor subsidence and fill the lake, bringing the water level to shallow depths and periodically drying out of the lake. The new subsidence would permit the lake to change. After some time, this cycle of filling and subsidence caused the collection and safeguarding of the thick grouping of algae-rich lake silt which is now called oil shale. This oil shale contains kerogen, an organic chemical compound from which oil can be generated.

In this area, just west of Gladstone, around 500 meters of oil shale amassed in the lake bowl known as The Narrows Graben. This happened over a 10 million-year period during the late Eocene to early Oligocene periods (around 30 to 40 million years ago).

Fig. 7.2 also demonstrates the resultant oil shale as a succession of layer groupings. There are around 100 climatically controlled layers from 1 to 7 m thick. It should be noted that the shale is generally around 3 m thick. It is assessed that each layer took around 75,000 years to gather on the lake floor [4].

3.2 The Green River Formation

One specific underground structure, known as the Green River Formation, contains a considerable amount of oil shale. The Green River Formation was made around 48 million years ago in the middle of the Eocene Era. The district contained an arrangement of intermountain lakes where fine sediment deposits and organic matter were deposited over roughly 6 million years. Over the long haul, the organic matter was converted into the oil shale reservoirs which are seen today. There are different formations in the strata, deposited in the same way, that also contain oil shale assets. These reservoirs likewise possibly contain reasonable oil shale reserves.

In March 2009, the United States Geologic Survey (USGS) finished a reassessment of the present oil shale deposits in this area and expanded their evaluation from around one trillion barrels of shale oil to 1.525 trillion barrels of shale oil. The Department of Energy (DOE) reports that around 1.8 trillion barrels of shale oil possibly underlie Colorado, Utah, and Wyoming in reserves of more than 15 barrels/ton [5].

Fig. 7.3 shows the oil shale deposits in the three states [5].

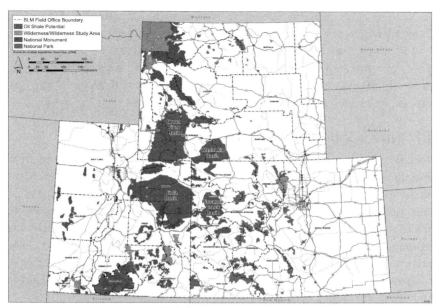

Figure 7.3 Oil shale deposits in three USA states.

4. KEROGEN: TYPES, STRUCTURE, AND HISTORY

Around 10 years ago, laborers at the French Petroleum Institute (FPI) developed a valuable plan for portraying kerogens, which is still considered as the standard nowadays. Kerogen is the portion of solid organic matter occurring naturally in source rocks and can yield oil upon further thermal maturation. Kerogen is insoluble and nonextractable using organic solvents [6]. It is mostly insoluble in most organic solvents because its components have a high molecular weight (more than 1000 Daltons; 1 Da = 1 atomic mass unit).

Every kerogen particle is exceptional. It has patchwork structures framed by the irregular mixture of many smaller molecular parts. The physical and chemical characteristics of a kerogen are considerably impacted by the type of biogenic particles, resulting in formation of the kerogen and transformations of those constituent molecules.

Components of kerogen are influenced by the process of thermal maturation (such as catagenesis and metagenesis) that changes the initial kerogen. Heating of kerogen below the surface results in chemical reactions which in turn lead to breaking off of fragments of kerogen into gas or oil molecules. The leftover kerogens additionally experience paramount changes, which are reflected in their chemical and physical properties.

FPI distinguished three principal sorts of kerogen (types I, II, and III) and have contemplated the compound attributes and the way of the creatures from which different kinds of kerogens were determined. Type IV kerogen has also been identified after subsequent experiments.

4.1 Type I: Sapropelic

This type of kerogen is very uncommon as it is essentially created from growth of lacustrine algae. The best-known sample is the Green River Shale of the Eocene age in Wyoming, Utah, and Colorado reserves. These oil shale reserves have attracted a lot of attention and have prompted numerous investigations conducted on the Green River Shale kerogens so that the reputation of type I kerogens has been substantially enhanced. The occurrence of type I kerogens is restricted to anoxic lakes and a couple of abnormal marine environments. Type I kerogens have high-generation limitations for liquid hydrocarbons.

4.2 Type II: Planktonic

This type of kerogens emerges from a few very different sources, including leaf waxes, marine algae, dust and spores, and fossil resins. They likewise incorporate contributions from cell lipids of bacterial species. Various species of type II kerogens are grouped together, in spite of their extremely divergent origins that have incredible abilities to produce liquid hydrocarbons. Most type II kerogens are found in marine sediments under anaerobic conditions.

4.3 Type III: Humic

This type of kerogens is made of land-based organic matter that is deficient in fatty or waxy parts. Cellulose and lignin are the main components. This type of kerogens has much lower ability to generate hydrocarbons than do type II kerogens and, unless they have small incorporations of type II materials they normally generate gas.

4.4 Type IV: Residue

These kerogens are mainly constituted of restructured organic debris. They contain oxidized matters of various origins. Normally, they do not have any ability to generate hydrocarbons. This type of kerogen is mainly correlated with the amount of hydrogen content present in it (H/C ratio). Type I kerogens contain much less rings or aromatic structures. Hence, they have the highest hydrogen content (H/C > 1.25). The type II kerogens also have higher H/C ratios (<1.25). The type III kerogens contain much less hydrogen content (H/C < 1) because of heavy aromatic structures present in them. Type IV kerogens contain the lowest amount of hydrogen (H/C < 0.5) as they have polycyclic aromatic content. In the case of oxygen content, type IV contains the maximum amount of oxygen as they are highly oxidized, followed by type III, type II, and type I [7].

Fig. 7.4 demonstrates the different types of kerogens.

Since every kerogen molecule has its own type, it does not seem useful to know the composition of all kerogen molecules. Regardless of the possibility that such a portrayal of kerogen can be achieved, it would not be of immediate importance to exploration geologists. However, it is important to create a general technique for depicting kerogen composition and relating it to its hydrocarbon generation ability. One possible way is characterizing kerogens into some general types [8].

The composition of kerogens differs from place to place. No specific/efficient techniques are available to determine the composition of kerogens.

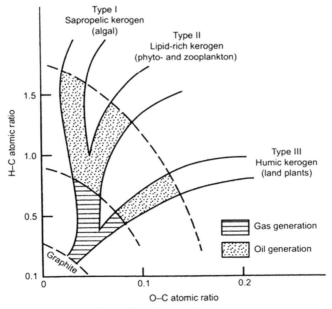

Figure 7.4 Types of kerogens.

A study conducted by Menzela et al. discusses about the molecular composition of kerogen in Pliocene Mediterranean sapropels and associated homogeneous calcareous ooze. They used compound identification based on mass spectral and retention time data. Determination of relative measures of individual components was envisaged by the integration of individual peak zones from summed mass chromatograms utilizing applicable m/z values. These measured peak zones were further multiplied by respective correction factors as these m/z values show much less ion count than is present. These correction factors were taken from Hartgers et al. [7a]. Table 7.1 shows the obtained results from their work [8]:

There is another example of kerogen composition introduced by Collerus and Prien. The Green River Formation constitutes about 16% of insoluble organic matter which is known as kerogen. This accounts for around 80% of total organic matter present and the remaining 20% of the organic matter is soluble bitumen. Generally, two techniques and their combination were used in this experimental work, namely micropyrochromatography and mass spectrometry. The apparatus used is illustrated in Fig. 7.5 [9].

Table 7.2 lists the results obtained from the above experiments. Figs. 7.6 and 7.7 show the generalized structure of kerogen and its matrix, respectively.

5. CHARACTERIZATION METHODS FOR SHALE OIL

There are several techniques for the characterization of shale oil, such as Induced Coupled Argon Plasma (ICAP), X-Ray Diffraction (XRD), Gas Chromatography (GC), Infrared Spectroscopy (IR), Nuclear Magnetic Resonance (NMR) spectroscopy, and High-Resolution Mass Spectrometry (HRMS).

In general, three main analytical methods are employed for determination of shale oil properties, namely: distillation, spectroscopic and spectrometric analysis, gas chromatography or/and mass spectrometric analysis.

Some of the experimental tests and equipment used for characterization of shale oil are given in Table 7.3.

5.1 X-Ray CT (Computer Tomography)

Hounsfield [9a] developed the X-ray computed tomography which is a nondestructive radiological imaging method that uses computer-processed

Table 7.1 Characteristic Mass Fragment and Correction Factor for Kerogens Based on Hartgers' Work

Compound Class	Characteristic Mass Fragment (m/z)	Correction Factor
n-Alkanes	55 + 57	2.9[a]
n-Alkenes	55 + 57	4.9[a]
Unsaturated isoprenoids	55 + 57	4.9
Saturated isoprenoids	55 + 57	2.9
Methyl ketones	58	3.6[b]
Alkyl benzenes	78 + 91 + 92 + 105 + 106 + 119 + 120 + 133 + 134	1.6[a]
Alkyl thiophenes	84 + 97 + 98 + 111 + 112 + 125 + 126 + 139 + 140	2.5[a]
Alkyl phenols	94 + 107 + 108 + 121 + 122	2.2[a]
Alkyl pyrroles	80 + 81 + 94 + 95 + 108 + 109 + 122 + 123 + 136 + 137	1.5
Oxygenated aromatics	78 + 91 + 92 + 105 + 106 + 119 + 120 + 133 + 134	2.6
Alkylindenes	115 + 116 + 129 + 130 + 143 + 144	1.8[a]
Indole	117 + 132	7.0
Phytadiene/phytene	55 + 57 + 68 + 70 + 82 + 95 + 97 + 123	2.4
Prist-1-ene/prist-2-ene	55 + 57	4.9
3-Ethyl-4-methyl-1H-pyrrole-2,5-dione	53 + 67 + 96 + 110 + 124 + 139	2.9
6,10,14-Trimethyl-2-pentadecan-2-one	58	7.5
C_{20} isoprenoid thiophenes	98 + 111 + 125	2.6

Characteristic m/z values and correction factors used to calculate relative abundances and internal distribution in pyrolysates.
[a] From Hartgers et al. [7a] and Hartgers et al. [9b].
[b] From Hoefs et al. (1996).

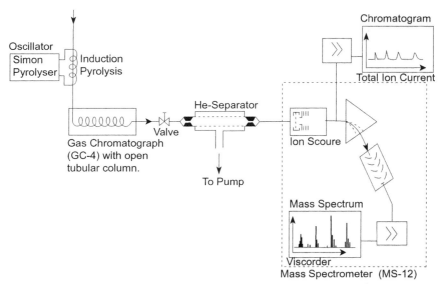

Figure 7.5 Schematic of the apparatus which is a combination of micropyrochromatography and mass spectrometry to determine kerogen composition.

combinations of many cross-sectional X-ray images of a sample to create a 3D data-set. This method has successfully been applied in geoscience since the 1980s (e.g., [9c,d]). CT-scan imagery involves a 2D or 3D linear X-ray attenuation pixel matrix. This attenuation depends on atomic number, density, and thickness of the sample. In studies related to shales, the CT method can be used to determine the presence of nodules and/or fractures as well as the orientation of the core relative to bedding by viewing full-diameter sections of core samples. CT-scanning is also used for selecting undamaged full-diameter sections of the core for further sampling and detailed density studies for intervals which exhibit a highly interbedded nature.

5.2 High-Resolution Micro-CT

The principles based on which high-resolution micro-CT is performed are the same as for conventional CT; however, a much higher resolution is achieved by using smaller samples and reducing the source-to-detector distance. The sample is illuminated by a microfocus X-ray source and magnified projection images are collected by a planar X-ray detector. The sample is rotated and numerous angular views are acquired. Based on these views, a stack of virtual cross-sectional slices of the sample is generated.

Properties of Shale Oil

Table 7.2 Fragmentation Products of Kerogen

No.	Name	Formula	Identified in Fraction
1	Aliphatic Hydrocarbons	n-C_{10} to n-C_{34}	85-7
		b-C_{10} to b-C_{36}	122-1
2	Alicyclic Hydrocarbons		
	Cyclohexanes	C_{10-13} H_{21-27}	123-1
	Decalins	C_{5-8} H_{11-17}	123-1
3	Hydroaromatic Hydrocarbons		
	Dialkyltetralins	C_{2-5} H_{5-11}	122-1, 123-1
		C_{8-12} H_{17-25}	122-1
	Hexahydro-phenanthrenes	C_{1-3} H_{3-7} (+6H)	123-1
4	Dialkylbenzenes	C_{8-13} H_{17-27}	123-1
5	Dialkylnaphthalenes	C_{3-4} H_{7-9}	123-1, 123-4
6	Alkylphenanthrenes	C_{1-3} H_{3-7}	121-4, 123-2

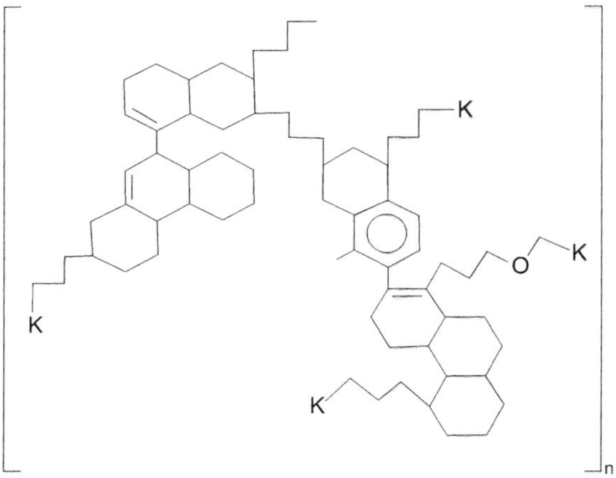

Figure 7.6 Generalized structure of kerogen of the Green River Formation.

5.3 Scanning Electron Microscope (SEM)

The most direct method to determine the porosity is scanning electron microscopy (SEM). However, this approach suffers from the following main limitations:

- Surfaces are prepared mechanically, hence their quality is poor.
- Resolution that can be achieved in a traditional filament instrument is very low.
- Visualization of the rock surface is limited to 2D.

These issues have recently been resolved by development of field emission microscopes which are coupled with ion milling tools (Focused Ion Beam, FIB) for producing in situ polished cross-sections with very high quality. This advancement has made it possible to generate SEM images of the pores down to nano-scale as well as to visualize the sample in 3D volumes through excavation of the sample surface.

5.4 Mercury Injection Porosimetry

One of the standard techniques of assessing the distribution of pore through size in samples with porosities from micron-scale to nano-scale is mercury injection capillary pressure (MICP). Mercury can penetrate between the course rigid grains of shales and areas between clay grains and secondary minerals.

Properties of Shale Oil 243

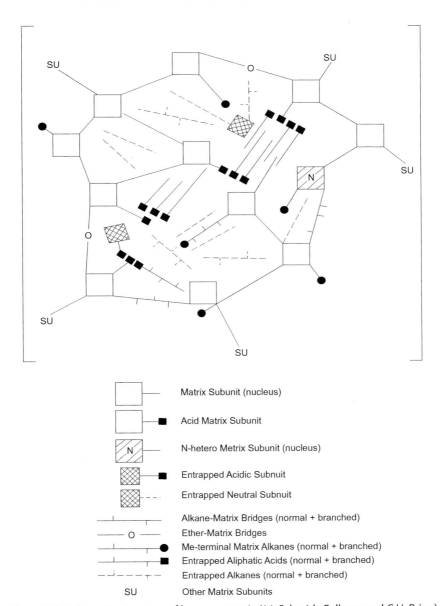

Figure 7.7 Schematic structure of kerogen matrix (J.J. Schmidt Collerus and C.H. Prien).

Table 7.3 Experimental Tests and Laboratory Apparatus Employed (B.A. Akashj and O. Jaber)

Test	Standard Testing Method	Name of Apparatus Used
Distillation	ASTM D-110/IP-24	Vacuum distillation
Initial boiling point		Gecil process
Density	ASTM D-1298	Hydrometer
Water content	ASTM D-1796	Distillation
Sediment	ASTM D-1796	Centrifuge
Flash point	ASTM D-93	PMC tester
Total sulfur	ASTM D-4294	XR-F MiniPal
Kinematic viscosity	ASTM D-2170	Viscometers
Pour point	ASTM D-2170	Viscometers
Salt content	IP-77	Separation flask
Heavy metals		
Fe	CMM-82	ASS/vario-6
Na, K, V, and Ni	ASTM D-5863	ASS/vario-6
Ca and Mg	ASTM D-4628	ASS/vario-6
Si	ASTM D-5184	ASS/vario-6
Heating value	ASTM D-240	IKA C-5003

5.5 Ultrasonic

For conventional reservoirs, the geophysical studies often utilize ultrasonic methods, often through wave-velocity-based approaches, to generate information related to stress fields and pore fluid saturations and pressures.

5.6 MICP and NMR

MICP and NMR tools are capable of providing direct measurements of shale permeabilities to aqueous fluids in the presence of applied stress conditions. Such measurements are done through the simulated effect of a pore fluid on a disc-shaped sample of shale being subjected to confining and pore pressures [10].

5.7 Raman Spectroscopic Analysis

Raman spectroscopy is mostly used for spectroscopic analysis of shale oils. The kerogen structure of shale oil is characterized by spectroscopy. This measurement tool is utilized to understand the mechanisms and nature of chemical processes used for the production of petrochemical products from sedimentary shales. As the shale oils contain carbon, their Raman spectra often resembles that of amorphous elemental carbon. Fig. 7.8 shows the spectroscopy of a particular shale oil sample. As Fig. 7.8 implies, the sample mainly consists of kerogen and calcium carbonate. The broad peak which is seen at 1354 cm^{-1} is normally symbolized by "D" band which

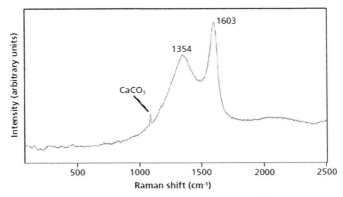

Figure 7.8 Raman spectrum of black shale having calcite and kerogen.

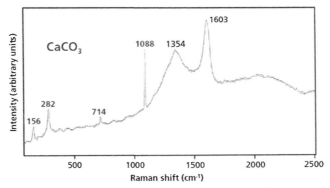

Figure 7.9 Intensity versus Raman shift to identify $CaCO_3$ and kerogen in a shale oil sample. (The Raman bands at 1354 and 1603 cm^{-1} show the presence of kerogen.)

refers to the region of disorder. The sharp peak at 1603 cm^{-1} is referred as "G" band that stands for graphite [11].

Fig. 7.9 demonstrates the presence of $CaCO_3$ in a shale oil sample that is specified with different peaks.

Raman spectra of black shales with various amounts of calcium carbonate from different areas of shale are presented in Fig. 7.10.

5.8 Distillation

This method is performed to obtain the distillation of the shale oil. The first fractions of distillate are normally obtained at higher pressures and lower (to moderate) temperatures. For instance, at a pressure of about 760 mm of mercury, the first seven distillates are obtained, and the remainder is taken at lower pressures. It could be observed that thermal decomposition before

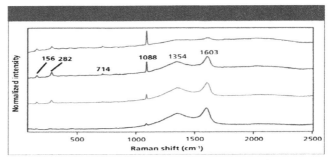

Raman spectra from different areas of black shale indicating the varying amounts of calcium carbonate relative to kerogen.

Figure 7.10 Normalized intensity versus Raman shift, indicating calcium carbonate with respect to kerogen.

a vapor temperature (the cutoff point of the atmospheric distillation in the crude petroleum method) is achieved when the shale oil is distilled.

5.9 Gas Chromatography Analysis

This method is used to determine the shale oil composition. For further understanding, a case study is presented here.

The analysis of Paraho shale oil naphtha 260°C (500°F) was performed using a 60-m DB1 fused silica capillary column in a 5700A Hewlett Packard gas chromatograph. The column was coupled directly to the source of a VG 7070H double-focusing mass spectrometer tuned to a resolution of 1000.

The column was programmed from 30 to 200°C at 1°C/min. The sample was splitless injected with the injection port at 300°C. Helium was used as the carrier gas at a pressure of 1.4 kg/cm. Optimum resolution was attained at an approximate linear flow velocity of 20 cm. Data were collected using a magnet scan cycle of 1.5 s from mass 12 to 300 using an INCOS instrument control and data collection computer system. The multiple ion detection (MID) data were acquired at a mass spectrometric (MS) resolution MAM = 9600 at a repetition rate of 1.5 s.

A 0.05-μL sample was injected using a 1-μL syringe. The injection port is designed so that the needle is received by a 0.76-mm (0.03-in.) sleeve which couples to the fused silica capillary column [12].

The Paraho shale oil naphtha 260°C (500°F) distillate sample was obtained from the Laramie Energy Research Center specifically for method development and use by ASTM Committee E−14 on Mass Spectrometry.

Fig. 7.11 shows the flame ionization detector (FID) gas chromatogram of the Paraho shale oil naphtha using a 60-m DB1 fused silica capillary column. This is compared to that obtained using a second 60-m DB1 column on the gas chromatographic mass spectrometric (GCMS) system. A comparable

Properties of Shale Oil

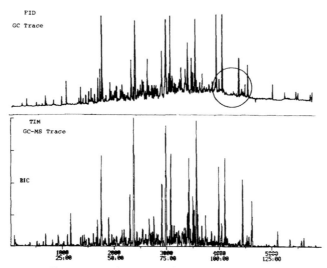

Figure 7.11 GC traces of the shale oil sample.

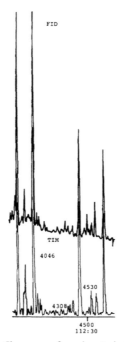

Figure 7.12 Close-up of peaks *circled* in Fig. 7.11.

resolution of the traces is noticed. Fig. 7.12 is a close-up of the peaks circled in Fig. 7.11. There is some shifting of peak positions; however, the resolution is the same [12].

5.10 Gas Chromatography-Mass Spectrometry (GC—MS)

The GC—MS method is an integrated technique which gives us the exact condition of the sample, such as its origin, environmental deposition, and the level of thermal maturity of the sample to be analyzed. It is a real technique which gives us an insight into the qualitative and quantitative information for analysis. To obtain the component composition of reservoir oils, a reservoir sample is flashed into gas and liquid phases at room temperature. The volume of the flashed gas, and the mass, molar mass, and density of the flashed liquid are noted. Then, compositions of the gas and liquid phases are analyzed using a gas chromatograph.

There are several analytical errors, which may include those due to contamination, measurement errors, mechanical/instrumental errors, fractionation errors, and loading errors on the GC. In order to minimize the errors, the analyst should take precautionary measures. For example, if a weak signal from the GC—MS analysis of a fraction is received, the sample should be prepared at higher concentration and be re-rinsed using selected ion monitoring (SIM) mode in order to improve the signal to noise ratio. The results obtained from GC—MS should be processed with GC—MS software with biomarker definitions to obtain more authenticated and vigorous interpretations.

Normally, the reservoir crudes are composed of:
- Hydrocarbon compounds that are made exclusively from carbon and hydrogen;
- Nonhydrocarbon but still organic compounds that contain, in addition to carbon and hydrogen, heteroatoms including sulfur, nitrogen, and oxygen;
- Organometallic compounds: organic compounds, normally molecules of porphyrin type that have a metal atom (Ni, V, or Fe) attached to them.

The hydrocarbons are mainly formed from the following constituents:
- Linear (or normal) alkanes (paraffins);
- Branched alkanes (paraffins);
- Cyclic alkanes or cycloparaffins (naphthenes);
- Aromatic alkanes (aromatics).

From the gas chromatography perspective, the analysis of alkanes is performedusing a nonpolar column and boiling point is the basis of

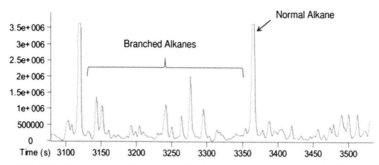

Figure 7.13 GC-MC peaks to differentiate the branched alkanes from the linear alkanes.

separation. The boiling point of normal alkanes is higher than their respective branched ones. Fig. 7.13 presents the chromatography results of a hydrocarbon.

In another experiment, GC/MS chromatography was utilized for seven oils (A—G) with known physical and chemical properties. The results are given as in Fig. 7.14 [13].

6. EXTRACTION PROCESSES FOR SHALE OIL

There are two common extraction processes for shale oil including ultrasonic and supercritical.

6.1 Ultrasonic Method

One known ultrasonic method is the DCM/MeOH ultrasonic extraction. The chief reagents in DCM/MeOH ultrasonic extraction are dichloromethane (CH_2Cl_2) and methanol (CH_3OH). Other reagents in use are internal and external standards. Common internal standards are adamantane, squalane, hexamethyl benzene, and p-terphenyl. The method is briefly explained here.

6.1.1 Sample Preparation

The dichloromethane (CH_2Cl_2) and methanol (CH_3OH) ultrasonic extraction method consists of cleaning of the dirt from the rock sample collected from the field by washing or its surface is scraped to get rid of the dirty portions or materials. The samples can then be wrapped in tinfoil and put in a drying oven at 110°C to dry properly. By means of mechanical machines, such as pestle and mortar, the sample is disaggregated/pulverized and homogenized. 2—3 g of the sample are then put into a centrifuge which

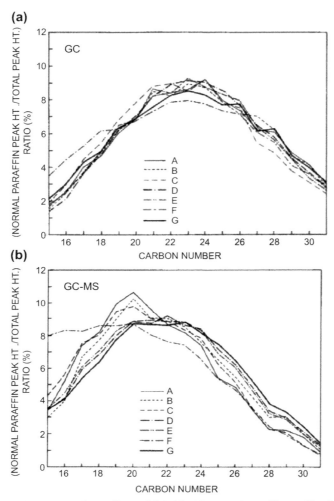

Figure 7.14 (a) GC normal paraffin and (b) GC-MC normal paraffin profiles for a heavy shale oil.

works on the principles of sedimentation where the centripetal acceleration is used to separate substances based on the density difference sand settling velocities as introduced by Stokes' law. Thus, the denser solid particles settle at the bottom of the test tube due to gravity while the supernatant in the solution having organic matter remains on top. Then, the supernatant solution is filtered off from the test pipette in the form of a rubber dropper from pellet without disturbance. For the second time, the solvent mixture is added to the pellet (residue) in the test tube to the half-way mark as before. The procedure should be repeated once again, and then repeated in the same order for the third time for each sample. It is concentrated either by evaporation using a rotary evaporator or under a stream of nitrogen. At this stage, the concentrated fractions are transferred into clean, empty, preweighed vials, in order to evaporate. The volume of the fraction becomes lower than 1 mL, and the leftover is completely evaporated overnight to obtain the mass of dry extracted organic matter (rock powder). The units for the acquiesce, the mass of Extractable Organic Matter (EOM) are mg of extract/g of rock or its percentile value can be calculated as follows:

$$\text{Extract Weight} = \text{Vial} + \text{Extract} - \text{clean vial} \quad (7.1)$$

Percentage extractable of Organic matter (% EOM)

$$= [\text{Extract Weight/Mass of Sample before extraction}] \times 100 \quad (7.2)$$

The rocks which are obtained from the sea contain type II kerogen. In this case, Sulfur removal from the extract might be very necessary. The elementary state of Sulfur is very important to be removed before analysis. Lab analytical instruments such as GC fractionation (column chromatographic separation) and gas chromatography-mass spectrometry (GC—MS) need to not have Sulfur, to avoid interference with the other compounds of interest.

6.1.2 Sulfur Removal
As per Imperial Lab Manual (2009) the following steps must be taken to eliminate Sulfur from the sample:
- For copper activation, we need to place copper turnings in 10% HCl for a few hours;
- In the second step, the removal of the activated copper takes place with the RO water wash and decant. Then, the activated copper is stored under solvent (usually hexane). It is reactivate when the shiny surface becomes dull;
- Sample treatment.
  The following steps are involved in sample treatment:
- Dilute extract with solvent (93:7 v/v DCM/MeOH) in small vial;
- Place activated copper turnings in vial and leave overnight;
- If copper turns black replace with new activated copper. If any copper remains shiny reaction is complete;
- Thus, shininess of the copper grains is an indication that the sample is sculpture free.
  Fig. 7.15 shows a simple schematic of the steps described above.

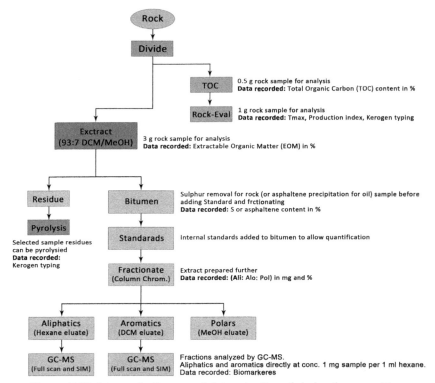

Figure 7.15 Schematic diagram of determination of shale oil composition.

6.1.3 Column Chromatographic Separation (Fractionation) and GC–MS

After DCM/MeOH ultrasonic extraction, the samples pass through the fractionating column of the chromatography as shown in Fig. 7.16. The silica gel/alumina was used as stationary phase or solid phase which was activated at a temperature of 110°C for an hour where the CH mobile phase comprises of three different solvents: hexane (C_6H_{14}), dichloromethane (DCM), and methanol (MeOH). The column is prepared by rinsing with nonpolar solvent mixture and by plugging the lower part of the column with a small amount of glass wool. Then, in the next step, the activated silica gel (or alumina) was put into a column to the required volume. Lastly, the sample extractable organic matter is placed on top of the column for fractionation.

The elution solution for fractionating column comprises of three different fractions: 3 mL of hexane (C_6H_{14}), 3 mL of dichloromethane ($CHCl_3$) methanol (CH_3OH) respectively, and the fractions are collected in three separate clean, preweighed test tubes. The first fraction comes out in the order of aliphatic, aromatic and polar fraction, respectively. In the next step, all three fractions are concentrated under nitrogen and left to obtain the dry masses.

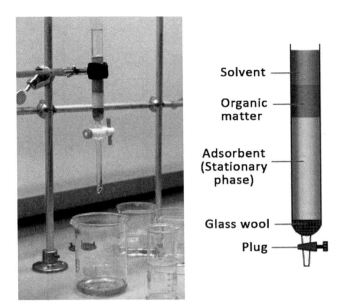

Figure 7.16 Column chromatography for separation of aliphatic, aromatic, and polar fractions.

The normalized percentages composition of three fractions can be displayed on an aliphatic-aromatic polar fraction ternary diagram.

6.1.4 Qualitative Analysis of GC and MS

The quantitative analysis of the analyzed compound includes nomenclature and identification of the compound on the basis of classification. It is done on the basis of their mass spectrum, which is basically a unique fingerprint for each molecule. This fingerprint can then be utilized to identify unknown compounds. For example, the mass chromatogram is illustrated in Fig. 7.17, showing the first peak of dodecane at a retention time of 4.97 min.

The mass spectrum generated by MS is also presented in Fig. 7.18.

The GC–MS samples are identified by comparing the sample spectrum with the reference spectrum.

6.2 Supercritical Extraction

Supercritical gas (SCG) extraction can possibly make the vitality in oil shale accessible in a more concentrated form, due to the high solvent extraction capacity of supercritical gases. SCG extraction has been the subject of various research papers and surveys. A broad study has been made of its uses in coal.

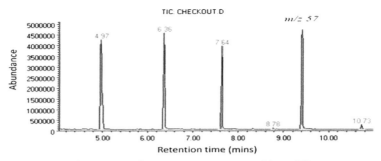

Figure 7.17 Chromatogram generated by a GC.

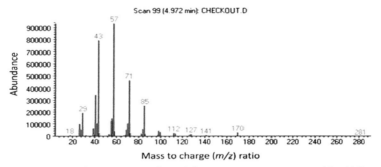

Figure 7.18 Chromatogram by MS (mass spectrum generated by MS).

The SCG extraction capacity is because of close fluid density solvent power consolidated with gas phase mass exchange. A remarkable feature of the extracts is their high molecular weight, a property not normal for matter which is essentially volatilized from coal. These results propose that SCG extraction may be effectively connected to oil shale. Regardless of the disadvantages connected with customary oil shale retorting and the high SCG coal extraction results, examination of SCG oil shale extraction is apparently not present in the literature.

Table 7.4 and Fig. 7.19 show that around 80% of the organic matter can be extracted by using a supercritical gas [14].

7. CHARACTERISTICS OF SHALE OIL

Physical properties [15]: like other types of oils, shale oils have various physical properties. The typical values of basic characteristics of the crude shale oil are provided in Table 7.5.

Table 7.4 Extraction Results Using Supercritical Method

Solvent	Temperature (°C)	Pressure (MPa)[b]	SCG Density (g/mL)	TOM[c] Removed (wt%)
Toluene[d]	330	6.9	0.53	33.8
	340	7.2	0.53	40.5
	357	8.3	0.53	54.7
	385	13.8	0.53	80.1
	393	15.2	0.53	87.5
	450	22.4	0.53	92.5
Pyridine[d]	363	7.6	0.55	63.3
Nitrogen	385	0.21	0.0013	16.9

Extraction Conditions and Results[a]

[a]Material Extracted: 12.9 wt% organic matter raw oil shale (21 gallon/ton) produced at Anvil Points, Colorado. Organic carbon and hydrogen contents: 10.3 and 1.53 wt%. Kjeldahl nitrogen content: 0.27 wt%.
[b]0.101 MPa/Atm.
[c]Total organic matter.
[d]Critical parameters: toluene, 319°C, 4.11 MPa; pyridine, 347°C, 5.63 MPa.

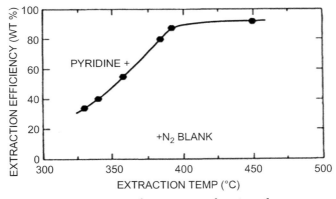

Figure 7.19 Extraction performance as a function of temperature.

Table 7.5 Common Properties of Shale Oils

Characteristic	Typical Range
Viscosity (at 100°F)	120–256
Gravity, API	19–28
Pour point, °F	80–90
Nitrogen, wt%	1.7–2.2
Sulfur, wt%	0.7–0.8

Table 7.6 lists the physical properties of some important shale oil deposits. Table 7.7 also presents the physical properties of Estonian shale oil.

The properties of shale oil from in situ retorting are shown in Table 7.8.

7.1 Composition

Naturally occurring shale oil is stored in shales. Shale is similar to other naturally occurring organic sediments. It has diverse composition with changeable properties according to the site of deposit and type of formation. Oil shale, peat, and tar sands have significant ordinary individuality so that shale oil is less combustible than natural gas. Therefore, the weight basis energy content is less than natural gas, petroleum, and coal. Its organic content is 50% by weight of a sample and the inorganic content is 30% of very rich shale and 95% for very lean shale. Its physical properties show that shales are brittle and rigid enough to maintain open fractures. It is mostly composed of solid organic materials of mean molecular weight of 3000. Table 7.9 presents a comparison between elemental composition of various samples of shale oils, crude oil, and coal, while elemental compositions of various shale oils are listed in Table 7.10.

Table 7.11 also demonstrates a basic comparison of crude oil and shale oil in terms of elemental composition.

Composition, which varies from region to region, strongly affects the processing and production technologies of shale oil.

It is important to note that sulfur and nitrogen must be removed through hydro treating, hydro cracking, and delayed coking.

Shale oil contains higher levels of nitrogen, sulfur, ash, and some toxic inorganic matter, compared to conventional oil samples. The presence of these kinds of biologically active compounds in shale oils with high concentrations may restrict the use of shale oil instead of petroleum-derived products. It can act as a possible health hazard. Therefore, shale oil requires more extensive refining than is needed for crude oil.

In a study, shale oil was used as a fuel for a water-cooled furnace to investigate the impact of heat transfer and environmental emissions [15a]. It was found that the rate of heat transfer was enhanced when shale oil was utilized instead of diesel fuel. Also, there was a positive effect on exhaust emissions. In another study, shale oil was used instead of diesel fuel in a single-cylinder, direct-injection diesel engine [15b]. It was found that shale oil increases the thermal efficiency of the engine, compared to the diesel fuel [16].

Table 7.6 Properties of Some Important Oil Shale Deposits [4]

Country	Location	Age	Oil Shale Organic Carbon (%)	Kerogen (Atomic Ratio) H/C	Kerogen (Atomic Ratio) O/C	Retorting Oil Yield (%)	Retorting Conversion Ratio[a] (%)	Retorting Density (15°C)	Shale Oil H/C (Atomic)	Shale Oil N (%)	Shale Oil S (%)
Australia	Glen Davis	Permian	40	1.6	0.03	31	66	0.89	1.7	0.5	0.6
Australia	Tasmania	Permian	81	1.5	0.09	75.0	78				
Brazil	Irati	Permian		1.2	0.05	7.4		0.94	1.6	0.8	1.0–1.7
Brazil	Tremembé-Taubaté	Permian	13–16.5	1.6		6.8–11.5	45–59	0.92	1.7	1.1	0.7
Canada	Nova Scotia	Permian	8–26	1.2		3.6–19.0	40–60	0.88	1.5		
China	Fushun	Tertiary	7.9			3	33	0.92			
Estonia	Estonia deposit	Ordovician	77	1.4–1.5	0.16–0.20	22	66	0.97	1.4	0.1	1.1
France	Autun, st. Hilarie	Permian	8–22	1.4–1.5	0.03	5–10	45–55	0.89–0.93	1.6	0.6–0.9	0.5–0.6
France	Crevenay, Severac	Toarcian	5–10	1.3	0.08–0.10	4–5	60	0.91–0.95	1.4–1.5	0.5–1.0	3.0–3.5
S. Africa	Ermelo	Permian	44–52	1.35		18–35	34–60	0.93	1.6		0.6
Spain	Puertollano	Permian	26	1.4		18	57	0.90		0.7	0.4
Sweden	Kvarntorp	Lower Paleozic	19			6	26	0.98	1.3	0.7	1.7
UK	Scotland		12	1.5	0.05	8	56	0.88			
USA	Alaska	Jurassic	25–55	1.6	0.10	0.4–0.5	28–57	0.80		0.8	0.4
USA	Colorado	Eocene	11–16	1.55	0.05–0.10	9–13	70	0.90–0.94	1.65	1.8–2.1	0.6–0.8

[a] Conversion to oil based on organic carbon.

Table 7.7 Common Properties of Estonian Oil Shale

	GGS	SHC
Density at 20°C, g/cm^3	0.9998	0.9685
Viscosity at 75°C, mm^2/s	18.7	3.5
Flash point, °C	104	2.8
Initial boiling point, °C	170	80
Calorific value, MJ/kg	39.4	40.4
Phenolic compounds, %	28.1	11.5
Fraction boiling up to 200°C, %	3.9	15.7
Fraction boiling at 200—350°C	28.3	33.7
Molecular mass, M	287	275

Table 7.8 Properties of Shale Oils From In Situ Retorting Process

Characteristic	Typical Range
Viscosity (at 100°F),	40—100
Gravity, API	30.6—54.2
Pour point, °F	−15 to +35
Nitrogen, wt%	0.35—1.35
Sulfur, wt%	0.6—1.2

Table 7.9 Elemental Composition of Oil Shale, Crude Oil, and Coal

Type/Location	% Organic C	% S	% N	% Ash
Oil shale/Piceance, Colorado	12.4	0.63	0.41	65.7
Oil shale/Dunnet, Scotland	12.3	0.73	0.46	77.8
Oil shale/Alaska	53.9	1.50	0.30	34.1
Coal, sub-bituminous/ C. Wyoming	61.7	3.40	1.30	14.9
Coal, bituminous/ C. Colorado	73.1	0.60	1.50	7.2
Coal, anthracite/ Pennsylvania	76.8	0.80	1.80	16.7
Crude oil/Texas	85.0	0.40	0.10	0.5
Shale oil/Colorado	84.0	0.70	1.80	0.7

The samples of shale oil used in this experimental work were mined from Ellujjun deposit in Jordan. The elemental and proximate analysis of shale oil is presented in Table 7.12 [16]. It can be clearly seen from Table 7.12 that the sample has a high sulfur content of around 3—4%, while the H/C ratio is 8.9.

Table 7.10 Elemental Composition of Various Oil Shales

Location of Sample	% Organic C	% S	% N	% Ash	Fischer Assay
Piceance, Creek, CO	12.4	0.63	0.41	65.7	28
Elko, NÉE	8.6	1.10	0.48	81.6	8.4
Lone, CA	62.9	2.10	0.42	23.0	52
Soldier Summit, UT	13.5	0.28	0.39	66.1	17
Rifle, CO	11.3	0.54	0.35	66.8	26.2
Cieveland, OH	11.3	0.84	0.40	72.3	7.9
Condor, Australia	15.9	0.22	0.39	64.1	32

Table 7.11 Chemical Composition of Crude Oil to Synthetic Crude Oil

	Crude Shale Oil	Crude Petroleum
C/H ratio	1.2–1.7	~1.25
Nitrogen	1.1–2.2%	<0.5%
Sulfur	0.4–3.5%	1–3%

Table 7.12 Analysis of Jordanian Oil Shale

Proximate Analysis

Volatile matter, %	43.96
Fixed carbon, %	0.42
Carbonate, %	40.60
Ash, %	54.51
Moisture, %	1.11

Elemental Analysis

Organic carbon, %	17.93
Total carbon, %	22.80
Sulfur, %	4.54
Hydrogen, %	2.57
Nitrogen, %	0.40
Oxygen, %	—
Higher heating value, MJ/kg	6.0

Fig. 7.20 shows the volumetric distribution of various fractions obtained from shale oil and crude oil in the form of a bar chart.

7.2 Pyrolysis Method for Composition Determination

Shale oil is heated at different temperatures (e.g., 300, 400, 500, and 550°C) at a constant heating rate, such as 10°C/min. Shale oil is also heated at a certain temperature, but at a variety of heating rates (e.g., 2.5, 5, 10, and

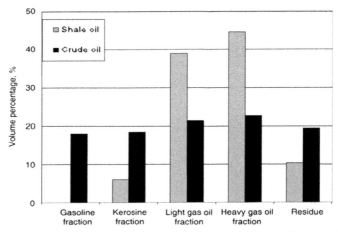

Figure 7.20 Volume percentage versus various fractions obtained from shale oil and crude oil.

20°C/min) in order to study the effect of heating rate. As an example, Table 7.13 shows the proximate analysis and ultimate analysis of Hudian Oil Shale. Table 7.14 presents the retorting products at different retorting conditions and proximate results of the semi cokes, respectively.

7.3 Boiling Range of Oil Shale

The initial boiling point for shale oil is around 200°C, which is too high compared to that of petroleum crude. However, it is close to the boiling point of crude residue. Around 90% of shale oil can be recovered up to 530°C. Fig. 7.21 exhibits the comparison of volume percentage recovered between heavy gas oil, light gas oil, and shale oil.

7.4 Self-Ignition Temperature of Shale Oil

The self-ignition temperature is the temperature at which oil shale samples spontaneously ignite. There is no particular method to make this

Table 7.13 Proximate Analysis % and Ultimate Analysis % of Hudian Oil Shale [8]

Proximate Analysis (%)		Ultimate Analysis (%)	
M_{ad}	4.46	C_{ad}	30.01
V_{ad}	37.88	H_{ad}	3.69
A_{ad}	52.60	N_{ad}	1.02
FC_{ad}	5.06	O_{ad}	8.04
$Q_{ar,net}$ (J/g)	13,145.21	S_{ad}	0.18

Table 7.14 Effect of Retorting Condition on Each Product, %

Retorting Condition	Oil	Water	Semicoke	Gas
350°C	1.29	7.00	91.23	0.48
400°C	3.73	7.20	87.09	1.98
450°C	13.32	7.57	73.90	5.22
500°C	18.09	8.63	67.43	5.85
550°C	17.50	9.00	66.19	7.32
2.5°C/min	17.99	8.89	66.59	6.53
5°C/min	18.10	8.77	66.61	6.53
10°C/min	18.07	8.80	67.13	6.00
20°C/min	17.90	8.79	67.18	6.12

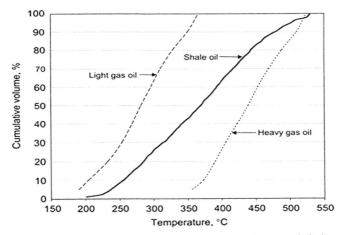

Figure 7.21 Boiling temperature ranges for heavy, light, and shale oils.

measurement. However, a consistent measurement of this temperature can help obtain invaluable information about the oil shale fuel characteristics. Measurement of oil shale spontaneous ignition temperature has been considered under various conditions. The significance of the information related to oil shale self-ignition temperature is due to the fact that this temperature controls the onset of the combustion retorting process as well as the oil shale retorting dynamics by the advancing oxidation zone. The importance of the retorting process in countercurrent combustion was highlighted by Branch; the combustion front moves toward the injected oxidizer, whereas in the cocurrent process, the front moves in the same direction as the oxidizer. Therefore, in the countercurrent process, for the raw oil shale, the spontaneous ignition temperature should be less than the retorting temperature of the shale oil. The char that remains after retorting in the cocurrent combustion process is burned in order to maintain the retort. For Colorado oil shale, the self-ignition temperature measurements have been performed for a wide total pressure and oxygen partial pressure range. The results showed that, in presence of nitrogen as a diluent, the ignition temperature is a function of oxygen partial pressure, but is not affected remarkably by the total pressure. Moreover, a close connection was also observed between the ignition temperature and the temperature at which lighter hydrocarbons, such as methane, were devolatilized from oil shale. For lower oxygen partial pressures, higher values of ignition temperature were needed, the lowest ignition temperature (450K) was much less than the temperature at which oil was being produced (640K). Figs. 7.22 and 7.23 show that the behavior of Colorado shale oil ignition in air at the pressure range of 0.3 to 14×10^5 Pa was similar to that of cool flame oxidation of light hydrocarbons. It can be concluded from such an experimental observation that the ignition of oil shale may be related to oxidation of gaseous hydrocarbons that evolve from the oil shale [17]. In Fig. 7.23, the temperature profile of self-ignition as a function of Fischer assay is shown.

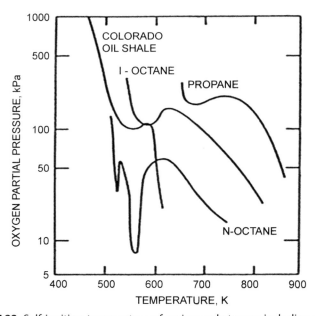

Figure 7.22 Self-ignition temperature of various substances including shale oil.

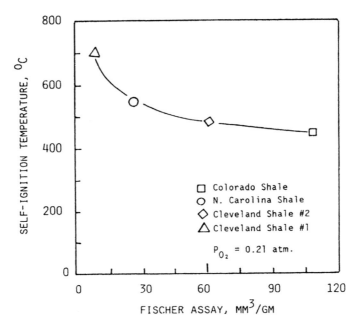

Figure 7.23 Self-ignition temperature versus Fisher assay.

It should be noted that the empirical correlation of Fischer assay of oil shale is given by Cook's equation as follows:

$$F = 2.216 \, wp - .7714 \text{ gal/ton} \qquad (7.3)$$

In Eq. (7.3), F is the Fischer assay estimated in a gallon of oil recoverable per ton of shale and wp stands for the weight percentage of kerogen in the shale. Using this equation, it should be noted that the mass fraction of kerogen in the shale is dependent on the measurement technique. For some shales, the maximum recoverable oil amount via supercritical extraction or CO, retorting is significantly higher than the Fischer assay value.

It is also conceivable that such a correlation can be sensitive to the type of oil shale retorted as well as dependent on the employed extraction process. There is no doubt that the Fischer assay of any oil shale is strongly correlated with the kerogen content of the shale [17].

Spontaneous ignition temperatures of Colorado oil shales in oxygen-containing gas streams of varying concentration have also been determined for the pressure range of atmospheric to 1000 psig. These self-ignition temperatures were found to be primarily a function of oxygen partial pressure and largely independent of the total pressure. Ignition temperatures were determined in a flow-type system so that conditions similar to those in combustion processes would be simulated. Ignition is characterized by a rapid temperature rise; therefore, a differential thermopile can be used to detect the temperature at which combustion occurred. Advantage can also be taken of the fact that ignition is also accompanied by the simultaneous release of considerable carbon dioxide. In this case, the gas stream can be continually monitored as a function of temperature with a differential thermal conductivity cell. Both techniques have been used in experiments and proved equally effective. In several research and engineering works, determinations have been made in air and in gas mixtures of oxygen and nitrogen over a pressure range from atmospheric to 1000 psig. The gas mixtures contained 6%, 13%, 21%, and 55%, by volume, of oxygen. Ignition temperatures as a function of oxygen partial pressure for data covering several oxygen concentrations in the gas stream are shown in Fig. 7.24 [18].

7.5 Diffusivity Parameter in Oil Shale

Understanding and proper characterization of thermo-physical behavior of materials like shale oils which are of technological and industrial significance is a vital step in their efficient utilization. For designing in situ and surface retorting facilities that are currently envisaged for extraction of oil from oil shale, it is crucial to develop solid knowledge of thermal transport

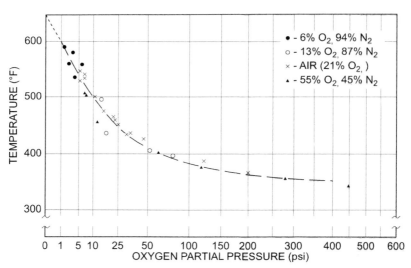

Figure 7.24 Self-ignition temperature of Colorado shale oil versus oxygen partial pressure.

parameters such as thermal diffusivity and conductivity. Thermal diffusivity, an important property for determining the diffusion of heat in different thermal excitation operations, specifies the history of temperature variations versus time for a material considering a certain set of boundary conditions. Moreover, thermal conductivity, which is another crucially important thermo-physical parameter, can be calculated from thermal diffusivity using the following relationship:

$$\alpha = \frac{k}{\rho c} \tag{7.4}$$

where α is the thermal diffusivity, k is the thermal conductivity, ρ is the density, and c is the specific heat of the material.

The laser flash technique is one of the new and successful methods for the measurement of thermal diffusivity. This method has been used to measure the thermal diffusivity of Green River oil shale in different shale grades from 6 to 100 gal/ton with a temperature range of 25 to 350°C. The results show that the thermal diffusivity decreases with increased temperature, particularly for shales which are not rich in organic matter. On the contrary, for shales with higher levels of organic matter content, thermal diffusivity is not very sensitive to grade and/or temperature. Thermal diffusivities measured for the shale oils of Green River were in the range of $0.1-0.9 \times 10^{-2}$ m^2/s and showed evidence of the effect anisotropy, as the cores with their axes parallel to the stratigraphic plane exhibited up to 30% higher α values than the samples cored perpendicular to bedding. Moreover, thermal diffusivity is observed to be severely influenced by the pore water content of the shale [19].

The main technique applied for measurement of thermal conductivity is transient line heat source in which a probe with a long electric heater and a thermocouple is inserted into a sample and the temperature rise near the center of the sample is measured. The heater is turned on, thus producing heat at an axially uniform constant rate. The temperature rise versus time is then measured by the thermocouple and recorded. After the heater is turned on (time zero), depending on the probe diameter, its construction, and thermal contact resistance, a log-linear temperature rise is observed versus time, typically starting after about 10–40 s. Such a log-linear increase of temperature with time usually continues for about 1 minute for samples with a thermal conductivity of approximately 1.5 Btu/ft-h-°F and longer for samples exhibiting lower values of thermal conductivity. Such a time interval is sufficient for determining the thermal conductivity of all oil shale samples. This measurement method is analogous to measurement of the product of permeability and thickness (k×h) based on the pressure drop observed versus logarithm of time, resulting from well drawdown tests [20]. The literature indicates that the thermal conductivity of most shales at ambient conditions is within the range of 0.5−2.2 W/m·K. A number of researchers have concluded that parameters that affect thermal conductivity are dependent on the type of shale sample. No systematic studies have been found which address variables affecting the thermal characteristics of several shale types. These influential parameters can be composition, temperature, porosity, pressure, and anisotropy [21].

Diment and Robertson reported a relationship between thermal conductivity and composition of shale from the Conasauga group as:

$$X = 2761 - 15R \tag{7.5}$$

where X is the thermal conductivity (in W/m·K) and R is the weight percent of shale that is insoluble in dilute hydrochloric acid.

In another study, Tihen, Carpenter, and Sohn obtained the relationship between thermal conductivity and composition of oil shales from the Green River Formation which assay between 0.04 and 0.24 L/kg as [21]:

$$X = C_1 + C_2 F + C_3 T + C_4 F^2 + C_5 T^2 + C_6 FT \tag{7.6}$$

where X is the thermal conductivity, W/m·K, C_1−C_6 are constants; F stands for the Fisher assay of the shale, L/kg; and T is the temperature, K.

This equation shows that for some shale oils, the effects of composition and temperature are interrelated. Thermal conductivity versus temperature is depicted in Fig. 7.25.

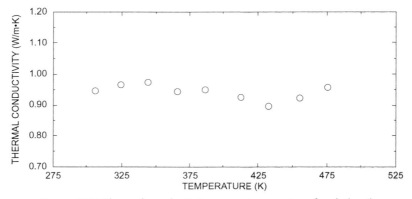

Figure 7.25 Thermal conductivity versus temperature for shale oil.

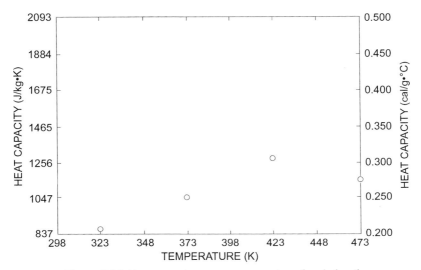

Figure 7.26 Heat capacity versus temperature for shale oil.

Similarly, for heat capacity, there is no adequate literature available. Fig. 7.26 demonstrates heat capacity versus temperature.

7.6 Time—Temperature Index (TTI) of Maturity

TTI is a great technique by which we can theoretically determine the maturity and oil generation. By modeling by geological burial history of the area (depth vs. time), the geothermal history can be determined (see Fig. 7.27). One should also make an estimation of uplift and erosion as well.

The thermal contact was calculated by multiplying each time interval by a temperature factor which is on the basis of an old chemical rule that

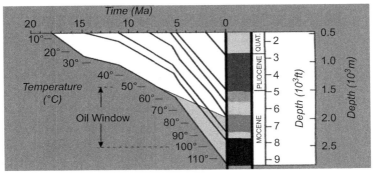

Figure 7.27 Time—temperature index of maturity based on the modeling investigation.

reaction rates double for each 10°C rise in temperature. This is the only reason for having isotherms at 10°C intervals.

In fact, TTI is an illustrative diagram to best determine the level of thermal maturity of the organic matter and degree of biodegradation of the sample with respect to time. The straight chain and the ring compound's samples are further investigated by GC—MS in order to achieve their respective chemical composition by concentration peaks in spectra.

7.7 Electrical Properties of Oil Shale

It is observed from measurements performed on different oil shales that the resistivity values decrease exponentially with time under the effect of the electric field of the Direct Current (DC); a typically characteristic trend of ionic solids in which conductance occurs through a transport mechanism which is enabled by heat. It is, however, not very easy to exactly identify the ions responsible for carrying the current in the oil shale as these rocks possess different minerals. Yet, it might be possible to argue that most of the current is carried by carbonate ions based on the comparison of activation energies at high temperatures (e.g. 380°C) typically associated with carbonate minerals with those of oil shale samples. Such estimations are merely based on speculation though, and should be considered cautiously. Heating can also impose changes on the chemical nature of oil shale, thus causing the conduction behavior of the shale to change. For example, changes in the resistivity values of Russian shales (from 10 ohm-cm at room temperature to the same value at 900°C) might be due to thermal decomposition of the kerogen of the oil shale. In Fig. 7.27, the frequency-dependent behavior of electric resistivity versus the reciprocal of temperature for a 117 L/ton sample of raw Green River oil shales is shown. In the resistivity curves, the minimum resistivity values fall in temperature ranges of 40—210°C. Such minimum values occur as the free moisture and bonded water molecules are gradually lost to clay particles in the matrix of shale. Fig. 7.28 illustrates the same behavior of the reheated materials, although the curves show a completely different trend from raw shales. For these experiments, the temperature of the shales was reduced by cooling to room temperature. They were then reheated and their temperatures were raised to about 500°C. In Fig. 7.28, the typical behavior of ionic solids is depicted. The resistivity does not show a minimum peak and the results correspond to thermally activated conduction [17].

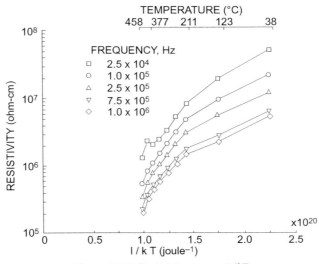

Figure 7.28 Resistivity versus 1/kT.

7.7.1 Dielectric Constant

The dielectric constant of oil shales is controlled by temperature and frequency. At low temperatures, oil shales have shown high dielectric constant anomalies which, according to some researchers, might be due to polarization effects on electrodes. This can be explained by the fact that the presence of moisture and accumulation of charges at the shale sedimentary gives rise to interfacial polarization in these materials. In Fig. 7.29, which shows the

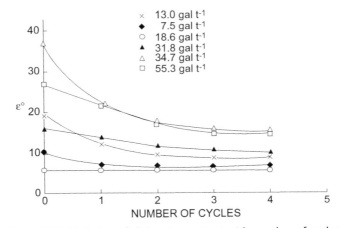

Figure 7.29 Variation of dielectric constant with number of cycles.

changes in dielectric constant versus the number of heating cycles for Green River oil shales with different grades, it can be seen that the dielectric constant decreases significantly with each subsequent drying cycle. It should be noted that the samples were heated at 110°C for 24 h and then cooled to room temperature before being tested.

8. CHARACTERISTICS OF OIL SHALES

Petrophysical analysis is the starting point, combining laboratory measurements, core data, and well logs. Rock physics then establishes the relationship between petrophysical and elastic properties of the formation and enables the creation of synthetics for missing and bad log data from drilling and invasion effects. In order to characterize the petrophysical properties of the rocks, i.e., rock capability to accumulate and transport reservoir fluids, it is necessary to determine the values of two key parameters including porosity and permeability.

8.1 Porosity of Oil Shales

Porosity is generally defined as the fraction of void spaces in the material relative to the total volume. Porosity is usually expressed as a fraction (between 0 and 1) or as a percentage out of 100. Based on the methods used for measurement of void spaces volume and the specific pores which are to be investigated, the porosity of the material can be estimated by different methods: intraparticle porosity, packing porosity, interparticle porosity, bed porosity, superficial porosity, total open porosity, porosity by liquid absorption, internal porosity, porosity by saturation, and porosity by liquid penetration. Due to the presence of organic matter as a solid and insoluble portion in the oil shale, it is not possible to estimate the porosity of the mineral matrix of the oil shales by the same techniques used for determining the porosity of reservoir rocks. The inorganic particles, however, contain a micropore structure measuring about 2.36—2.66 % vol/vol. The porosity measurement of oil shales might be limited to the external surface characteristics instead of the internal pore structure itself, despite the noticeable surface area of minerals (4.24—4.73 m^2/g) which enables the oil shale to produce 29—75 g/t in the Fischer assay. Generally, naturally occurring porosities in the raw oil shales are extremely low. It is, however, possible for some structural phenomena such as folding or faulting to cause new pores to be created and/or some disconnected pores to be connected. Otherwise, a fairly good portion of the oil shale porosity is known to consist of closed or blind pores. Normally, in porosimetry using mercury injection, disconnected or blind pores are not filled with mercury, not even at substantially high pressures; this leaves such pores out of porosity calculations. Currently, porosimetry using pressurized mercury injection is not performed anymore because mercury is highly poisonous. To efficiently produce oil from the oil shales, both physical and chemical properties of these rocks need to be taken into consideration. The efficiency of oil extraction from the matrix of oil shales may be reduced because of the low-porosity and low-permeability nature of these rocks as well as their high mechanical strength, as these factors all tend to make the mass transport of the reactants and products much more difficult.

In general, two main types of porosity are distinguished as follows:
- Total porosity—calculated as the total pore volume divided by bulk volume of the rock; and
- Effective porosity—calculated as the volume of interconnected (permeable) pores divided by bulk volume of the rock.

Shale rock is built with micro- and nano-sized space pores with varying degrees of water saturation and partly of residual organic matter. Void spaces also occur between rock grains (inorganic pores and micropores), but their volume is minimal. Effective porosity appears as a result of fracturing [23].

In the case of conventional oil and gas reservoirs (sandstones, carbonate rocks), porosity was defined as the void space between rock grains (inorganic pores and micropores). In that space and within laminae enriched in silica, as well as in the system of natural fractures and microfractures, the gas is accumulated in the form of free gas.

In shale rocks, natural gas occurs as:
- Free gas within rock particles;
- Free gas within the dispersed organic matter;
- Gas adsorbed by the dispersed organic matter;
- Gas adsorbed by certain clay minerals.

In addition to the aforementioned various accumulation spaces, free gas is present in clay-mud shale complexes and also in laminae that are enriched in organic matter. However, a significant amount of natural gas is present in organic pores located within insoluble organic matter which is called kerogen [19].

8.2 Permeability and Fracturability of Shales

Permeability is associated with the presence of natural cracks/fractures in the rock which enable the flow of reservoir fluids between pore spaces.

Permeability enables the flow of natural gas or oil into the borehole and their production. Permeability coefficient is dependent on:
- Size of pores;
- Relative configuration of the rock-building grains;
- Grain grading and cementation; and
- Rock fracturing patterns.

In the case of shale rocks, both permeability and porosity are highly dependent on:
- Mineral composition;
- Organic matter distribution;
- Quantitative (%) content of organic matter; and
- Thermal maturity of organic matter.

Shale rocks characterized by low permeability normally prevent any unrestrained flow of hydrocarbons. Accordingly, stimulation jobs (such as fracturing operations) should be performed in order to connect the pores to the borehole and allow for an unrestrained flow of gas and reservoir fluids [19].

Table 7.15 presents the permeability charts of various basins.

The conventional and unconventional reservoirs have the same petro physical data analysis techniques for all of the shale formations (e.g., gamma rays, resistivity, porosity, and acoustic, along with addition of neutron capture spectroscopy data). The petro physical analysis of shale oil starts with a gamma ray log. This indicates the presence of organic-rich shale. The organic matter contains higher levels of naturally occurring radioactive material than the ordinary mineral reservoirs. Petro physicists use gamma ray count to identify organic-rich shale formations. A triple combo tool provides the measurement of resistivity and porosity. The resistivity measurement of the shale that has the potential of gas is higher than that of the shale having no gas potential. Porosity measurement of the gas-bearing shales also has distinct characteristics. Organic-rich shales exhibit more variability, higher density porosity, and lower neutron porosity. This indicates the presence of gas in the shale. Lower neutron porosity may occur due to lower clay mineral contents in those shales.

The shales have higher bulk density than that of conventional reservoirs like sandstone or limestone because of the constituent material that plays an important role in the formation of shale. Kerogen has lower bulk density, which leads to a higher computed porosity. The grain density, derived from Elemental Capture Spectroscopy (ECS) must be known in order to

Table 7.15 Permeability Charts of Different Sediments

Permeability	Pervious		Semipervious			Impervious		
Unconsolidated sand and gravel	Well sorted gravel	Well sorted sand or sand and gravel		Very fine sand, silt, loess, loam				
Unconsolidated clay and organic			Peat		Layered clay	Unweathered clay		
Consolidated rocks	Highly fractured rocks		Oil reservoir rocks		Fresh sandstone	Fresh limestone, dolomite		Fresh granite
k (cm^2)	0.001 | 0.0001	10^{-5} | 10^{-6}	10^{-7}	10^{-8} | 10^{-9}	10^{-10} | 10^{-11}	10^{-12}	10^{-13}	10^{-14} | 10^{-15}
k (millidarcy)	10^{+8} | 10^{+7}	10^{+6} | 10^{+5}	10,000	1,000 | 100	10 | 1	0.1	0.01	0.001 | 0.0001

Modified from Bear (1972).

compute density porosity of the shale. Silicon, calcium, iron, sulfur, titanium, gadolinium, and potassium are the primary outputs of spectroscopy.

The spectroscopy data also provide information about the clay type, so that engineers can predict the sensitivity of the fracking fluid and understand the fracturing characteristics of the formation using these data. When the clay is in contact with water, it swells and hence inhibits gas production, therefore a lot of operational issues can arise. Smectite is the most common form of swelling clay. It also indicates the rocks that are ductile.

For long-term productivity of a shale gas well, acoustic measurements are very important that provide the mechanical properties of anisotropic shale media. To enhance mechanical earth models and to optimize drilling, the sonic scanner acoustic scanning data are used. The mechanical properties may include bulk modulus, Poisson's ratio, Young's modulus, yield strength, shear modulus, and compressive strength which are computed from compressional shear and Stoneley wave measurement.

When the difference between the vertically and horizontally measured Young's modules becomes large, the closure stress will be higher than that in isotropic rock. These anisotropies are associated with the rock having a higher clay volume. As the proppant is more embedded into ductile formation, therefore it will be difficult to retain the conductivity of the fracture during production.

Another acoustic measurement that is beneficial in shale analysis is sonic porosity. Sonic porosity is much lower than neutron porosity in shale. That is a function of high clay-bound water which is common in shale. The high sonic porosity indicates the gas potential within the pore space. When both the sonic and neutron porosity values are similar, it indicates that shale may be oil-prone. To define the orientation and concentration, log analysts use wire line borehole image. It can be interpreted from these data that either the hole is open or closed.

The measurement from these various tools can be combined in an integrated display like a shale montage log. Geologists can directly compare the quality of the rock by the formation properties that are presented using a single platform. Free and absorbed gas has the units of SCF/ton.

Seismic data analysis moves the analysis beyond well control to the whole field. The work flow is described as follows:
- Determine TOC and mineralogy, including porosity and water saturation, using petrophysical and rock properties analysis. To do that, determine bulk density for each mineral, calculate TOC weight percentage, and convert this measure to bulk volume kerogen.

- Extend analysis beyond well control to visualize the entire area of interest by combining well log and seismic data. Characterize structural and stratigraphic complexities to identify high-value intervals and potential hazards like water conduits.
- Evaluate relative brittleness and ductility from well logs and seismic inversion to identify areas prone to fracturing.
- Analyze rock stresses, natural fracture networks, and fracture directionality by examining image logs, directional borehole acoustics, and azimuthal seismic inversion data to determine optimal horizontal well direction and fracturing strategy.
- Plan the well bore trajectory.
- At the conclusion of the workflow, there should be sufficient information about the reservoir characters to select optimal drilling locations, as well as orientation and placement of horizontal wells for the most effective production program.

  Shale properties can be described in different scales of observation:
- Mineral grains (nano- and microscale);
- Packages of lithologically uniform laminae (millimeter or centimeter scale);
- Higher-order sedimentary complexes displaying internal lithological variability and higher-order patterns of the sedimentary structure (scale of meters).

8.3 Pore Structure, Pore Size Distribution, and Surface Area in Oil Shales

During pyrolysis of oil shale, it was discovered that the temperature and the heating rates have a pronounced effect on the surface area as well as the pore structure. Specifically, these parameters can affect the specific surface area, specific pore volume, and development of mesopores. The specific surface area was calculated based on the Brunauer, Emmett, and Teller (BET) equation where the specific surface area and the pore size distribution were calculated by the Barrett-Joyner-Halenda (BJH) method.

During experiments, the pore structure of raw and semicoke samples can be measured by the method of low-temperature adsorption of nitrogen and thus produce adsorption desorption isotherms. It could be observed from the isotherms that with the final rise of temperature, the specific surface and the total pore volume decrease in the beginning then increase in large scale [3].

Fig. 7.30 shows the variation in pore sizes of different oil shale samples [24].

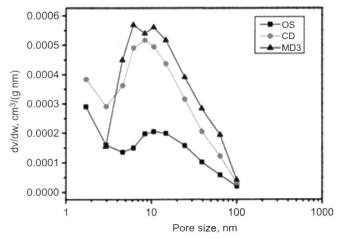

Figure 7.30 Pore size distribution of oil shale samples.

Table 7.16 Specific Surface Areas and Pore Properties of Oil Shale Samples

Sample	BET Specific Surface Area (m^2/g)	Correlation Coefficient of BET Equation (R)	Single Point Specific Pore Volume at $P/P_0 = 0.9846$ (cm^3/g)	Average Pore Size (nm)
OS	3.2478	0.9989	0.0112	14.3630
CD	5.5659	0.9996	0.0233	14.6622
MD3	5.3853	0.9997	0.0303	19.2616

The variations in specific surface areas for different oil shale samples are presented in Table 7.16 [24].

8.3.1 Pore Structure Measurement Using Adsorption/ Desorption Method

The low-temperature nitrogen adsorption/desorption method was conducted in a Gemini 2380 automated surface area and pore size analyzer at a temperature of 150°C and a pressure of 0.334 MPa for 2 h in an atmosphere of nitrogen in order to degas the sample. At a temperature of 77.3K and a relative pressure P/P_0 of 0.05—0.986, the nitrogen was adsorbed on the degassed sample.

By applying $0.05 < P/P_0 < 0.25$ of the adsorption branch to the BET equation, the specific surface area can be determined. The desorption branch

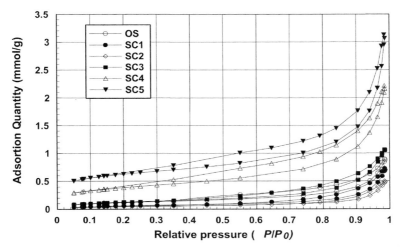

Figure 7.31 Adsorption/desorption isotherm for final retorting temperature SC1—SC5.

of the isotherms can be used to calculate the pore size distribution by BJH method [3].

Figs. 7.31 and 7.32 represent the adsorption/desorption isotherms formed by semicokes produced at different final retorting temperatures (SC1—SC5) and different heating rates (SC6—SC9).

It is clearly observed that the raw oil shale and its semicoke (SC1—SC9) form the same inverted "S" shape adsorption/desorption isotherms, having type II classification by the BET method, which shows that the micropores and the mesopores are prominent in samples. At low pressure, the isotherm branch prudes slowly, where at P/P_0 greater than 0.9 the adsorption quality

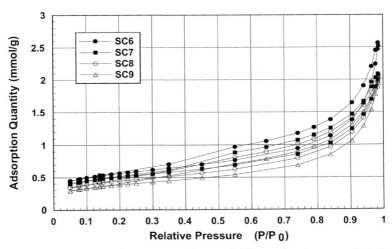

Figure 7.32 Adsorption/desorption isotherm of different heating rates SC6—SC9.

Table 7.17 Surface Area by BET

Samples	S_{RRT} (m²/g)	r	Error (m²/g)
OS	9.581 9	0.999,962 6	0.0276
SC1	5.325 8	0.999,725 3	0.0407
SC2	4.237 3	0.999,346 6	0.0500
SC3	10.735 9	0.999,945 0	0.0371
SC4	30.639 7	0.999,891 7	0.1512
SC5	48.142 6	0.999,483 9	0.5226

increases sharply because of capillary condensation in mesopores and micropores.

It could be concluded from the isotherms that increasing the temperature from room temperature to 400°C, the pore content decreases but with the rise of temperature from 400 to 550°C, there is a gradual increase in pore content. It is also observed that the hysteresis loops of SC6—SC9 detach in the same scale, meaning the heating rate has little impact on pore content. Table 7.17 shows the surface area of OS and SC1—SC5 by BET.

Fig. 7.33 indicates the specific surface area distribution of OS and SC1—SC5.

Fig. 7.34 illustrates that with the rise of temperature from room temperature to 400°C, the surface area decreases mainly due to the formation of less porous thermoplast and due to the deformation of the kerogens below 400°C.

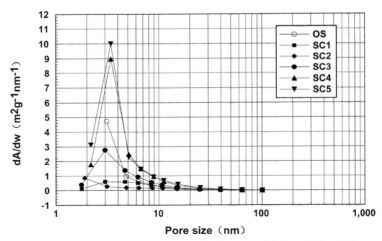

Figure 7.33 Specific surface area distribution of OS and SC1—SC5.

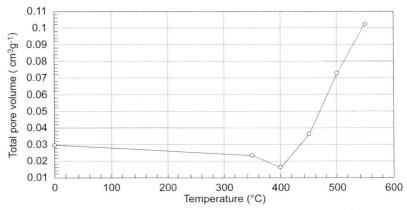

Figure 7.34 Change in total pore volume during oil shale pyrolysis.

Fig. 7.34 indicates the relation between total pore volume and final temperature, where the pore size distribution of OS and SC1—SC5 is given. When the heat is provided from room temperature to 400°C, the total pore volume decreases gradually, that is why most of the mesopores and the micropores are closed and the size distribution becomes uniform. The pore volume increases gradually at a temperature $> 400°C$, producing 3 nm pore size [3].

8.4 Petrology and Geochemistry of Oil Shale

Shales are tough and dense sedimentary rocks formed over very lengthy periods of geological time possessing various sediments. Shales are found in a range of colors from light tan to black and based on these colors are generally called black shale or brown shale. Other names may be used for shales in different regions. For example, it has been referred to as "the rock that burns" by Ute Indians who observed the properties of oil shale outcrops that burst into flames when hit by lightning. Therefore, it seems logical that different definitions of oil shale types are available. However, the source of each definition and the exact type of shale which is being defined by each specific definition should be qualified. For example, in one definition, oil shales are categorized based on their mineral content as: (1) Carbonate-rich oil shales, that are rich in carbonate minerals such as calcite and dolomite, with the organic-rich layers pressed as thin layers between carbonate-rich layers. These shales are high-strength formations with noticeable resistance to weathering, and due to their hardness, it is not easy to process them by mining (ex situ). (2) Siliceous oil shales, which do not contain much carbonate minerals, but are very rich in siliceous minerals such as quartz, feldspar, clay, chert, and opal.

These shales, which exhibit dark brown or black colors, are usually not as hard as the carbonate shale. Nor do they show such resistance to weathering. As a result, ex situ mining might be considered for this category. (3) Channel oil shales, which are rich in organic matter that totally surrounds other minerals in the rock. These shales are also dark brown or black and since they are not very hard, they are considered suitable for extraction by ex situ mining methods. Shales are also more commonly categorized based on their origin and formation as well as the type and nature of the organic matter included in the rock as: (1) terrestrial origin; (2) marine origin; (3) and lacustrine origin (Hutton, 1987, 1991). The type of organic matter and its composition forms the basis of this classification and the shales are categorized based on the environment in which sedimentation of the shale has occurred and the organisms that have formed the organic matter content of the rock. It also considers the distillable products the shale can produce.

9. PRESSURE ON CONFIGURATION OF OIL AND GAS MARKETS

One of the first questions raised by the shale "revolution" will be about the impact of rising shale oil (and gas) production on markets and prices. In a normal market, prices should fall and the only issue for an economist would be to calculate how fast and by how much. However, the oil market is not "normal": It features a producer cartel with a long history of trying to manage prices and production. In oil markets, the question of the consequences of higher (shale) oil supplies seamlessly translates into the question of how OPEC is likely to react.

There are good reasons to presume that OPEC members are willing and capable of reacting with production cuts to rising non-OPEC supplies.

Production cuts, in turn, lead to spare capacity. The build-up of spare capacity in OPEC countries required to neutralize the additional supplies (including biofuels and oil sand production) is substantial indeed. Taking our conservative production profile as a guide, spare capacity to accommodate the new supplies will have to exceed 6 million barrels per day (Mb/day) within this decade—the highest since the late 1980s, as shown in Fig. 7.35.

This will be no easy task for OPEC. The cohesion of the organization is the key uncertainty, especially in the current decade. It will be a world in which Russia, the US, and Saudi Arabia account for a third of global production. Of the three, only OPEC member Saudi Arabia is likely to incur the cost of maintaining large amounts of spare production capacity. And Saudi Arabia is

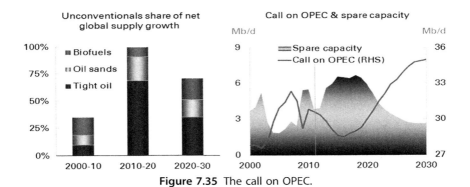
Figure 7.35 The call on OPEC.

situated between Iraq, an OPEC member with steadily rising output but without a quota agreement, and Iran, for which no-one can predict how long the current sanctions will limit production.

The role of "above-ground" political discussions is likely to reach new heights in determining whether, where, and when shale resources will be accessed.

On the other hand, we are unlikely to witness the return of the very low oil prices last seen in the 1990 or 1980s (assuming OPEC stays put). Tight oil production is not only relatively expensive; with its high well density, it is also eminently scalable (different from the conventional supplies in Alaska or the North Sea added in the 1980s). Supplies will react if prices fall substantially.

In the first instance, the advent of shale oil (and gas) is likely to put pressure on prices by stressing the configuration of markets as we know them. In the case of oil, OPEC is the lynch pin of these markets. In the case of gas, it is the oil indexation of gas prices.

REFERENCES
[1] Oil shale & tar sands programmatic. EIS; 2012.
[2] www.slb.com [Schlumberger].
[3] www.spectroscopyonline.com.
[4] http://www.eccos.us/oil-shale-in-co-ut-wy.
[5] Akashj BA, Jaber O. Characterization of shale oil as compared to crude oil and some refined petroleum products. 2003.
[6] Menzela D, van Bergena PF, Veldb H, Brinkhuisc H, Sinninghe Damstéa JS. The molecular composition of kerogen in Pliocene Mediterranean sapropels and associated homogeneous calcareous ooze. 2005.
[7] http://www.qer.com.au/understanding/oil-shale-z/oil-shale-formation.
[7a] Hartgers WA, Sinninghe Damste JS, de Leeuw JW. Identification of C2-C4 alkylated benzenes in flash pyrolyzates of kerogens, coals and asphaltenes. Journal of Chromatography 1992;606:211–20.

[8]   Schmidt Collerus JJ, Prien CH. Investigations of the hydrocarbon structure of kerogen from oil shale of the green river formation. 1974.
[9]   Altun NE, Hiçyilmaz C, Hwang J-Y, Suat Bağci A, Kök MV. Oil shales in the world and Turkey; reserves, current situation and future prospects: a review. 2006.
[9a] Hounsfield GN. Computerized transverse axial scanning (tomography): Part I. Description of system. The British Journal of Radiology 1973;46:1016−22.
[9b] Hartgers WA, Sinninghe Damste JS, Requejo AG, Allan J, Hayes JM, Ling Y, Xie T-M, Primack J, de Leeuw JW. A molecular and carbon isotopic study towards the origin and diagenetic fate of diaromatic carotenoids. In: Telnaes N, van Graas G, Oygard K, editors. Advances in organic geochemistry 1993. Pergmon: Oxford. Organic Geochemistry 1994;22:703−25.
[9c] Colletta B, Letouzey J, Pinedo R, Ballard JF, Balé P. Computerized X-ray tomography analysis of sandbox models: examples of thin-skinned thrust systems. Geology 1991;19:1063−7.
[9d] Wellington SL, Vinegar HJ. X-ray computerized tomography. Journal of Petroleum Technology 1987;39:885−98.
[10] Dean Allred V. Some characteristic properties of Colorado oil shale which may influence in situ processing.
[11] Aczel T, editor. Mass spectrometric characterization of shale oils: a symposium; 1986. Issue 902.
[12] Zeng H, Zou F, Lehne E, Zuo JY, Zhang D. Gas chromatograph applications in petroleum hydrocarbon fluids.
[13] Compton LE, Supercritical gas extraction of oil shale.
[14] Oil shales in the world and Turkey; reserves, current situation and future prospects: a review.
[15] Gilliam TM, Morgan IL. Shale: measurement and thermal properties.
[15a] Abu-Qudais M. Performance and emissions characteristics of a cylindrical water cooled furnace using non-petroleum fuel. Energy Conversion and Management 2002; 43(5):683−91.
[15b] Abu-Qudais M, Al-Widyan MI. Performance and emissions characteristics of a diesel engine operating on shale oil. Energy Conversion and Management 2002;43(5): 673−82.
[16] www.popularmechanics.com.
[17] Geology and resources of some world oil-shale deposits − Scientific investigations report 2005−5294.
[18] Prats SM, O'Brien SM. The thermal conductivity and diffusivity of Green River oil. SPE-AIME, Shell Development Co; 1975.
[19] Qing W, Liang Z, Jingru B, Hongpeng L, Shaohua L. The influence of microwave drying on physicochemical properties of Liushuhe oil shale.
[20] Speight J.G. Shale oil production processes.
[21] Lee S. Oil shale technology.
[22] Wang Y, Dubow J. Thermal diffusivity of green river oil shale by the laser-flash technique. Thermochimica Acta 1979;28(1):23−35.
[23] Wang Y, Dubow J, Rajeshwar K, Nottenburg R. Thermal diffusivity of Green River oil shale by laser-flash technique.
[24] Josh M, Esteban L, Delle Piane C, Sarout J, Dewhurst DN, Clennell MB. Laboratory characterisation of shale properties.

FURTHER READING
[1]   http://infolupki.pgi.gov.pl/en/gas/petrophysical-properties-shale-rocks.
[2]   [Online] http://www.geomore.com/porosity-and-permeability-2/.

[3] Bai J, Wang Q, Jiao G. Study on the pore structure of oil shale during low-temperature pyrolysis. International Conference on Future Electrical Power and Energy Systems. Energy Procedia 2012;17:1689—96.
[4] Sandrea R. Evaluating production potential of mature US oil, gas shale plays [Online] Oil and Gas Journal March 12, 2012. http://www.ogj.com/articles/print/vol-110/issue-12/exploration-development/evaluating-production-potential-of-mature-us-oil.html.
[5] Ruhi C. The five important implication of Shale oil and gas [Online] EP Energy Post January 10, 2014. http://www.energypost.eu/five-global-implications-shale-revolution/.
[6] A primer for understanding Canadian shale gas; energy briefing note. National Energy Board Canada November 2009.
[7] Natural Resources Canada, http://www.nrcan.gc.ca/energy/sources/crude/2114.
[8] Understanding shale gas in Canada; Canadian Society for unconventional gas (CSUG) brochure.
[9] Understanding hydraulic fracturing; Canadian Society for unconventional gas (CSUG) Brochure.
[10] http://en.wikipedia.org/wiki/History_of_the_oil_shale_industry.
[11] http://en.wikipedia.org/wiki/Oil_shale_reserves#cite_note-wec-4.
[12] Rice University, news and media relations. Shale Gas and US National Security July 21, 2011.
[13] The economic and employment contributions of shale gas in the United States. IHS Global Insight December 2011.
[14] Canada's energy future: energy supply and demand projections to 2035. The National Energy Board November 2011.
[15] IEA. Golden rules for a golden age of gas: world energy outlook- special reports on unconventional gas. 2012.
[16] Statics Canada. Report on energy supply and demand in Canada-2008. Ottawa, ON: Statistics Canada; 2010. p. 29.
[17] Worldwide Geochemistry. Review of data from the Elm worth energy Corp. Kennetcook #1 #2 Wells Windsor Basin 2008. Canada. 19p.
[18] US Office of Technology Assessment. An assessment of oil shale technologies. 1980.
[19] Dawson FM. Cross Canada check up unconventional gas emerging opportunities and status of activity. Paper presented at the CSUG Technical Luncheon, Calgary, AB. 2010.
[20] Gillan C, Boone S, LeBlanc M, Picard R, Fox T. Applying computer based precision drill pipe rotation and oscillation to automate slide drilling steering control. In: Canadian unconventional resources conference. Alberta, Canada: Society of Petroleum Engineers; 2011.

CHAPTER EIGHT

Production Methods in Shale Oil Reservoirs

1. INTRODUCTION

Shale oil production is a nonconventional oil production method in which hydrocarbons are extracted from tight shale rock and brought to surface. Unlike conventional oil production from permeable and porous reservoirs, shale oil does not freely flow under normal conditions due to the very low permeability of the shale that holds the oil. This often requires production techniques such as hydraulic fracturing to open the rocks and allow fluid to flow. Another form of oil found in shales is known as oil shale. This type of production often uses mining to obtain the oil shale, and a number of processes (such as heating or the addition of chemicals) are employed to convert kerogen in the rock into shale oil, which is a form of synthetic crude [1]. A picture of oil shale is shown in Fig. 8.1.

Shale oil is produced both at the surface through mining or using in situ methods that directly extract the fluid from the reservoir without ever bringing any shale to surface for processing.

Figure 8.1 Picture of oil shale [2].

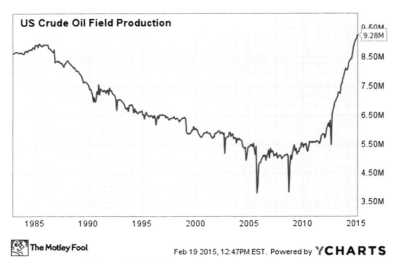

Figure 8.2 US oil production versus time [4].

Since the early 2000s, advances in drilling technology, particularly in horizontal drilling and horizontal fracturing, have made the production of large reserves of tight oil and shale oil possible. This, combined with high oil prices from 2010 to 2015, led to a sharp increase in the oil output of the United States (termed the shale revolution), as shown by the plot in Fig. 8.2 [3].

Potential exists for the development of shale and tight oil reserves in Eastern Canada as well: There are large quantities of hydrocarbon-bearing shales in the rocks under Anticosti Island in Quebec, and the West Coast of Newfoundland in what is known as the Green Point Formation of the Cow Head Group in the Port au Port Region. Political pressure by environmental groups and members of the public to ban hydraulic fracturing on the island has created a standstill in development of the region, but perhaps in the future as technology proves itself and if the current economic climate allows it, the area will see its first commercial developments.

This chapter will examine topics including the history of shale oil production with a section on Eastern Canada and Newfoundland. There is also a critical analysis of the production methods involved with producing shale oils compared to conventional crude, the effects of reservoir and rock properties, the production facilities used, and some operational issues experienced. A discussion of the theory involved with shale oil production and modeling is included, as well as some information on economic, practicality, and environmental concerns related to shale and tight oil production.

2. HISTORY OF SHALE OIL PRODUCTION DEVELOPMENT

In very early history, oil shales have been used as fuel since they are usually capable of burning without any kind of processing, like charcoal or wood [5]. It was also used as a decoration and in rock sculpting, as well as for construction purposes [6]. In Britain in 1684, a group was awarded a patent for their method to "extract and make great quantities of pitch, tarr, and oyle out of a sort of stone" [7].

The first real commercial production of shale oil from oil shales started in France in the 1830s. Large quantities of shale were mined and processed by heating it in specialized ovens called retorts. The production of shale oil spread around other parts of the world by the late 19th century.

In North America, early pioneers learned about the flammable properties of oil shales the hard way when their chimneys made of oil shale bricks caught fire along with the logs inside. Native Americans also understood the flammable properties of the rock, referring to it as "the rock that burns." The first shale oil operations started up around the state of Utah [7].

In the 20th century, production of shale oils actually declined, as conventional crude was much easier to produce, cost less, and was more readily available, so producing oil from oil shales was not economically feasible. In 2016, there is limited oil production from oil shales, but producing tight oil from shale formations through horizontal drilling and hydraulic fracturing has seen a huge increase in adoption in recent years, especially in areas such as North Dakota in the Bakken Shale formations. In the last year or two, low oil prices have made many of these new projects economically unfeasible due to their high break-even oil prices. There exists plenty of potential for future shale and tight oil projects in the United States and Canada if the economic climate becomes friendlier to their development.

In Newfoundland, naturally occurring hydrocarbon seeps have been observed in the western half of the province along coastlines and waterways for the last two centuries. All along the western section of the island up the Northern Peninsula show evidence of oil in the form of gaseous emission at surface, live and dead oil shows in rocks, and oily odors coming from rock that has been freshly cracked open. The main regions of interest are shown in Fig. 8.3.

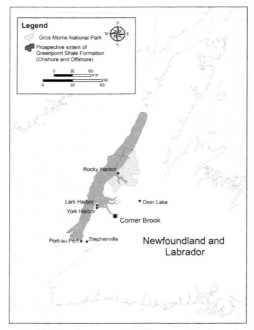

Figure 8.3 Shale oil reserves in Western Newfoundland [8].

The first well drilled in Western Newfoundland was at Parson's Pond in 1867, with a total of 64 wells being drilled from this time until 1991, none of which used seismic data for placement. Many of these wells did not successfully reach total depth, but most of those that did encountered hydrocarbons. It is estimated that a total of 5000 to 10,000 barrels of oil may have been produced in the region during this time. Since 1994, 40 onshore wells, 9 onshore-to-offshore wells, and 1 offshore well have been drilled using seismic data for the purposes of exploration, delineation, or stratigraphic testing. Production during this period has been approximately 40,000 barrels of oil [9]. Currently, understanding of the oil reserves in the area is in its fairly early stages due to the limited amount of exploration and delineation work completed, and at the current time, there is a government-ordered pause in development in the region to better understand the geology of the region and its implications for unconventional oil production. As well, pressure from environmental groups to not allow hydraulic fracturing or "fracking" in the province has contributed to the government's decision to put a stop to current development for the time being. As such, the future of development in the area is unknown at this point, pending further review by the provincial government.

3. PRODUCTION METHODS FOR OIL RESERVOIRS

Various production methods are used to extract petroleum from the earth. There are three types of methods used for the recovery of oil: primary recovery, secondary recovery, and tertiary recovery. Generally, shale oil reservoirs require more advanced production techniques to be applied in order to recover oil.

3.1 Primary Recovery

The first stage in producing life of almost all conventional reservoirs is primary production where the reservoir fluids are produced by natural depletion. The reservoir energy during this production stage is, in fact, the pressure gradient that causes the fluids to move to the surface through the producers. The reservoir energy, however, decreases gradually as the reservoir pressure declines due to production and the wells will cease to produce unless the reservoir energy is augmented by secondary recovery methods.

3.2 Secondary Recovery

Secondary recovery includes methods of petroleum production that are based on the use of man-made energy to produce oil. This means injecting fluids to increase the pressure of the reservoir and creating an artificial drive. This includes water injection and natural gas injection.

3.2.1 Water Injection

Water injection is water-flooding a reservoir using man-made systems to increase the production from oil reservoirs and is injected directly into the production zone. Since the density of water is greater than the density of oil, the water injected will cause the oil to rise and flow toward the production well. Water flooding can increase the amount of oil recovered from a reservoir but is sometimes not the best method to use for oil recovery as it can have complicating factors. Water flooding is the most commonly used secondary oil recovery method because it is inexpensive—water is cheap and usually available in large volumes. The mobility ratio of the oil and water and the geology of the oil reservoir are the key factors to determining how effective the water flooding will be.

3.2.2 Gas Injection

Gas injection is similar to water injection, whereby gas is injected to maintain pressure in the reservoir for the production of oil. Gas injection is achieved by injecting gas through dedicated injection wells. Gas is usually injected into the gas cap of the formation instead of directly into the production zone like water injection.

3.3 Tertiary Recovery

Tertiary recovery includes methods of petroleum production that are used to increase the mobility of the oil in order to increase extraction. This process would be completed through the use of steam and chemicals. This would include steam flooding, surfactant flooding, and CO_2 flooding.

3.3.1 Steam Flooding

Steam flooding is the process whereby steam is pumped into the well and eventually that steam will condense to hot water. In the hot water zone the oil will expand and in the steam zone the oil will evaporate, causing the viscosity drops to expand and increasing the permeability of the reservoir. This is a cycling process and is a commonly enhanced oil recovery method that is used today [10].

3.3.2 CO_2 Flooding

Gas injection is the most commonly used method for enhanced oil recovery in today's industry [11]. Gas injection is similar to water injection, whereby gas is injected into the reservoir to maintain pressure in the reservoir for the production of oil. This is because the interfacial tension between the oil and the water is reduced. Gas injection is achieved by injecting gas through its own injection wells. A commonly used gas is CO_2 because it reduces the oil viscosity and is less expensive than liquefied petroleum gas.

3.3.3 Polymer Flooding

Polymer flooding is the process whereby the viscosity of the water is increased by mixing long chain polymer molecules with the injected water. From this process, the mobility ratio of the oil and the water increases. To reduce the residual oil saturation (the remaining oil in a reservoir after a flood) surfactants can be added with the polymers as they decrease the surface tension between the oil and the water.

4. PRODUCTION TECHNIQUES FOR SHALE OIL RESERVOIRS

Shale oil, often grouped with other types of tight oil, is petroleum that consists of light crude oil contained in petroleum-bearing formations of low permeability, in this case, shale.

When organic hydrocarbon is heated, it yields combustible gases, crude shale oil, and retorted shale. This shale oil is obtained by various techniques such as in situ processing or underground processing. In situ processing, fracturing, and retorting of the deposit is done, whereas in underground processing, shale is mined and heated in retorting vessels. Oil shale mining is further classified as underground mining and surface mining on the basis of topographical features, accessibility, overburden thickness, and presence of groundwater in the mining zone. There are exploited either by surface processing techniques or by in situ technology. Surface processing is mainly done using three steps such as mining, thermal processing or retorting, and processing to obtain refinery feedstock [12]. Fig. 8.4 shows various stages to obtain shale oil [12].

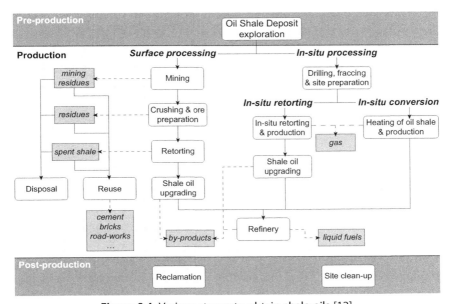

Figure 8.4 Various stages to obtain shale oils [12].

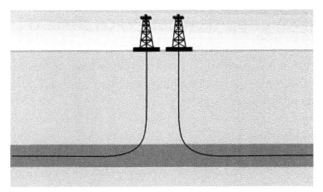

Figure 8.5 Horizontal drilling [13].

As noted before, different types of technologies are used for the production of shale oil, but in today's industry the most common methods used for shale oil production are horizontal drilling and hydraulic fracturing.

4.1 Horizontal Drilling

Horizontal drilling is a commonly used technology because drilling at an angle other than vertically can stimulate reservoirs and obtain information that cannot be done by drilling vertically. Horizontal drilling can increase the contact between the reservoir and the wellbore. As displayed in Fig. 8.5, the wells are drilled vertically until a depth that is above the shale oil reservoir is reached. This depth is usually 1000–3000 m below the surface [12]. Once this calculated depth is reached, the well begins to turn at an angle that is steadily increased until the well becomes parallel with the reservoir. Once the well is parallel with the reservoir, it is drilled until the desired length is reached.

4.2 Multiple Fracturing

Multiple fracturing is implemented in most of the horizontal wells because it is necessary to stimulate the reservoir rock. It consists of dividing the horizontal leg into sections, which are then fractured independently (see Fig. 8.6). During the multiple fracturing operation, each stage is isolated from the rest of the wellbore. It is isolated with the help of various types of plugs or packers. After the completion of all the fracture stages, this plug is removed and all the stages of the wellbore are allowed to flow back to the surface [14].

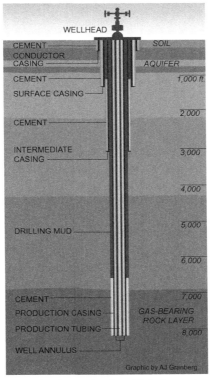

Figure 8.6 A schematic of multiple fracturing [14].

4.3 Hydraulic Fracturing

When a well is drilled in tight oil reservoirs, they require some sort of well stimulation. A common type of stimulation used in today's oil and gas industry is hydraulic fracturing [14]. In unconventional oil, the reservoir permeability is typically low and additional pathways must be developed to allow the flows of hydrocarbons [14]. As displayed in Fig. 8.7, hydraulic fracturing applies pressure by pumping fluids into the wellbore, which develops pathways in the reservoir so oil can flow to the wellbore. Hydraulic fracturing (also known as fracking) also improves the permeability of the reservoir by pumping these fluids into the reservoir. Once these fractures in the rock have been created, sand is usually pumped into these small fractures to keep them in place. As shown in Fig. 8.7, the fracking process usually involves many stages. Each stage is isolated using packers or plugs to ensure that the fracture grows in the desired direction and distance.

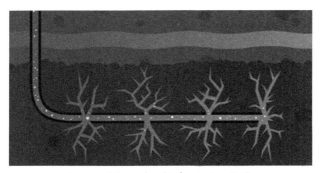

Figure 8.7 Hydraulic fracturing [15].

Fracking uses a large amount of water (millions of gallons) along with sand and chemicals that are pumped underground. Environmental and health problems have developed due to hydraulic fracturing, mainly caused by the chemicals being injected underground. Water contamination has become a problem due to the wells being drilled beyond qualifiers, and the lack of disclosure of the chemicals by several gas industries is keeping people from knowing what they could potentially be drinking. Also, earthquakes, blowouts, and the method of waste disposal of produced waters could also have added concerns to the adverse effects of hydraulic fracturing.

4.4 Oil Shale Production

The modern industrial use of oil shale for oil extraction dates back to the mid 19th century and began to decline after World War II due to the accessibility of conventional crude oil. The extraction of oil from the oil shale was completed by a process called retorting. Retorting is the conversion of the solid hydrocarbons in rock to liquid form when extracting oil from oil shale so that it can be processed. This process is completed by heating the rock to very high temperatures and separating the resultant liquid. There are two main methods to oil shale production, surface retorting and in situ retorting.

4.4.1 Surface Retorting

Surface retorting usually follows the standard process of mining the oil shale, retorting aboveground, and the processing of the shale oil. Once the mining process is completed, the oil shale is taken to a facility for retorting, which is the process explained in the previous sections. Once this heating process is completed, the oil is upgraded by further processing before being sent to a refinery.

There are two types of mining methods, surface mining and underground mining.

4.4.1.1 Surface Mining

Surface mining is a process whereby soil and rock overlying the mineral deposit are removed. There are two principle types of surface mining, strip mining and open-pit mining (see Fig. 8.8). Strip mining is the process of mining a seam of mineral by removing a long strip of overlying soil and rock. Strip mining is a practical type of mining when the ore body that is to be extracted is near the surface. Open-pit mining is the process of extracting rock or minerals from the earth through their removal from an open pit or borrow.

Surface mining is usually used to develop coal seams and deposits of many other minerals, but their feasibilities vary with the nature of the body ore. Large, low-grade ore deposits are usually economically attractive because it permits high recovery of the resource and allows sufficient space for large and efficient mining equipment. An open pit mine could recover almost 90% of the oil shale in a very thick deposit [17]. It is said that strip mining could provide an even higher recovery. Shown in Fig. 8.8 is an open-pit mine used to explain the process of mining the oil shale. The overburden is drilled and blasted loose over an area above the oil shale zone and the load is then transported to a disposal area. Once the shale beds are exposed from the blasting, the shale is then drilled and blasted and then extracted from the pit.

4.4.1.2 Underground Mining

Underground mining is the process of extracting oil shale and other minerals from sedimentary rocks through various underground mining techniques. Underground mining techniques are vastly different than surface mining

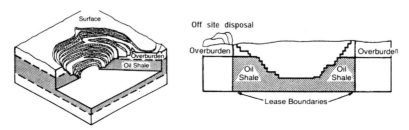

Figure 8.8 Open-pit mine [16].

techniques. Some underground mining techniques include room and pillar mining, and long wall mining. Room and pillar mining is usually completed in flat dipping bedded ores. Pillars are placed in a particular pattern and the rooms are mined out. This method is used in deposits such as oil shale and coal. Long wall mining is a mostly automated process whereby long-wall mining machines are used to extract blocks of deposits as wide as the face of the equipment, and potentially kilometers long. Cutters are then used to cut the deposits from the face of the equipment. Room and pillar mining is a common underground mining method for mining oil shale.

4.4.2 Aboveground Retorting

There are many different types of aboveground retorting, which tend to differ in terms of operating characteristics and technical details. In general, there are four commonly used aboveground retorts. The first type of retort is the process whereby heat is transferred by conduction through the retort wall [17]. Since the method of conduction heating tends to be very slow, this type of retort is not commonly used. The next type of retort is the process whereby heat is transferred by flowing gases generated within the retort by combustion of carbonaceous retorted shale and pyrolysis gases [17]. The next type of retort is the process whereby heat is transferred by gases that are heated outside of the retort vessel [17]. The final type of retort is a process whereby heat is transferred by mixing hot solid particles with the oil shale [17].

4.4.2.1 In Situ Retorting

In situ retorting is the processing of oil shale underground. The oil shale is slowly heated underground and the resultant liquids and gas are extracted directly from the reservoir. True in situ retorting uses injection and production wells that are drilled in a particular pattern. One pattern that is commonly used is the five-spot pattern, whereby four wells are drilled at the corners of a square and the injection well is drilled in the center [15]. The deposit is then heated through the injection wells and to be efficient, the deposit must be highly permeable. In situ retorting does not generate the amount of waste disposal that accumulates from mining in surface retorting, making it an attractive option.

4.4.2.2 In Situ Method

One method for in situ retorting is the wall conduction method. This method uses heating pipes that are placed in the formation of the oil shale. This uses electricity within the heating elements to heat the oil shale layer between 650 and 700° over a time frame of approximately 4 years [18]. The area in which processing occurs is isolated from the surrounding groundwater by a freeze wall consisting of wells filled with a circulating super chilled fluid. This method has many disadvantages, including the risk of groundwater pollution, the extensive use of water, and large electrical power consumption [18].

5. EFFECT OF ROCK PROPERTIES ON OIL PRODUCTION FROM OIL SHALES

One of the characteristics of tight shale oil reservoirs is that they have very low natural porosities and permeability. This means that unlike a conventional reservoir, the pores are so small and poorly connected that the hydrocarbons are unable to flow using standard production techniques. The porosity in fractured shales is generally less than 5% [19]. The permeability in oil-rich shales is often extremely low—in the range of 0.001 to 0.0001 millidarcies, compared to tens to hundreds of millidarcies for a conventional sandstone reservoir [20]. As such, these types of reservoirs require some form of stimulation to achieve producible flow, typically through the use of hydraulic fracturing whereby water and proppant is pumped into a well at high pressures to create microfractures, which allow fluid to flow.

Because porosity is the open space in rocks where hydrocarbons can accumulate, and permeability is what allows the hydrocarbons to flow, it is clear that higher permeability and higher porosity lead to higher production rates and more oil in place in most cases. As such, shales are not naturally good oil reservoirs unless they are fractured to artificially increase permeability and porosity. Shales are sedimentary rocks that form from some of the smallest grains, usually silt and clay-sized materials. Shale is essentially mud that has been compressed after being buried and transformed into rock. It is because of its tiny particle makeup that the rocks are so densely packed with little pore space and poor connectivity between spaces [13]. Therefore, within the classification of shale rocks, those that have formed from the smallest grains of silt would be expected to be the worst candidates for oil production with all other things kept equal.

Another important aspect of the rock formation for shale oil is the deposition environment and whether it allowed organic material to accumulate. Organic material is one of the requirements for hydrocarbon formation, and is important to consider when determining whether a shale bed may have oil production potential. Shale that contains organic material that can turn into hydrocarbons in the right conditions is called black shale [21].

One other component to consider is the structure of the reservoir itself, especially the thickness of the beds and the amount of faulting present. In general, a thicker reservoir with less faulting will have better potential for oil production than one that is thin and has been broken up over time into many smaller blocks separated by faults, requiring more wells to reach all the target zones.

6. SHALE OIL WELLHEAD AND GATHERING

Because shale oil reservoirs often need large numbers of wells or sidetracks to access all the target zones, along with the fact that shale oil reservoirs are often much more spread out than traditional reservoirs that are concentrated in smaller areas, developing a shale oil reservoir may require larger numbers of spaced out well sites with relatively low individual production rates and short economic well lives. This kind of setup with many drill sites is shown below in Fig. 8.9.

This kind of site often requires pipe networks to gather production fluids from the many wells to transport the crude oil to a central processing location, which may be a fair distance away from the well site itself.

As well, the production equipment for a shale reservoir well, including the wellhead, casing, Christmas tree, and such, must be rated for high

Figure 8.9 Tight/shale reservoir wellhead sites [22].

pressure operation in order to facilitate hydraulic fracturing operations, which require pumping fluid at very high pressure to fracture the reservoir rock (see Fig. 8.10).

As can be seen in Fig. 8.10, there are many components in a wellhead and Christmas tree assembly. The wellhead is the equipment at surface that connects to the production tubing and casing, sealing these components from the external environment. From the wellhead, there is also the Christmas tree, which is essentially a large assembly of valves and lines to direct and contain the flow of fluid coming from the well. Some of the most important components are the production and annulus master/wing valves, crossover valves, production choke, gages, tubing hanger, equipment for chemical/gas lift/methanol injection, and any required safety equipment such as pressure safety valves. The production master and wing valves are used to shut in the production tubing if required, and the annulus valves are used to gain access to the annular space between the tubing and casing. The production choke can be used to choke back the production coming from the well for reservoir management purposes. There are often pressure, temperature, density, and other gages such as

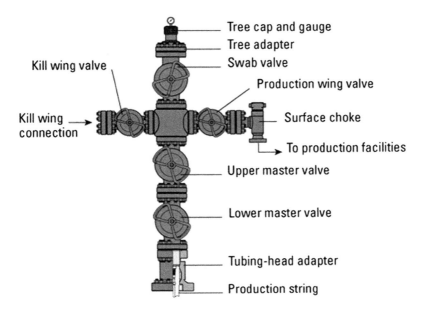

Figure 8.10 Wellhead and Christmas tree diagram [23].

those for measuring sand production or resistivity. The tubing hanger supports the tubing string as it hangs in the wellbore. The flow leaving each Christmas tree is collected at a central manifold where it is sent to the production plant to be separated, treated, and sent to storage.

7. LIMITATIONS FOR PRODUCTION FROM OIL SHALE: OPERATIONAL PROBLEMS

7.1 Price of Oil

The greatest limitation for oil shale production is the price of oil. Up until the last decade or so, oil shale production could only produce small amounts of oil, making production not economically feasible. Before newer technology came around, such as the developments in horizontal wells, the main process of extracting shale oil was surface mining like open-pit mining and strip mining. With these methods, they have a high processing cost and depending on the price of oil may not produce enough oil to be economical especially when there is an availability of cheaper petroleum. After World War II, the oil shale production around the world continued to grow until the 1973 oil crisis. The world production of oil shale peaked at 46 million tonnes in 1980 and fell to about 16 million tonnes in 2000 due to the availability of cheap conventional petroleum.

In the start of the 21st century a boom in oil shale production started due to multiple factors. The main factors were the advancement in technologies that allowed for more oil to be extracted and the continually rising price of the barrel. This started a revolution in the United States for oil production. The amount of oil shale production kept growing at a fast rate until the collapse of the oil industry in 2015. When the price of the barrel plummeted, so did the production of all methods but especially oil shale. Since the collapse the production rate fell drastically and still hasn't recovered. This is just another example to show that oil shale production is limited on the price of the barrel.

7.2 Public Opposition

Many oil shale operations have been paused or canceled altogether due to public opposition. Due to a list of environmental issues that come with oil shale production, many people do not wish for operations to be done in their area. This list includes gas emissions, water consumption, water contamination, increased seismic activity, and radioactive materials being released into water. There have been many studies done and still being conducted on oil shale operations, all with differing results. For example, there are currently plans to start production on the west coast of Newfoundland that has been put on hold due to the overwhelming opposition from the general public.

8. EXAMPLES OF PRODUCTION TECHNIQUES IMPLICATION IN REAL OIL SHALE CASES

One of the biggest conventional oil shale production sites is the Athabasca oil sands in northern Alberta. Currently about 1.3 million barrels of oil is produced daily, mainly from surface mining. Although most of the operations are surface mining, in situ methods are also used such as steam-assisted gravity drainage.

With the downturn the amount of in situ oil production sites has dropped but there are still many sites across the world, especially in North America. In the United States, many sites are still in production and new ones are being created regularly, such as the Eagle Ford Shale sites.

9. GOVERNING EQUATIONS TO MODEL SHALE OIL PRODUCTION METHODS

When considering shale oil production methods, like all other fluid flow through porous media, the primary governing equation to model most production methods is Darcy's Law [24]:

$$q = -\frac{KA}{\mu}\frac{\partial P}{\partial L} \qquad (8.1)$$

where q is the volumetric flow rate in cm^3/s, K is the permeability in darcy, A is the cross-sectional area of the flow in cm^2, μ is the viscosity in centipoise, and $\partial P/\partial L$ is the pressure gradient over the length of the flow path in atm/cm.

This equation is a one-dimensional flow simplification where fluid flows in only one direction. Fluid flow through a porous media is generally far more complex than this, however Darcy's equation gives the user some bounds on what the potential fluid flow can become [24]. This equation demonstrates that permeability and cross-sectional area are directly proportional to flow rate while viscosity is inversely proportional. Additionally, as the pressure gradient becomes higher, the flow rate also becomes higher.

Lastly, the negative sign gives that for positive flow rate the pressure gradient must be negative, meaning that the fluid flows toward areas of lower pressure [24].

In primary shale oil production, primarily Darcy's Law governs the production rate. A good estimate of production rates can be obtained using the equation. The key difference between conventional oil and shale oil is that the permeability is significantly lower. Conventional oil typically sees a permeability of 0.5–50 darcy while shale oil can be up to 1000 times less ranging from 0.5 to 5 millidarcy [25,26].

In secondary shale oil production, material balance equations can be applied to determine recovery factors and original oil in place in the reservoir. The following equation describes material balance within a shale oil reservoir.

Change in oil phase volume + Change in gas phase volume
+ Change in water phase volume + Change in solid phase volume
= 0

(8.2)

Ultimately, knowing the changes in each of these volumes along with pressure changes and reservoir rock properties can determine the bulk volume of the reservoir and estimated recovery factors among other things.

Finally, for tertiary recovery, one major governing equation is that of crack propagation within the reservoir during hydraulic fracturing. Since hydraulic fracturing is able to improve near wellbore permeability and improve connectivity and production, it is frequently used in shale oil production to improve flow rates. The governing equation for hydraulic fracture propagation is as follows [26]:

$$w_{(x,t)} = 2.52 \left[\frac{(1-\nu)Q\mu L}{G} \right]^{1/4} \quad (8.3)$$

Here, w is the width of the fracture, ν is Poisson's ratio of the shale, Q is fracture liquid flow rate, μ is fluid viscosity, L is the length of the fracture propagation and G is the shear modulus of the shale [26]. Here it is evident that as flow rate, viscosity, and fracture length increase, the width increases. When Poisson's ratio or the shear modulus of the shale increase, the width of the fracture decreases. If the width is known, then length can be obtained.

Knowing these parameters can assist in estimating the improved wellbore connectivity and improved inflow oil rate from improved near wellbore porosity and permeability.

The fracture pressure is also a parameter often used in shale oil production. Knowing this parameter can assist an engineer in determining how high they must pressurize the well to achieve fracture. The governing equation is as follows [27]:

$$p_w^{frac} = 3\sigma_H - \sigma_v - p_f + T_o \qquad (8.4)$$

where p_w^{frac} is the wellbore pressure for fracture to occur, σ_H is the larger horizontal stress, σ_v is the vertical stress, p_f is the pore pressure, and T_o is the tensile strength of the formation.

Finally, fracture size can be related to the permeability of the near wellbore media to better predict the flow profile within the reservoir. When aperture data (length, width, and height) of the fracture are known, the relationship between said data and the permeability and porosity could be calculated by the following equations: [28]

$$\phi_{frac} = 0.001 * W_f * D_f * K_{fl} \qquad (8.5)$$

$$K_{frac} = 833 * 10^2 * W_f^3 * D_f * K_{fl} \qquad (8.6)$$

where ϕ_{frac} is the fracture porosity, K_{frac} is the fracture permeability, W_f is the fracture aperture, D_f is the fracture frequency, and K_{fl} is the number of the main fracture direction. Its value is 1 for subhorizontal or subvertical fractures, 2 for orthogonal subvertical fractures, and 3 for chaotic or brecciated fractures. Using this equation in conjunction with Eqs. (8.3) and (8.4) can give rise to many correlations depending on quality and type of data obtained from the production well [28].

10. MODELING AND OPTIMIZATION OF PRODUCTION TECHNIQUES IN SHALE OIL RESERVOIRS

Modeling and optimization of shale oil production techniques is often done in computer programs specifically designed for the modeling and simulation of oil fields. Once rock parameters, such as porosity and permeability as well as fluid properties such as viscosity, are known a computer model can be generated to simulate the production of the shale oil field [27].

For oil production wells, selection of the control variable to be incorporated into the optimization formulation is based on the following considerations:
- Choosing out of the variables in the well model equation the ones that are to be considered as control variables;
- If the valve setting is selected as a control variable, it can be either a continuous variable or an integer.

The use of integer variables is considered an initiative approach for modeling the shut-ins. It is also meaningful to use such integer variables to model on/off valves. Shut-in of a well results in a pressure build-up in the reservoir. This pressure build-up is triggered by the valve setting being set to zero, giving a step response in the grid pressure. This is an important property both in simulation and in optimization of the switching of a shale gas well. It is not possible to implement shut-ins only through controlling the wellhead pressure, and simultaneously avoiding integer variables, without assuming a specific structure of the optimal solution. Using integer values is an efficient way to handle the lower bound on the flow rate. Note, however, that the procedure of the solution and its implementation changes noticeably after the integer variables are introduced in the formulation of the problem..

A variety of reservoir modeling and production modeling software is available in the industry, but they all take advantage of the same governing equation, which is mass balance in three directions using a finite difference analysis. Some commercial software available includes Eclipse, EM Power, ExcSim, ReservoirGrail, and Merlin, just to name a few [29].

As previously mentioned, the governing equation for a finite difference model is the mass balance in three directions. This is so that the model can be discretized into small 3D elements and the mass balance can be applied at each face. The mass balance equation is as follows [30]:

$$(\text{mass accumulation rate}) = (\text{mass flux in}) - (\text{mass flux out}) + (\text{net rate of chemical production})$$

$$\frac{dm}{dt} = \dot{m}_{in} - \dot{m}_{out} + \dot{m}_{reaction} \tag{8.7}$$

One program frequently used to model shale oil reservoirs for simulation is Petrel and Eclipse, supplied by Schlumberger. Geomodelers generate the reservoir structure, lithography, and stratigraphy virtually within the program and assign porosity, permeability, and other rock properties based on core samples obtained from exploration well drilling. Next reservoir engineers fill the structure with oil, water, and gas based on well testing data and fluid analysis. Then the model is matched to production data by editing all reservoir parameters [31].

Once history matched, the model can be used to optimize production by identifying which wells to produce from and which wells to shut in. This is called the development strategy and can be constrained or bottlenecked by gas or water production capacity, injection capacity, or many other factors. Additionally, the model can identify reservoir locations that have not been swept to existing producers and provide potential infill opportunities.

In addition to optimizing using the reservoir model, injection fluid can be optimized to react best with the fluid and rock composition to stimulate injection, improve injection efficiency and sweep, loosen the oil from the shale with surfactants, and improve productivity index by fracturing or acidization [32].

When considering fracturing in the optimization of shale oil production, we must consider the composition of the fracture fluid. The fluid used for fracture generation and propagation must be incompressible such that pressure down hole can be sufficiently increased to the fracture pressure. In addition, a proppant is typically used to maintain the fracture width after pressure is released such that permeability near the wellbore remains high and the productivity index is improved. Fig. 8.11 shows the typical volumetric composition of a fracture fluid.

As seen in Fig. 8.11, 99.51% of the fluid volume is water or sand. Water is incompressible and nonreactive and therefore very useful in fracture processes. The sand acts as the proppant down hole and can be of varying grain size based on the anticipated fracture width. Sand that is too fine will not be effective, as it will allow the fracture to close, whereas too large of a grain

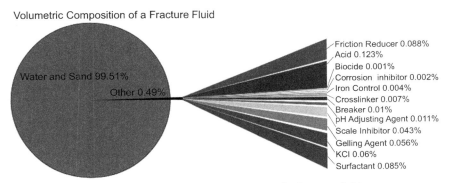

Figure 8.11 Volumetric composition of a fracture fluid.

will not enter the fracture at all [33]. The remaining 0.49% by volume are other compounds meant to reduce the friction of the slurry, inhibit corrosion, iron and scale, adjust pH, and even a surfactant to reduce the surface tension between the slurry and the reservoir fluid [33]. Altogether this fracture fluid effectively improves the flow performance of the well and optimizes the production of the shale oil.

11. SCREEN CRITERIA FOR OIL PRODUCTION IN OIL SHALE

When considering screening criteria for oil production in shale oil, there is one main type of production method to consider—hydraulic fracturing. When considering hydraulic fracturing as a method of oil production, the reservoir must first be screened to ensure it meets the proper criteria for the process. In the case of this process, primarily the rock strength properties should be considered, as well as flow capacity and fluid type. First, if the rock strength is too strong in the reservoir, a very high pressure will be required to fracture the rock, which may not be obtainable using available equipment or may become too expensive given the expected recovery. If the rock is too brittle, high fracture pressures could cause the rock to fissure too long and cause direct pathways of production, which is equally less favorable as the unfractured rock will not be produced at all. Second, if there is insufficient flow capacity at the surface facility, it may not be necessary to fracture a well if the uplift will not be able to be produced. Finally, for hydraulic fracture to work well, the fluid must be of light to volatile oil type. Medium to heavy oils will not able to be produced through the fractures and therefore other production methods, perhaps in line with fracturing should be applied.

An alternative, or in addition, to fracturing could be steam injection. Steam injection is more useful for heavy to medium oils where the viscosity is high and the molecules are much larger. When steam is injected, energy is added to the oil and it is loosened from the shale reservoir rock. As such, the fluid is able to flow much more freely within the reservoir and fluid flow is made easier based on Darcy's Law, decreasing viscosity.

A third alternative production method would be using gas lift to stimulate production by lightening the fluid column within the wellbore and reducing hydrostatic head. This method should be used preferably in undersaturated oil as the added gas is absorbed into solution and then separated at surface. If the gas is added below the bubble point, slugging can occur, which may hinder production slightly. In addition to this, gas injection can be used only if sufficient gas supply is available from the surface facility. If there is insufficient lift available, then production can become erratic or even stop [34].

12. ADVANTAGES AND LIMITATIONS OF OIL PRODUCTION TECHNOLOGY IN SHALE OIL RESERVOIRS

In recent years, the technology involved with producing hydrocarbons from shale reservoirs has seen a dramatic increase in sophistication due to the shale oil boom in North America. This technology has both advantages and limitations.

One advantage of current technology used in shale oil reservoirs is that horizontal drilling can be used to target long, thin sections of reservoir much more effectively than what could be done using only vertical wells. As well, the ability to sidetrack wells or drill multilaterals can help reduce the cost by reducing the number of new wells that have to be drilled. Many wells can be drilled from a single pad at surface, and mobile equipment on trucks can be used again and again to drill more wells, meaning that the additional cost per well can be quite low. As well, modern fracturing operations have the ability to bring most of the injected material back to surface and the ability to reuse the water for future jobs exists. Often, when a well has reached the end of its useful life, its equipment can also be reused on another well.

Some of the most significant disadvantages with shale oil production technology involve its public reputation, especially in the media. Many local governments have been pressured to ban hydraulic fracturing (fracking; including in Atlantic Canada) due to the public's fear of fracking being linked to a number of negative issues, including seismic events, groundwater pollution, health issues, and excessive fresh water use. For these reasons, developing a shale oil reservoir in a frontier-type location can be challenging and may require a lengthy process of expensive approvals. As well, shale oil wells often have high breakeven oil prices, as they are more expensive to drill than conventional wells (require horizontal drilling plus stimulation techniques), require precise drilling methods to hit thin beds, and often do not produce large quantities of hydrocarbons for long periods of time.

As such, in uncertain economic times, especially when oil prices are low or particularly volatile (such as the present time), justifying shale oil development may be difficult, especially in areas that are not already at least partially mature.

13. TECHNICAL AND ECONOMIC ASPECTS OF OIL PRODUCTION IN SHALE OIL RESERVOIRS

13.1 Technical Aspect

Shale oil reservoirs are usually developed through the following process.

The first step to the development process would be exploration. A potential reservoir must be assessed by studying its geology, leasing the mineral rights, then applying for licenses and permits.

Next, the site preparation and well construction stage is completed. Exploratory drilling is used to determine the host rock's characteristics and to assess the quality and quantity of the resource. To prepare the site, the well pad must be constructed and the wellbore must be drilled to determine if the well looks like it could be successful.

Furthermore, the horizontal drilling of the wells occurs. The well is drilled vertically until some distance above the reservoir and then begins to turn on an angle until the well is parallel with the reservoir. Horizontal drilling is favorable over vertical drilling because even though it costs more, the increased production from the larger body of rock is enough to offset the cheaper costs of vertical drilling.

After the directional drilling is completed the well must be stimulated. This is done by injecting water into the well at high pressures, creating fractures in the reservoir, allowing oil to flow to the wellbore.

Once the oil is able to flow to the wellbore, the oil is collected at the wellhead and is under operation and production. Producing wells are commonly monitored and inspected for leaks.

The final stage of the process is the end of the production stage, when the well is no longer productive and is abandoned. Before abandoning the well, the company must ensure that the well is properly sealed.

13.2 Economic Aspect

There are many benefits for the world in production of shale oils. If each barrel of imported oil is replaced by a barrel of domestic oil production, the trade deficit is reduced by the cost of that barrel. The average minimum

economic prices for a mature plant with a capacity of 100,000 Bbl/d are as follows:
- $38/Bbl: true in situ;
- $38/Bbl: surface mining;
- $57/Bbl: underground mining;
- $62/Bbl: modified in situ.

The estimates show that by 2030, it is possible for oil shale production of 3 MMBbl/d to cause the global oil prices to reduce by 3—10%. Boosting domestic production during the next 30 years can help avoid $325 billion of imports. Commercial, full-scale, in situ oil shale projects can be in the range of 10,000—100,000 Bbl/d for a surface retort of about 300,000 Bbl/d. The process technology and resource quality are two factors that control the capital and operating costs. Model estimates of the capital and operating costs are, respectively, in the ranges of $12—20/Bbl (CAPEX) and $40,000—55,000 per stream of daily capacity (OPEX), which include mining retorting and upgrading. However, the cost varies depending upon the mining, retorting, and upgrading. Gross revenue from the potential oil shale production is used to estimate the GDP as well as the influence on the trade deficit. The labor component of all major cost elements is isolated to calculate the labor cost. The minimum economic price, which is defined as the world crude oil price required to yield a 15% rate of return on the project, for oil shale projects is also under the influence of technology and quality of the resource. Estimates show that the production from shale oils is expected to reach 500,000 Bbl/d by 2020 and then remain steady for the next 15 years. Furthermore, the cumulative shale oil production is estimated to reach 12.8 billion barrels. Shale oil development is also expected to have a remarkable impact on employment by generating up to 300,000 high-paying jobs. Table 8.1 summarizes the typical costs [35].

The economic analysis of a tight oil reservoir has been based on the Permian Basin. The Permian Basin is a sedimentary basin largely contained in the western part of Texas and the southeastern part of New Mexico. According to [35], "Since the early 1990s the cost of oil production has nearly doubled and the break-even price for the Permian basin tight oil plays is about 61 USD per barrel," which can be seen in Table 8.2.

From Table 8.3 it can be seen that the estimated drilling and completions cost for the Permian basin is about $19 million. The operating cost is approximately $36 per barrel of oil. The severance tax is 4.6% and has a net revenue interest of 75% with an 8% discount.

Table 8.1 A Summary of Typical Cost for Shale Oil Production

Item	Unit	Business as Usual	With Targeted Tax Incentive	With RD&D
Production	Billion Bbls	3.2	7.4	12.8
Direct federal revenues	Billion $	15	27	48
Direct local/state revenues	Billion $	10	21	37
Direct public sector revenues	Billion $	25	48	85
Contribution to GDP	Billion $	310	770	1300
Value of imports avoided	Billion $	70	170	325
New jobs	FTE (thousand)	60	190	300

Table 8.2 Break-Even Oil Price [35]

Play	Number of Wells	Break-Even Oil Price
Trend Area–Spraberry	873	$54.75
Wolfcamp	992	$75.36
Bone Spring	692	$49.24
Total/weighted average	**2,557**	**$61.25**

Table 8.3 Economic Costs of Permian Basin [35]

Permian Basin Economic Assumptions			
	Trend Area-Spraberry	Wolfcamp	Bone Spring
Drilling and Completion cost	$6.4 million	$7 million	$6 million
Operating expenses	$12/BOE	$12/BOE	$12/BOE
Severance tax	4.6%	4.6%	4.6%
Net revenue interest	75%	75%	75%
Discount	3%	8%	8%

BOE, barrel of oil equivalent

14. TYPICAL COSTS TO CONDUCT OIL PRODUCTION PROCESSES IN OIL SHALE

A cost breakdown was completed in a report released by the University of Pittsburgh showing a general breakdown cost of drilling an in situ well. Table 8.4 lists the cost.

This cost breakdown would not cover every well as certain types of operations will cost more or less depending on the project, but this is a good estimate to the total cost.

Table 8.4 Economic Costs of In Situ Well

Type of Operation	Cost ($)
Land acquisition and leasing	2,100,000
Permitting	663,000
Horizontal drilling	1,200,000
Hydraulic fracturing	2,500,000
Completion	200,000
Production to gathering	472,000
Total Cost	**7,135,000**

15. ENVIRONMENTAL/PUBLIC SUPPORT ISSUES WITH SHALE OIL

The environmental impact has always been a big issue with oil shale production. Many environmental groups have fought against oil shale.

There are a series of federal laws that govern most of the environmental aspects of shale gas and oil developments in the United States and Canada. The federal laws can address the regional- and state-specific character of the activities such as geology, hydrology, climate, topography, industry characteristics, and local economics. The state agency not only imposes federal laws but also its own sets of federal NEPA law, which requires environmental assessments at the state level. The federal laws have the right to regulate, permit, and enforce all the activities from drilling to fracturing of the well. Every state requires developers to obtain a permit before drilling and fracturing the well. This permit involves the data about the well location, construction, operation, and reclamation. There is always site inspection before the permit is approved [36].

A project planned for the west coast of Newfoundland has been held up, waiting for an environmental impact review before it can go ahead. The oil shale projects have an impact upon the land usage, water management, air pollution, and greenhouse gas emissions. Surface mining, retorting, and fracking both have their unique set of impacts but surface mining has the greatest impact on the environment.

15.1 Surface Mining

15.1.1 Area Usage

To carry out surface mining in oil shale, a large area of land is required. This area is needed for mining, processing, and waste disposal, which will leave the land scarred and unusable for a long time. For this reason, production should not be done near highly populated areas. After any operation is complete, the operators have to reclaim the land by trying to replace the original biodiversity in the area, but reclaiming takes time and usually does not make it 100% back to the way it was.

15.1.2 Waste Materials

The waste materials created during production such as mining waste, combustion ash, and used oil shale also use up land to be disposed of. The waste material may be polluted with sulfates, polycyclic aromatic hydrocarbons, and heavy metals that are known to be toxic and carcinogenic. For this reason, the waste should not be disposed of underground but put in a landfill to slow or stop the contamination of the groundwater.

These open pits and landfills also bring up air pollution concerns. Depositing the waste material in a landfill has been studied to prove the distribution of pollutants by air. The harmful chemicals that can be spread are known carcinogens and may be a possible link to the rising of asthma and lung cancer in oil shale areas [36].

15.1.3 Gas Emissions

Another environmental concern is the greenhouse gases produced during production. The gas emissions for oil shale production are higher than the emissions produced during conventional methods. These emissions enter into the conversation of global warming and are a leading reason for opposition from the general public who want to shut down oil shale production.

15.1.4 Water Usage

The biggest factor that gains so much opposition from the general public is water contamination and water usage from the oil shale production. The surface mining heavily affects the water runoff in the greater area around it. Many times, water must be pumped to lower the groundwater levels to allow for production. The lowering of the water level can be extremely harmful to the surrounding environment by reducing the amount of water normally in the area. This effect can kill off forest or other biometric materials. A large amount of water is also used throughout production. Water is need for quenching hot materials and needed to help control the dust kicked up by the machinery. The amount of water used gained a lot of opposition when oil shale was booming across America, specifically the west coast of the United States, where there has been a drought for the last 10 years [37].

15.2 In Situ Oil Shale

Similar to the surface mining, in situ oil shale production has many concerns regarding its effect on the water supply in the surrounding area. There have been many studies done on groundwater and surface water contamination, water usage, and radioactive contamination.

15.2.1 Water Contamination

For in situ production, the fluid that is injected into the ground is composed of many chemicals and sometimes radioactive tracers. Most of the chemicals injected are harmless to humans, but in some projects throughout the United States, carcinogenic chemicals have been used. The use of these carcinogens have sparked a lot of controversy and led some companies to go as far as to hide their use of the chemicals by listing them as trade secrets. All these chemicals pumped into the ground have led to groundwater contamination in many cases. Many studies have been completed in detail to investigate the contamination with varying results. Sometimes during production, a fracture may create a pathway that can allow for the fracture fluid to seep into the groundwater. Many times, companies drill separate wells to act as disposal wells. They pump all unnecessary fluids back into the ground, which has also led to groundwater contamination [38].

In addition to groundwater contamination, in situ oil shale has been known for contamination at the surface level as well. Of the fluid that's injected, a low percentage of the fluid flows back to surface but needs to be dealt with properly. In some cases, recovery is less than 30% and includes a mixture of the injection fluid, formation material, and brine water. There are surface spills that occur more often than thought. Most of these spills happen due to equipment failure or engineering mistakes.

15.2.2 Radionuclides

Another concern to the fluid that flows back to surface is the radionuclides it can bring. In oil shale, there are naturally occurring radioactive materials. These materials can be radium, radon, uranium, and thorium. In many cases, especially in the United States, the local sewage plants filter the flow-back fluid. These sewage plants are not built to screen or deal with the radioactive materials, and allow the materials back into the other filtered water. This has become a big issue and is yet another fighting point against oil shale. The US Environmental Protection Agency considers it a risk to both the workers on site and the other people living in the nearby area [39].

15.2.3 Seismic Effects

Much opposition and many studies have been sparked due to the idea that in situ oil shale production increases seismic activity. These studies have produced evidence that the oil shale operations can cause increased seismicity that can lead to microearthquakes. These microearthquakes are usually too small to be felt but in some cases, such as an event that occurred in England in 2011, the earthquake was felt by local residents.

The reinjecting of waste fluid in the disposal wells has had the greatest impact on seismic activity. Similar to the seismic activity caused by operations, many of the cases are too small to be felt but these cases have a higher amount felt by the public and are of greater magnitude. The severity of these disposal wells has been said to be due to location near existing faults but research is still ongoing.

15.2.4 Air Emissions

In situ operations pose a high risk to climate change due to the gas emissions that happen during operations. The main difference as compared to conventional oil methods is the amount of methane gas released. There is an ongoing very controversial debate about whether oil shale is worse than any other method when contributing to global warming. Some studies show that the oil shale production emissions are 3.5—12% higher than conventional oil. A report by Robert W. Howarth, is a professor at Cornell University, completed a study showing how oil shale emissions are significantly worse than conventional oil or coal operations. His study has sparked many debates and arguments criticizing him and comparing his report to another that has the oil shale emissions similar to that of conventional oil and coal [40].

16. OPPOSING VIEWS REGARDING SHALE OIL

As previously mentioned, there are two main viewpoints on the development of shale oil: those who are for it, and those who are against it. Proponents of shale oil development argue that it has helped the United States reduce its dependency on foreign oil imports, meaning that oil imports from OPEC member countries could be reduced. This takes some of the power away from OPEC in driving oil prices up. As well, oil shale developments can be a huge driver of local and even large-scale economic growth, as new jobs are created in not only the upstream production of the oil, but also the downstream refining and processing of the oil into useful products. A booming oil industry also leads to the growth of spin-off businesses, including oil service companies that specialize in things like drilling, completions, fracturing, and manufacturing of oilfield equipment. The growth in well-paying jobs also has positive effects on other businesses as well, because more people with higher paying jobs in a region leads to more spending at local restaurants, stores, and entertainment destinations. Oil boom locations also often see a growth in new home construction and business development, creating further spin-off effects. In general, oil shale supporters cite the economic benefits of development as the prime reasons for progressing shale oil developments [41].

There is another group of people who believe that the negative consequences of shale oil projects outweigh the economic benefits incurred. While the industry creates jobs, many people are opposed to the increase in truck traffic and heavy industry work that comes along with an oil boom. More oil production and pipelines naturally lead to a higher potential for spills and mishaps to occur. There is evidence to show that hydraulic fracturing operations could be linked to increased seismic activity in an area [42]. Many people believe that shale oil development could be the cause of groundwater contamination, and that frack fluid or actual petroleum could seep upward and reach the water table after fracturing has taken place. As well, the construction of oilfield sites required clearing of forest for wellpads and roads, leading to protest from those against deforestation [41].

Clearly, there is debate over whether or not the pros of oil shale development outweigh the cons. Right now, the high cost of shale oil development and low price of oil are hindering the startup of new shale oil projects, but if the economic environment should become more favorable to production, the economic benefits versus environmental concerns debate will once again become a prominent issue.

17. FUTURE PROSPECTS OF OIL SHALE PRODUCTION

For proper shale oil production forecasts, the potential well drilling should also be considered. The productive area of land which is suitable for drilling as well as optimal well spacing are the two factors that control the maximum number of wells to be drilled. Current estimates show that it is possible to drill 33,000—39,000 wells at the Bakken formation in North Dakota.

Another very important point to consider when forecasting the future shale oil production is that ultimate recovery depends on the well spacing and decrease as the well spacing density increases. In other words, increased rates of drilling new wells to produce the shale oils will result in fast depletion of the reserves. For example, estimates show that daily oil production of 2000 barrels at the Bakken formation may not be sustainable since the drilling potential will be depleted by 2022—2025. A maximum production level of 1500 barrel/d by year 2020 is considered as the most likely scenario for the Bakken formation. Another issue is that production companies need to constantly increase the number of active drilling rigs in order to be able to produce at consistently high levels. Such a continuous increase in the number of active wells will make the companies need to hire more drilling crews. However, according to the statistics, the availability of qualified drilling crew is considered a challenge in near future. Shortage of hydrofracturing crews may also prove to be a critical limitation for future shale oil production globally.

18. RESEARCH AND TECHNOLOGY DEVELOPMENT IN SHALE OIL PRODUCTION

In the 1976 fiscal year, the Energy Research and Development Administration (ERDA) initiated a research program in the area of unconventional natural gas which was then continued by the Department of Energy (DOE) in 1978. This research program consisted of three elements:
- Eastern Gas Shale Program;
- Western Gas Sands Program;
- Methane Recovery from Coal-beds Program.

In 2001, the Nuclear Regulatory Commission (NRC) assessed the benefits and costs of many DOE research and development programs including the above three programs. In 2007, the history of the DOE's unconventional gas R D programs were documented in a report by National Energy Laboratory. The most significant technological innovations related to energy in the 1980s and 1990s, as well as the role of the DOE in development of the technologies were assessed by NRC.

The NRC identified the three technologies of horizontal drilling, 3D seismic imaging, and fracturing among the most important technological innovations. These three technologies are considered critical in the development of shale gas. The role of the DOE in improving horizontal drilling and 3D seismic imaging was concluded by the NRC as "absent or minimal," in "fracture technology for tight gas" as "influential." The NRC, however, did not include fracture technology for shale gas that had developed under the great influence of fracture technology for tight gas. The reason for such an exclusion was probably that by the time of the NRC report, the shale gas boom had not arrived. Microseismic frac mapping is another technology which might be critical to development of shale gas, but has not been fully developed or applied.

19. CONCLUSIONS AND RECOMMENDATIONS

Based on the research conducted, several recommendations and conclusions have been drawn. Shale oil is a type of unconventional crude oil reserve that requires more expensive and technologically advanced techniques to produce. There is also significant public controversy surrounding shale oil development in some areas. To go about developing a new shale oil project (such as on the west coast of Newfoundland), several important considerations should be made. It is important to be aware of the different production techniques that are employed for shale oil production, and based on an understanding of the reservoir properties, the optimal method or combination of methods should be selected. An environment analysis should be conducted, demonstrating how operations can proceed safely with minimal environmental disturbance. Public information sessions and consultation should be considered to inform the public of the actions taken by the developer to create a safe and efficient operation. An economic analysis must be completed, considering the price of oil, cost of development, transportation and refining costs, among other things. Work should progress only if there is a high likelihood of achieving a profitable project. As well, it is important to understand the theory involved in shale oil production, and create a high-quality reservoir model for simulation in order to optimize well placement and completion design. Following these recommendations should increase the likelihood of the development being a success.

REFERENCES

[1] AMSO, Oil shale extraction methods, [Online]. Available from: http://amso.net/about-oil-shale/oil-shale-extraction-methods/.
[2] HowStuffWorks, What's oil shale? [Online]. Available from: http://s.hswstatic.com/gif/oil-shale-2a.jpg.
[3] Council on Foreign Relations, The shale gas and tight oil boom: U.S. States' economic gains and vulnerabilities, [Online]. Available from: http://www.cfr.org/united-states/shale-gas-tight-oil-boom-us-states-economic-gains-vulnerabilities/p31568.
[4] yCharts, Us crude oil field production, [Online]. Available from: https://media.ycharts.com/charts/7f20a289fbe422ef0287d59f3c9c0c96.png.
[5] Knutson CF, Russell PL, Dana GF. Non-synfuel uses of oil shale. 1987. Golden, Colorado.

[6] National Geographic, Oil shale, [Online]. Available: http://nationalgeographic.org/encyclopedia/oil-shale/.
[7] Redleaf Resources, Inc. History of oil shale, 2013. [Online]. Available from: http://www.redleafinc.com/history-of-oil-shale.
[8] Bird L. Fracking not 'a game changer' for N.L., says independent report. CBC; May 31, 2016 [Online]. Available from: http://www.cbc.ca/news/canada/newfoundland-labrador/western-nl-hydraulic-fracturing-report-released-1.3607408.
[9] Natural Resources Canada, Newfoundland and labrador's shale and tight resources, [Online]. Available from: http://www.nrcan.gc.ca/energy/sources/shale-tight-resources/17700.
[10] Prats M. Thermal recovery, Richardson, Texas: monograph series. SPE; 1982.
[11] Hobson GD, Tiratsoo EN. Introduction to petroleum geology. Scientific Press; 1975.
[12] Pan Y, Mu J, Ning J, Yang S. Research on in-situ oil shale mining technology. Fushun City, Liaoning Province, China: Liaoning Shihua University; Allouche E, Ariaratnam S, Lueke J. Horizontal directional drilling: profile of an emerging industry; 2000.
[13] www.Geology.com, Shale, [Online]. Available from: http://geology.com/rocks/shale.shtml; Arogundade O, Sohrabi M. A review of recent developments and challenges in shale gas recovery. Heriot Watt University.
[14] Arthur JD, Uretsky M, Wilson P. Water resources and use for hydraulic fracturing in the marcellus shale region. Pittsburgh: ALL Consulting; 2010.
[15] World Wide Metric, The pros and cons of fracking for oil, [Online]. Available from: http://blog.worldwidemetric.com/trade-talk/the-pros-and-cons-of-fracking-for-oil/.
[16] Hearings on oil shale leasing. In: Subcommittee on minerals, materials and fuels; 1976.
[17] Daood A. Princeton.edu. 2014 [Online]. Available from: www.princeton.edu/~ota/disk3/1980/8004/800407.PDF.
[18] USOT Assessment. An assessment of oil shale technologies. Diane Publishing; 2007.
[19] Zendehboudi S. D2L. 2016 [Online]. Available from: https://online.mun.ca/d2l/le/content/218361/viewContent/1964867/View?ou=218361.
[20] Natural Resources Canada, Geology of shale and tight resources Canada, [Online]. Available from: http://www.nrcan.gc.ca/energy/sources/shale-tight-resources/17675.
[21] U.S. Energy Information Administration. Shale in the United States. July 20, 2016 [Online]. Available from: https://www.eia.gov/energy_in_brief/article/shale_in_the_united_states.cfm.
[22] E. &. S. S. News. Hydraulic fracturing water use is tied to environmental impact. November 4, 2015 [Online]. Available from: https://eos.org/research-spotlights/hydraulic-fracturing-water-use-is-tied-to-environmental-impact.
[23] Well head & Christmas tree components, [Online]. Available from: https://www.croftsystems.net/hs-fs/hub/367855/file-1535415665-png/xmas_tree_diagram.png?t=1477924376337.
[24] Lund L. Decline curve analysis of shale oil production. Uppsala University; 2014.
[25] CSUR. Understanding tight oil. 2014 [Online]. Available from: http://www.csur.com/sites/default/files/Understanding_TightOil_FINAL.pdf.
[26] SPE. Fracture propagation models. May 12, 2016 [Online]. Available from: http://petrowiki.org/Fracture_propagation_models.
[27] Holme A. Optimization of liquid-rich shale wells. 2013. Trondheim.
[28] Crain E. Crain's petrophysical handbook. 2016 [Online]. Available: https://www.spec2000.net/15-permfrac.htm.
[29] Reservoir simulation. August 8, 2016 [Online]. Available: https://en.wikipedia.org/wiki/Reservoir_simulation.
[30] MTU, Mass and energy balances, [Online]. Available from: http://www.cee.mtu.edu/~reh/courses/ce251/251_notes_dir/node3.html.

[31] Schlumberger. Eclipse. 2016 [Online]. Available from: https://www.software.slb.com/products/eclipse.
[32] Earthworks. Acidizing. 2016 [Online]. Available from: https://www.earthworksaction.org/issues/detail/acidizing#.V6Elr7y9Cp5.
[33] CSUR. Understanding hydraulic fracturing. 2015 [Online]. Available from: http://www.csur.com/sites/default/files/Hydr_Frac_FINAL_CSUR.pdf.
[34] SPE, Gas lift; July 1, 2015. [Online]. Available: http://petrowiki.org/Gas_lift; A study on the EU oil shale industry — viewed in the light of the Estonian experience: a report by EASAC to the committee on industry, research and energy of the European Parliament.
[35] Berman A. Artberman.com, June 19, 2016. [Online]. Available: Artberman.com/permian-basin-break-even-price-is-61-the-best-of-a-bad-lot/; J Daniel Arthur, Bruce Langhus, David Alleman. An overview of shale gas developments in USA.
[36] Lotman S. Op-ed: Don't let Estonian shale firm do to Utah what it has done to Estonia. The Salt Lake Tribune; June 12, 2016 [Online]. Available from: http://www.sltrib.com/opinion/3974651-155/op-ed-dont-let-estonian-shale-firm.
[37] Fischer PA. Hopes for shale oil are revived. WorldOil; August 2005 [Online]. Available from: http://web.archive.org/web/20061109140826/http://worldoil.com/magazine/MAGAZINE_DETAIL.asp?ART_ID=2658&MONTH_YEAR=Aug-2005.
[38] Jackson RB, Vengosh A, Carey JW, Davies RJ, Darrah TH, O'Sullivan F, Pétron G. The environmental costs and benefits of fracking. August 11, 2014 [Online]. Available from: http://www.annualreviews.org/doi/full/10.1146/annurev-environ-031113-144051.
[39] Pennsylvania Department of Environmental Protection. Dep study shows there is little potential for radiation exposure from oil. January 15, 2015 [Online]. Available from: http://files.dep.state.pa.us/OilGas/BOGM/BOGMPortalFiles/RadiationProtection/rls-DEP-TENORM-01xx15AW.pdf.
[40] Santoro R, Ingraffea A, Howarth RW. Methane and the greenhouse-gas footprint of natural gas from shale formations. April 12, 2011 [Online]. Available from: http://link.springer.com/article/10.1007%2Fs10584-011-0061-5.
[41] Lombardo C. Pros and cons of oil shale. February 6, 2015 [Online]. Available from: http://www.visionlaunch.com/pros-and-cons-of-oil-shale/.
[42] Hume M. Study confirms link between fracking, earthquakes in Western Canada. March 29, 2016 [Online]. Available from: http://www.theglobeandmail.com/news/british-columbia/study-confirms-link-between-fracking-earthquakes-in-western-canada/article29427905/.

CHAPTER NINE

Shale Oil Processing and Extraction Technologies

1. INTRODUCTION

The exploitation of oil shale involves mining, after which shale is directly burned to produce electricity or undergoes further processing. The two very common methods of surface mining, open-pit mining and strip mining, involve the removal of overlying material. In underground mining of oil shale, however, the removal of overlying material is very limited.

For the extraction of the oil shale both in situ or ex situ processes are involved (see Fig. 9.1). In either case, the pyrolysis converts the kerogen of the oil shale into the condensable vapors, which after condensation are turned into synthetic crude oil and noncondensable gas (shale gas). Pyrolysis includes heating in the absence of air at a high temperature. Usually, this takes place between 450 and 500°C. The decomposition starts at a very low temperature of 300°C, and then proceeds more rapidly at a higher temperature [1–3].

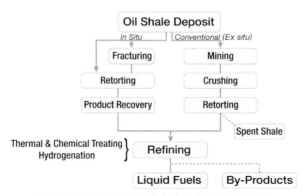

Figure 9.1 Overview of shale oil extraction [1].

Shale Oil and Gas Handbook
ISBN: 978-0-12-802100-2
http://dx.doi.org/10.1016/B978-0-12-802100-2.00009-5

© 2017 Elsevier Inc.
All rights reserved.

2. DESCRIPTION OF OIL SHALE PROCESSING

Shale oil processing is defined as an industrial process to which raw shale oil is subjected to extract oil from it—basically a process to produce unconventional oil [1]. Because shale oil exists as solid sedimentary rocks, its extraction is more complex than extraction of conventional oil [1]. It cannot be directly pumped out of the ground via wellbores [1]. It has to be mined first, followed by heating at high temperatures using chemical processes like pyrolysis, hydrogenation, or thermal dissolution [1], called retorting [1]. The resulting petroleum-like liquid has to be separated and collected [12]. Alternatively, raw shale oil can be heated underground and the resulting petroleum-like liquid can then be pumped out [1]. The resultant shale oil extracted from either of the methods is treated in processing facilities to become usable as a fuel or other feedstock specifications [1].

3. WHAT IS OIL SHALE RETORTING?

Once an oil shale deposit is identified, shale oil is mined, crushed, and transported to a processing facility for retorting—a heating process to separate oil and mineral fractions from raw shale oil [1]. This oil fraction separation process takes place in a vessel referred to as retort [1]. Upon completion of the retorting process, further processing is required to upgrade the oil as per the required feedstock specifications, and spent shale must be disposed of Ref. [1]. Previously mined areas are often leveraged to dispose of spent shale, reclaiming the mine land [1]. Two methods can be utilized for extraction of hydrocarbons from shale oil deposits—mining of shale oil followed by retorting at the surface (surface retorting), or in situ retorting, which involves underground heating of shale oil deposits [1]. Process steps involved in surface retorting as well as in situ retorting are shown in Fig. 9.2 [1].

4. CHEMISTRY OF OIL SHALE RETORTING

Common minerals typically associated with raw and retorted oil shale are shown in Table 9.1 [2]. Carbonate minerals which are originally present in raw oil decompose endothermically during the retorting process, thereby increasing the energy requirement of the process [2]. These carbonates also liberate CO_2, and form acids that end up leaching into surface waters or underground [2]. In addition to common mineral matters, retorted shale oil can also yield valuable metals in low concentrations (see Table 9.2 for estimated content of trace metals).

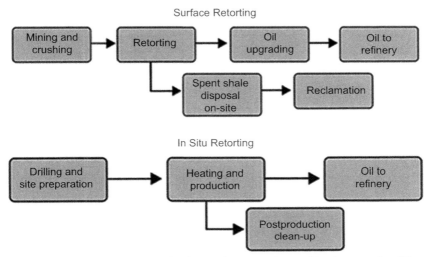

Figure 9.2 Process steps involved in surface retorting and in situ retorting [1].

Table 9.1 Raw Oil Versus Burnt Oil (Retorted) Minerals [2]

Raw Oil Shale		
Dolomite	$CaCO_3MgCO_3$	~35
Calcite	$CaCO_3$	~15%
Ferroan	$Mg_{1-x}Fe_x(CO_3)$	
Dawsonite	$NaAl(OH)_2CO_3$	
Nahcolite	$NaHCO_3$	
Spurrite	$Ca_3Si_2O_8CO_3$	
Potash feldspar	$KAlSi_3O_8$	
Soda feldspar	$NaAlSi_3O_8$	~10
Quartz	SiO_2	~30
Halite	$NaCl$	
Analcime	$NaAlSi_2O_6H_2O$	
Nordstrandite	$Al(OH)_3$	
Burnt Oil Shale		
Silica	SiO_2	40−60%
Alumina	Al_2O_3	10−15%
Iron oxide	Fe_2O_3	5−10%
Calcium oxide	CaO	10−25%
Magnesium oxide	MgO	5−10%

Table 9.2 Composition of Trace Metals [2]

Metal	Estimated Average Content of Shale		Approximate Value Per Ton of Shale (Assuming Complete Recovery)
	ppm	lb/Ton	
Uranium	30	0.06	$4.00
Molybdenum	100	0.2	$2.60
Vanadium	500	1.0	$8.50
Chromium	100	0.2	
Cobalt	100	0.2	$7.80
Nickel	400	0.8	$3.60
Total			$26.50

5. CHEMISTRY OF KEROGEN DECOMPOSITION

The dominant form of organic matter in the raw shale oil is kerogen [2]. It is insoluble in benzene or cyclohexane [2]. The organic matter of this raw shale oil also contains natural bitumen, less than 15%, also insoluble in benzene or cyclohexane [2]. Upon heating without oxygen, kerogen decomposes into bitumen (also referred to as pyrolytic bitumen) [2]. Additional heating converts this bitumen into oil, gas, and a carbonaceous residue that remains in the oil shale matrix [2].

$$\begin{array}{ccc} \text{Kerogen} \rightarrow & \text{Pyrolytic Bitumen} \rightarrow & \text{Residues} \\ \searrow \text{Volatiles} & \searrow \text{Volatiles} & \end{array} \quad (9.1)$$

Experimental studies conducted on the kerogen decomposition of Baltic oil shale yielded a considerable linear relation when a first-order rate constant for kerogen composition was plotted against the by-products of decomposition (see Fig. 9.3) [2].

6. CHEMISTRY OF CARBONATE DECOMPOSITION

The inorganic constituents in shale oil also go through chemical reactions [2]. One of the most significant reactions in the oil shale retorting process involves the endothermic decomposition of various carbonate minerals [2].

$$MCO_3 \overset{\text{heat}}{\rightleftharpoons} MO + CO_2 \quad (9.2)$$

As we see in Eq. (9.2), this reaction requires thermal energy to be able to decompose the carbonate mineral and liberate carbon dioxide [2]. M represents divalent, metallic atom species [2]. It is this reaction that causes an

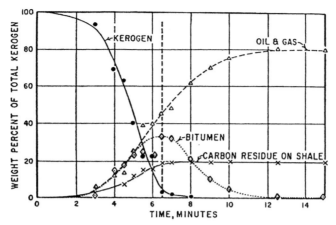

Figure 9.3 Experimental results of decomposition of kerogen [2].

additional energy requirement in the pyrolysis or any other process utilization of oil shale [2]. The raw shale oil may contain roughly up to 50% of carbonate minerals [2]. Out of these carbonate minerals, the most abundant minerals are dolomite, calcite, magnesium carbonate, and ferroan [2].

Some of the important carbonate decomposition reactions are shown in the following equations [2]. Although, these reactions are shown as equilibrium reactions, they experience forward direction when the system is subjected to elevated temperatures [2].

$$CaCO_3 \cdot MgCO_3 \rightleftharpoons CaCO_3 + MgO + CO_2$$
$$\rightleftharpoons CaO + MgO + 2CO_2$$
$$CaCO_3 \rightleftharpoons CaO + CO_2$$
$$MgCO_3 \rightleftharpoons MgO + CO_2$$
$$MgFe(CO_3)_2 \rightleftharpoons MgO \cdot FeCO_3 + CO_2$$

7. PYROLYSIS OR RETORTING OF OIL SHALE: EXPERIMENTS, APPARATUS, METHODOLOGY

Pyrolysis of shale rocks can be conducted above the ground (ex situ) in retorts. Retorts are specially designed vessels where shale rocks can be rapidly heated in an oxygen-free environment [3]. These pyrolytic reactions occur at temperatures ranging from 480 to 550°C [3]. Surface retort by-products typically contain high proportions of olefin, di-olefins, sulfur, and nitrogen compounds [3]. It should be noted that shale rocks have to be mined first, and crushed to a size that can be handled by the retort, prior to being subjected to pyrolysis temperatures [3].

Pyrolysis of shale rocks can also be conducted by heating shale rocks underground (in situ) [3]. Considering rock is a good insulator, heating shale rocks underground is a slow process, and can take months or even years [3]. Pyrolytic reactions at slow heating conditions occur at lower temperatures ranging from 325 to 400°C. This produces lighter oil and a higher gas-to-oil ratio [3]. Two of the most promising systems to perform in situ heating are (1) using electric heat to pyrolyze the rock by installing a large array of vertical or horizontal wells with electrical heaters within the formation and (2) drilling parallel horizontal wells by hydraulically fracturing the rock and injecting an electrical conductive propping medium into the fractured system [3]. A single horizontal well is then drilled at a right angle to the parallel wells, to connect them and establish a plate-like heating element [3]. Electric current would be passed through this to heat the shale rock underground [3].

The third pyrolysis approach involves mining of oil shale, and creating large surface capsules of tailored earth materials with it [3]. An excavated pit is leveraged and insulated with some type of engineered material to prevent escape of by-products [3]. This pit is filled with oil shale [3]. Heating and drainage pipes and sensors are installed inside the pit, and the filled capsule is capped with impermeable material and soil [3]. Hot gases are circulated through these pipes, and product is collected as vapor (see Fig. 9.4) [3]. This approach produces oil similar to in situ processes but offers comparatively shorter production time [3].

8. OPTIMAL RETORTING CONDITIONS

Results obtained from an experiment conducted on shale deposits in Jordan are shown in Fig. 9.5 [5]. The figure shows the variation in

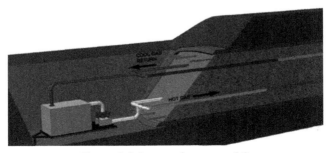

Figure 9.4 In capsule process [4].

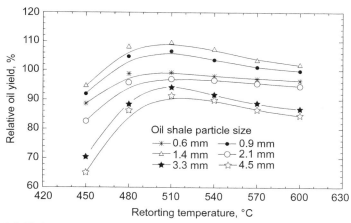

Figure 9.5 Variation in percentage oil yield with temperatures and particle size [5].

percentage oil yield with temperatures ranging from 450 to 600°C [5]. Different particle sizes were evaluated, 0.6–4.5 mm, and thermal decomposition was conducted at a constant shale hold time of 40 min inside the reactor [5].

The percentage oil yield was observed to increase with increasing temperature for all particle sizes, from 450–510°C [5]. This increase in yield is attributed to the conversion and volatilization of shale bitumen compounds to oil, water, and off-gas (hydrocarbons) vapors [5]. The peak was achieved at 510°C, after which particles were found to be less sensitive to temperature, particularly the smaller size particles (0.6–2.1 mm) [5]. Thermal cracking, coking of oil compounds to volatile fragments, and increased off-gases are deemed to be responsible for this yield decline post 510°C of temperature [5]. Hence, the optimum retorting temperature can be regarded as 510°C [5].

With respect to grain sizes, at 510°C of retorting temperature, a significant increase in oil yield was obtained with small shale grain sizes, 0.6–1.4 mm, as seen in Fig. 9.6 [5]. A relative oil yield of 99.5% was reported at a 0.6 mm shale grain size, and this was increased to about 110% at a 1.4 mm size [5]. The yield reduced considerably with large shale particle sizes, 2.1–4.5 mm, fell to around 105% at a 4.5 mm grain size [5]. Low yield at small shale size particles is attributed to particles having large specific surface area and pore volume, allowing the evolved oil (during the first decomposition process) to be retained onto the particle surface and inside the pores [5]. Additional retention time requires secondary decomposition (further cracking), and converts into noncondensable off-gases [5]. Hence, the optimum shale grain size can be regarded as 1.4 mm [5].

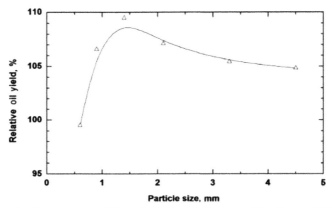

Figure 9.6 Percentage relative oil yield with shale particle size at 510°C [5].

9. KINETICS OF PYROLYSIS OR RETORTING OF OIL SHALE

Kinetics of retorting of oil shale (thermal degradation of oil shale) has been studied by various investigators for many years [6]. However, an agreement on the details of reaction mechanism does not quite exist yet [6]. Therefore, field experiments are given much higher importance when it comes to making any pyrolysis-related determination [6].

Under isothermal conditions, Hubbard and Robinson use a two-step first-order reaction mechanism to describe kerogen decomposition to bitumen, and then bitumen to oil [7]. In the following reaction, k is greater than k', and T is less than 300°C [7].

$$\text{Kerogen} \xrightarrow{k} \text{bitumen} \xrightarrow{k'} \text{oil}$$

Braun and Rothman have studied oil shale kinetics more in depth compared to Hubbard and Robinson [7]. They claimed that oil production kinetics can be described better by additionally considering an initial thermal induction period [7]. Later, Khraisha investigated the isothermal decomposition of oil shale as two consecutive first-order reactions, where activation energies were determined for each stage [7]. Similarly, many other theories or models per se were developed by investigators.

A new analytical methodology, a more general analytical approach to modeling pyrolysis kinetics of oil shale, was proposed by Xiaoshu, Youhong, Tao, and Martti [8]. It offered a closed-form expression of shale decomposition kinetics [8]. It is based on kinetic parameters and, unlike other models, does not make any assumptions on temperature integrals [8]. This model describes kerogen decomposition kinetics in two steps and the retorting process is modeled as an n-order reaction [8].

$$\text{kerogen} \rightarrow \text{bitumen} + \text{gas} \rightarrow \text{oil} + \text{gas} + \text{residue}$$

$$\frac{dX}{dt} = k(1-X)^n \tag{9.3}$$

where X is the weight loss or mass fraction or conversion (%) defined as

$$X = \frac{w_0 - w_t}{w_0 - w_f} \tag{9.4}$$

w_0: initial weight; w_t: weight at time t; w_f: final weight; k: rate coefficient given by Arrhenius equation:

$$k = A\exp\left(-\frac{E}{RT}\right) \tag{9.5}$$

A: preexponential factor; E: activation energy; R: gas constant; T: temperature; Assume $n = 1$.

Since a constant heating rate $\beta = dT/dt$ is the common case in applications, Eq. (9.3) leads to

$$\frac{dX}{1-X} = \frac{A}{\beta}\exp\left(-\frac{E}{RT}\right)dT \tag{9.6}$$

Integration of Eq. (9.6) gives

$$\int_{X_0}^{X} \frac{dX}{1-X} = \int_{T_0}^{T} \frac{A}{\beta}\exp\left(-\frac{E}{RT}\right)dT;$$

$$-\ln(1-X)\bigg|_{X_0}^{X} = \frac{A}{\beta}\int_{T_0}^{T} \exp\left(-\frac{E}{RT}\right)dT \tag{9.7}$$

The right-hand integral cannot be calculated analytically. We propose its approximation as follows:

$$\int_{T_0}^{T} \exp\left(-\frac{E}{RT}\right)dT = \frac{RT^2}{E}\exp\left(-\frac{E}{RT}\right)\bigg|_{T_0}^{T} - \int_{T_0}^{T} \frac{2RT}{E}\exp\left(-\frac{E}{RT}\right)dT$$

$$\tag{9.8}$$

Rearrangement of the terms containing integrals together leads to

$$\int_{T_0}^{T} \left(1 + \frac{2RT}{E}\right) \exp\left(-\frac{E}{RT}\right) dT = \left.\frac{RT^2}{E} \exp\left(-\frac{E}{RT}\right)\right|_{T_0}^{T} \quad (9.9)$$

For a small temperature change ΔT, $1 + 2RT/E$ is approximately constant. Therefore, Eq. (9.9) can be approximated as

$$\int_{T_0}^{T} \exp\left(-\frac{E}{RT}\right) dT \approx \frac{\frac{RT^2}{E}\exp\left(-\frac{E}{RT}\right)}{1 + \frac{2RT}{E}} - \frac{RT_0^2 \exp\left(-\frac{E}{RT_0}\right)}{1 + \frac{2RT_0}{E}} \quad (9.10)$$

and Eq. (9.7) becomes

$$-\ln(1-X) = -\ln(1-X_0) + \gamma_T - \gamma_{T_0}$$

$$\gamma_T = \frac{A}{\beta} \frac{\frac{RT^2}{E}\exp\left(-\frac{E}{RT}\right)}{1 + \frac{2RT}{E}} \quad (9.11)$$

In summary, the weight loss X can be computed base on the following simple closed-form formula:

$$\begin{aligned}\ln(1-X_1) &= -\gamma_{T_1} + \gamma_{T_0} \\ \ln(1-X_i) &= \ln(1-X_{i-1}) - \gamma_{T_i} + \gamma_{T_{i-1}}, \quad i = 1, 2, \ldots\end{aligned} \quad (9.12)$$

It can be seen that the proposed model is a systematic approach that is simple, efficient, and fully automatic. Extensions of it will be presented later.

10. ISOTHERMAL AND NONISOTHERMAL KINETICS MEASUREMENT AND EXPRESSIONS FOR SHALE OIL

The nonisothermal pyrolysis analysis has certain advantages over the isothermal method [7]. Errors caused by the thermal induction period are eliminated [7]. As well, it offers a rapid scan of the whole temperature range of interest [7]. Hence, nonisothermal technique is more commonly used to study oil shale pyrolysis [7]. A comparison of both isothermal and nonisothermal decomposition of oil shale experiments is discussed later. Models covered have been proposed by Yongjiang, Huaquing, Hongyan, Zhiping, and Chaohe [9].

The fraction of the material pyrolyzed, α, was defined as

$$\alpha = \frac{W_t}{W_0}, \quad (9.13)$$

where W_t is weight loss quantity after time t (minutes), and W_0 is total weight loss quantity after complete pyrolysis of oil shale kerogen [18].

Decomposition kinetic equation of solid matter is written as [9]

$$\frac{d\alpha}{dt} = k(1-\alpha)^n, \quad (9.14)$$

where $k = Ae^{-E/RT}$ is reaction rate constant, A is frequency factor, E is activation energy, R is gas constant, and T is temperature [9].

When assuming oil shale pyrolysis as a first-order reaction, the rate of kerogen decomposition can be expressed as [9]

$$\frac{d\alpha}{dt} = Ae^{-\frac{E}{RT}}(1-\alpha). \quad (9.15)$$

10.1 Nonisothermal Analysis

Taking constant heating rate as $\beta = dT/dt$, rate of kerogen decomposition can be expressed as [9]

$$\frac{d\alpha}{dT} = \frac{A}{\beta}e^{-\frac{E}{RT}}(1-\alpha). \quad (9.16)$$

Two models which were used to evaluate kinetic parameters are direct Arrhenius plot and integral methods [9].

Using the direct Arrhenius plot method, taking logarithm of both sides of Eq. (9.16) yields Eq. (9.17). Here, values of activation energy (E) and frequency factor (A) can be calculated by plotting the experimental value of $\ln(1/(1-\alpha)\cdot d\alpha/dT)$ against $1/T$ [9].

$$\ln\left(\frac{1}{1-\alpha}\cdot\frac{d\alpha}{dT}\right) = \ln\left(\frac{A}{\beta}\right) - \frac{E}{RT}. \quad (9.17)$$

Alternatively, using the integral method, the approximate integration of Eq. (9.16) yields Eq. (9.18) [9]. By using linear regression of $\ln[-\ln(1-\alpha)(E+2RT)/T^2]$ versus $1/T$ in Eq. (9.18), the activation energy and frequency factor can be obtained from the slope and intercept of regression line [9].

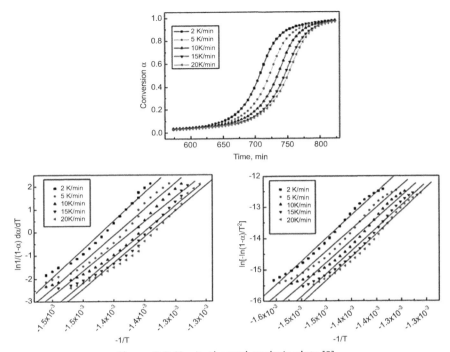

Figure 9.7 Nonisothermal analysis plots [9].

$$\ln\left[\frac{-\ln(1-\alpha)(E+2RT)}{T^2}\right] = \ln\frac{AR}{\beta} - \frac{E}{RT}. \quad (9.18)$$

Trends with different parameters versus time are shown in Fig. 9.7 [9].

10.2 Isothermal Analysis

Kerogen decomposition can be expressed as Eq. (9.19). A plot of $\ln(1-\alpha)$ against t will yield a straight line, with a slope equal to $-k$ (see Fig. 9.8) [9].

$$\ln(1-\alpha) = kt. \quad (9.19)$$

11. EX SITU RETORTING TECHNIQUES

There are various ex situ retorting techniques that are currently available [10]. These differ by reactor types (moving packed bed, solid mixers, fluidized bed, to name a few) as well as by operating conditions and technical

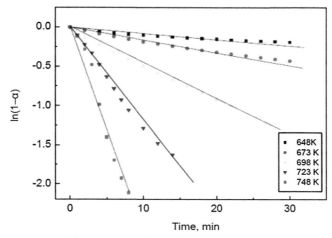

Figure 9.8 Isothermal analysis plots [9].

details such as drying of feed, feed distributor, spent shale discharge, and so on [10]. Therefore, a particular retorting technique includes various units of operations in plant, in addition to the pyrolysis reactor [10].

The main ex situ technologies are briefly described here.

11.1 Internal Combustion

In this method, the char and oil shale gas burn within a vertical shaft retort to supply heat for pyrolysis. Raw shale particles are introduced from the top of retort and heated by rising hot gases, which pass via descending oil shale. Hence decomposition of kerogen takes place at a temperature of 500°C. The shale oil evolved gases, and the cooled combustion gases are collected from the top. The condensable gas is converted into oil after condensation while noncondensable gases are recycled to provide heat energy to the retort. The air is injected from the bottom for combustion process. The spent oil shale and gases are heated to between 700 and 900°C. This technology has an efficiency of 80—90% Fischer assay yield. The only drawback is that combustible oil shale gas is diluted by combustion gases.

11.2 Hot Recycled Solids

In this method the heat is delivered to oil shale by hot solid particles, normally oil shale ash. Rotating kiln or fluidized bed retort is used for this process. The decomposition of oil shale has been completed at 500°C through

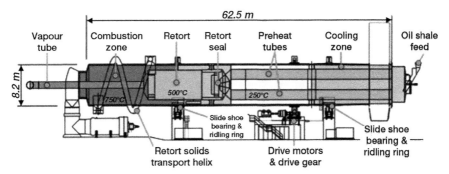

Figure 9.9 Alberta Taciuk processor.

recycled particles. Oil vapors are separated from the solids and condensed to collect the oil. Before mixing with the hot recycled solids, the heat is recovered from the shale ash and combustion gases to preheat the raw oil shale. As the hot recycled solids are heated in a separate furnace, therefore there is no way for oil shale gas to get diluted with combustion exhaust gas (see Fig. 9.9). The only disadvantage is that more water is used to handle the finer shale ash.

11.3 Conduction Through a Wall

In this method the heating transferred to the oil shale through conduction. The shale feed is converted into the fine particles. Different processes are used in this manner like combustion resources or oil-tech. In combustion resources, the rotating kiln is used in which hydrogen is used to fire the kiln while the hot gases are circulated through annulus. The process of conduction takes place through the wall of the retort. The only disadvantage in this method is that the kiln is most costly when scaled up.

11.4 Externally Generated Hot Gas

This process is similar to that of internal combustion technology. The processing of oil shale lumps takes place in vertical shaft kilns. The heat is transferred through the gases that circulated outside the retort vessel; hence the contamination of combustion gases with the retort vapors didn't take place. Due to less combustion of spent shale, the temperature of oil shale doesn't exceed 500°C. Hence, the decomposition of the carbonate mineral and CO_2 generation can be avoided for oil shales. These technologies are more stable and the control of internal combustion is more convenient.

11.5 Reactive Fluids

This technology is suited for processing of oil shale that has low hydrogen content. In this process, hydrogen or hydrogen donors react with coke precursors. The technology may include the IGT Hytort or Chattanooga fluidized bed reactor. The IGT Hytort method employed high-pressure hydrogen. In the Chattanooga process, a fluidized bed reactor is used with a heater, which is also fired by hydrogen for thermal cracking or hydrogenation of oil shale. Their result is better than the ordinary pyrolysis method, which is proved by laboratory scale tests. The only drawback is that a high-pressure retort vessel increases the cost and complexity.

11.6 Plasma Gasification

These technologies involve the bombardment of the radical (ions) on oil shale. The radical cracks the kerogen and hence is converted into synthetic oil and gas. The process is operated in plasma arc or plasma electrolysis mode and air, hydrogen, or nitrogen gas is used as plasma gas.

A few of the practical retorting techniques are discussed here.

11.7 Union Oil Retorting Process

Union Oil utilizes the conventional room and pillar method for production mining, with a mine portal designed open to a bench at 2100 m elevation [11]. Initial production mining takes place at a rate of 9900 tonnes per stream day [11]. Primary crushing (reduces ore to a particle size of 20 cm) and secondary crushing (reduces ore to a particle size of 5 cm) of shale ore is conducted underground [11]. Crushed shale is transferred to retort via solids pumps, which consist of a two-piston and cylinder assembly [11]. With the upstroke of the piston, the shale is moved upward through the retort, and comes in contact with a 510–538°C recycled gas [11]. The rising oil shale bed is heated to retorting temperature by countercurrent contact with the hot recycled gas [11]. This produces oil shale vapors and gas [11]. It is cooled by the cold incoming shale located at the lower section of retort cone. Union retort apparatus and flow diagram are shown in Fig. 9.10 [11].

11.8 US Bureau of Mines Gas Combustion Retort

This technique involves a vertical, refractory-lined vessel [2]. Crushed shale ore flows downward with the help of gravity, being countercurrent to the retorting gases [2]. Recycled gases enter the retort from the bottom [2].

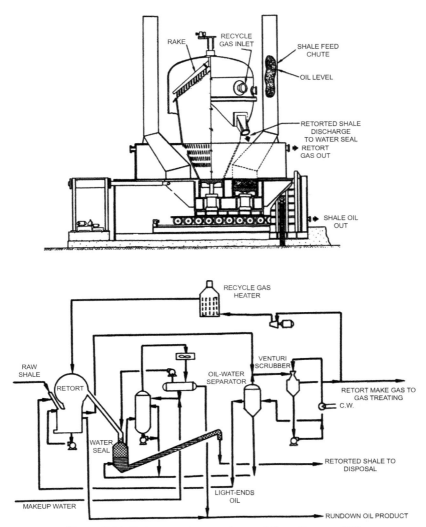

Figure 9.10 Union retort apparatus and flow diagram [11].

The hot retorted shale heats them up as they flow upward through the vessel [2]. The distributor system injects air and additional recycled gas (dilution gas) into the retort [2]. This takes place at a location about one-third of the way up from the bottom, where it mixes with rising hot recycled gases [2]. Schematic of the gas combustion retort is shown in Fig. 9.11 [2].

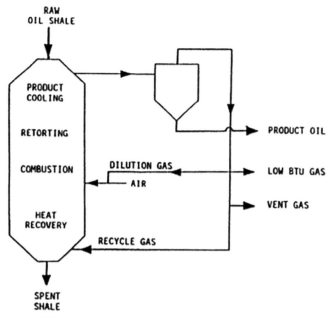

Figure 9.11 Gas combustion retort schematic [2].

11.9 Chevron Retort System

The chevron retort system utilizes a unit with shale feed capacity of 1 ton/day [6]. This technique uses a catalyst and a fractionation system [6]. The unit operates on a staged, turbulent flow bed process [6]. This technique is also referred to as the shale oil hydrofining process [6]. A schematic is shown in Fig. 9.12 [6]. It is above the ground and utilizes heat transfer to process mined and crushed shale ore to extract shale oil [12]. Heat transfer is achieved by mixing spent oil shale, heated in a separate combuster with fresh shale, allowing fresh shale to decompose and release shale oil [12].

12. ADVANTAGES AND DISADVANTAGES OF EX SITU PROCESSES

The advantages of the ex situ processes are:
- Offers high efficiency for organic matter recovery, up to 90% w/w of the total organic content [1].
- Offers better control of process operating variables [1].
- Process conditions that are undesired can be minimized [1].
- Simple product recovery [1].

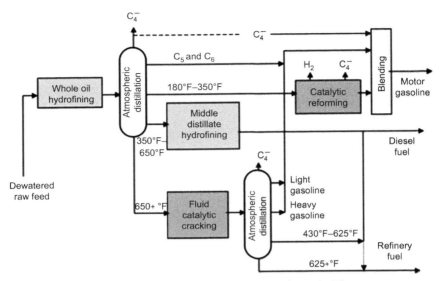

Figure 9.12 Chevron retort system schematic [6].

- Process units can be subjected to repetitive retorting operations [1].
  The disadvantages of the ex situ processes are as follows:
- High capital investment associated with large-scale units [1].
- High operating cost attributed to mining, crushing, transportation, and heating of oil shale; therefore, the process is more favorable to rich shale resources that are accessible to surface mining [1].
- Underground water contamination is always a possibility [1].
- Revegetation costs of site [1].

13. IN SITU RETORTING TECHNIQUES

Unlike ex situ oil shale retorting techniques, oil shale retorting can also be achieved underground; that is, shale formation does not have to be mined [6]. In situ techniques typically involve fracturing the shale deposits by either explosives or hydrostatic pressure [6]. Oil shale organic matter is burned in portion in order to achieve the necessary heat for retorting [6]. The retorted shale is extracted out of the production zone similar to crude petroleum [6]. In situ techniques have minimal environmental impact, and are economically more favorable [6].

The in situ techniques may include the following.

14. WALL CONDUCTION

In this method, a heating element or heating pipes are placed within the oil shale formation. The Shell ICP process utilizes the electrical heating elements for heating of oil shale. The temperature ranges between 340 and 370°C over a period of 4 years. The freeze wall, having circulated super-chilled fluid, isolates the groundwater of the processing area as depicted in Fig. 9.13. The main drawback of this process involves the large consumption of electric power and water. The risk of groundwater pollution is high in this scenario.

15. EXTERNALLY GENERATED HOT GAS

In this method, hot gases are heated above the ground and then injected into the oil shale formation. In the Chevron CRUSH process, the heated CO_2 is injected to heat the oil shale formation through the series of horizontal fractures from which the gas is circulated. A freeze wall is used to isolate the process of in situ shale oil production from its surroundings (Fig. 9.14).

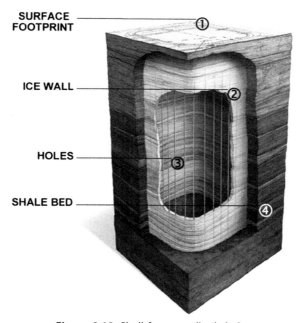

Figure 9.13 Shell freeze wall oil shale.

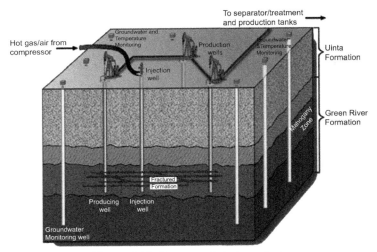

Figure 9.14 Chevron oil shale project.

16. EXXONMOBIL ELECTROFRAC

In this technology electrical heating with elements of both wall conduction and volumetric heating methods is used. In this method, the electrically conductive material is injected through the horizontal fractures, which then form a heating element. To allow opposing charges to be applied at either end, heating wells are placed in a parallel row with a horizontal well intersecting them in their toe.

17. VOLUMETRIC HEATING

In this method oil shale is heated by vertical electrode arrays. The deeper volume could be processed at slower heating rates by installation spaced at tens of meters. The only drawback in this case is that the electrical demand is very high and possibly char or groundwater would absorb an undue amount of energy. The microwave heating process can also be used that has the same principle as that of a radiowave heating system. The microwave heating system is tested by Global Resource Corporation. The radiowave has an advantage over microwave in that its energy can penetrate farther in the oil shale formation.

A few of these retorting techniques are discussed here.

17.1 Equity Oil Company Process

This process involves retorting the shale by injection of hot natural gas into the shale bed [6]. Taking Piceance Creek Basin as an example, one injection well and four producing wells are drilled into the shale formation [6]. Here, the natural gas was heated to 480°C of temperature, and compressed to about 85 atm [6]. The natural gas is delivered to the shale formation through insulated tubing [6]. Although this technique is economically more favorable, its economics is primarily dependent on the natural gas cost as well as the amount required for makeup of natural gas [6].

17.2 Dow Chemical Company's Process

A research study on feasibility of in situ recovery of low heat content gas was executed by Dow Chemical Company [6] in Michigan Antrim Shale where, using 19,000 kg of metalized ammonium nitrate slurry, a considerable part of shale was explosively fractured [6]. A 440-V electric heater and a 250,000-Btu/h propane burner were utilized to combust the shale [6]. Shale gasification and tolerance to severe operational conditions were the outstanding features of the process [6].

17.3 Talley Energy System's Process

A US DOE Industry Cooperative oil shale project was carried out by Talley Energy Systems [6]. This project was located 11 m west of Rock Springs, Wyoming [6]. The process utilizes a combination of hydraulic and explosive fracturing, involving no mining at all [6].

18. ADVANTAGES AND DISADVANTAGES OF IN SITU PROCESSES

The advantages of the in situ processes are:
- Offers oil recovery from deep oil shale formations [6].
- Considering all operations are conducted through wellbores, there is no solid waste disposal involved, making this process more desirable from an environmental perspective [6].
- Eliminates the need for transportation and crushing costs of the ore [6].
- Eliminates the need for mining and surface pyrolysis, considering the oil shale resource is heated in its natural depositional setting [6].

The disadvantages of the in situ processes are as follows:
- Insufficient permeability within the formation makes subsurface combustion difficult to control [6].
- High costs associated with drilling [6].
- Low recovery efficiencies [6].
- Extensive efforts required to prevent possible contamination of aquifers [6].
- Required permeability and porosity in the shale formation is difficult to establish [6].

19. SHALE OIL REFINING AND UPGRADING PROCESSES

Although, the composition of shale oil highly depends on the shale resource it has been derived from, as well as the method utilized to obtain it from the resource, it tends to possess certain characteristics such as high nitrogen content, oxygen compounds, carboxylic acids, and sulfur compounds [13]. The presence of these compounds can lead to stability issues in gasoline, jet, and diesel fuels. Not to mention, NO_x emissions will also be produced [13]. Therefore, further refining and processing is required to improve shale oil properties. A few of these techniques are discussed here.

19.1 Thermal Cracking Process

This process was developed for noncatalytic thermal cracking of shale oil in the presence of inert heat-carrying solids [2]. The process targets recovery of gaseous olefins as the desired cracked product [2]. It converts 15—20% of the feed shale oil to ethylene, a prevalent gaseous product. Remaining shale feed is converted into other gaseous and liquid products [2]. Gaseous products include propylene, ethane, 1-3 butadiene, C_4s, and hydrogen, whereas liquid products include benzene, toluene, xylene, and light and heavy oils [2]. Coke is produced as a solid product by polymerization of unsaturated materials [2]. It is removed as a deposit from the inert heat carrier solids [2]. The thermal cracking reactor does not require any hydrogen feed as such [2]. The thermal riser pushes through entrained solids concurrently at an average riser temperature of 700—1400°C [2].

19.2 Moving Bed Hydroprocessing Reactor

This process was developed for crude oil derived from oil shale as well as tar sands with large amounts of particulate matter (e.g., rock dust and ash) [2]. Hydroprocessing is carried out using the dual function moving bed reactor [2]. The catalyst bed offers a filtration action, which simultaneously removes particulate matter from oil shale [2]. Effluent coming from the moving bed reactor is separated, and further processed in fixed bed reactors. This function is carried out with the addition of fresh hydrogen [2]. Sulfurization is also promoted by the addition of this fresh hydrogen to the heavier hydrocarbon fractions. Preferably, shale oil treatment using a moving bed hydroprocessing reactor involves utilization of a moving bed reactor followed by a fractionation step to divide wide boiling range crude oil into separate fractions [2]. Lighter fraction is hydrotreated for residual removal such as metals, sulfur, and nitrogen [2], and heavier fractions are cracked in a second fixed bed reactor operating under high severity conditions [2].

19.3 Hydrocracking Process

This is a cracking process involving pyrolysis of high molecular weight hydrocarbons to low molecular weight paraffin and olefins in the presence of hydrogen [2]. The olefins formed are saturated by hydrogen during the cracking process [2]. The hydrocracking process is primarily utilized to process low value stocks possessing rich heavy metal content [2]. Additionally, high aromatic feeds that cannot be processed by conventional catalytic cracking are also subjected to a hydrocracking process [2].

20. ADVANTAGES AND DISADVANTAGES OF REFINING AND UPGRADING TECHNIQUES

The advantages of the refining and upgrading processes are:
- Allows production of product in line with end user requirements [14].
- Higher volumetric energy density liquid can be achieved [14].
- Refining or upgrading plants can be located inside or close to the existing refinery. Required process utilities and product distribution network will not have to be reinvented, and transportation costs of crude shale are avoided [14].
- Less organic matter (such as ash) can be returned to soil as fertilizer [14].
- Dry fractionation has the advantage of requiring only crystallizers and filters. This avoids operating costs [15,16].

The disadvantages of the refining and upgrading processes are as follows:
- Hydrogen addition (catalytic) and carbon rejection (noncatalytic) upgrading processes are very expensive. Hydrogen addition processes require excessive catalyst utilization and involve metals and carbon

deposition [15]. Noncatalytic processes yield large amounts of coke, and exhibit low liquid yield [15].
- Construction of refining or upgrading plants requires specialized equipment [15].
- Fractionation can also act as a disadvantage as some of the secondary products produced may not have any use in the market [16].

21. SUPERCRITICAL EXTRACTION OF OIL FROM SHALE

It was in the year 1977 when Williams and Martin developed a concept for shale oil extraction using various supercritical solvents [2]. They were able to successfully remove 92% of organic matter from Colorado oil shale utilizing supercritical methylcyclohexane as the solvent [2]. Following this, Poska and Warzel also came up with processes for shale oil extraction leveraging a variety of other solvents [2]. In 1981, Scinta and Hart disclosed a supercritical extraction technique involving 1 wt% concentration of hydrogen donor solvent in heptane [2]. A year later, Scinta and Classen developed a procedure using C_2-C_{20} alkanes or aromatics in 350–500°C temperature and 20–100 psi pressure range [2]. Another supercritical extraction technique was developed by Compton in 1983, where Compton developed selection criteria for solvents using solvent's Hildebrand solubility parameter comparison with that of oil shale bitumen [2]. Generally speaking, the supercritical extraction process is now used as a single stage extraction process for shale oil, as well as for heteroatom removal and residual conversion [2].

22. SUPERCRITICAL CO_2 EXTRACTION OF OIL SHALE: EXPERIMENTS, APPARATUS, AND PROCEDURE

The supercritical CO_2 extraction study described here utilizes a Stuart oil shale sample from Queensland, Australia [2]. Shale ore was manually crushed to small enough sizes to reduce mass transfer limitations as well as to promote random distribution of kerogen in particles [2]. Detailed schematic of the experimental setup of this supercritical extraction is shown in Fig. 9.15 [2]. It consists of an extraction unit that is heated externally, an ambient temperature gas liquid separator, as well as facilities to provide continuous flow of carbon dioxide [2]. An electric furnace is in place for heat supply to the system. A porous plate is used to support the bed of crushed shale [2]. The system used to separate the extracted shale oil from

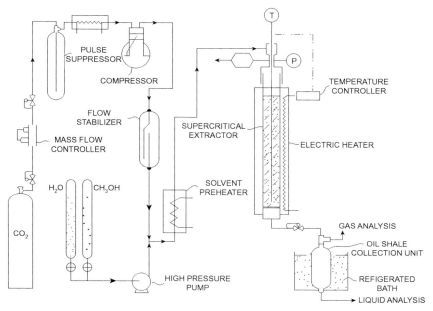

Figure 9.15 Supercritical CO_2 extraction system [2].

the carbon dioxide solvent consists of a 1-inch diameter stainless steel separation vessel with an inlet and outlet ports [2]. The separator is placed in a cold bath with a dry ice isopropanol mixture [2]. It condenses a waxy shale oil [2]. The inlet to this is located at the lower end of the extractor, and is short in length for minimal deposition of shale oil in the line [2]. Uncondensed gases are purged [2]. This technique uses a gas chromatograph to separate shale oil hydrocarbon constitutes according to their respective boiling points [2].

23. SUPERCRITICAL METHANOL/WATER EXTRACTION OF OIL SHALE

Supercritical methanol/water extraction methods typically produce higher yields than the yields obtained from retorting or subcritical solvent extractions [2]. This yield is applicable only to methanol and water mixtures, and cannot be duplicated with pure water or pure methanol [2]. A publication by Elington and Baugh yielded that supercritical extraction using a polar, protic, hydrogen bonding mixture of ammonia and water can liberate up to 87 wt% of organic matter [2]. The extraction yield is dependent on the type of shale as well as the water content [2]. Mixing speed also plays a significant role. Particles processed at higher speeds have higher extraction yields [2].

24. CONTINUOUS SUPERCRITICAL EXTRACTION

Continuous supercritical extraction of Colorado oil shale in the United States is discussed here [17]. A fixed bed reactor was used in this study [17]. Here, oil shale was placed in a reactor, and was extracted at temperature ranging from 370 to 400°C and 500—900 psi pressures, respectively, by flowing tolune at 20 mL/min. Spectrophotometry was used to monitor the performance of the effluent after it was depressurized and cooled to ambient conditions [17]. A change in absorption pattern was noticed with higher temperature extracts (400°C or higher) [17]. At lower temperatures, absorption patterns of isothermal runs stayed fairly unchanged, meaning extract composition did not change much during supercritical fluid extraction [17]. Additionally, the calculated activation energy was yielded much lower than that from pyrolysis reaction [17].

25. ADVANTAGES AND DISADVANTAGES OF SUPERCRITICAL EXTRACTION METHODS

The advantages of the supercritical extraction methods are:
- Energy efficient process is achieved for operations at lower temperatures (e.g., 380 vs. 500°C) [17]. This is applicable only in the presence of a fluid [17].
- Products obtained at lower temperature in the presence of fluid have better quality (e.g., oil contains less olefins and heteroatoms) [17].
- Supercritical fluids extraction yields higher H/C ratio [17].
- Supercritical extraction process works very well with richer shale [17].

The disadvantages of the supercritical extraction methods are as follows:
- With respect to solvent extraction, process derived fraction is regarded as an economic solution for solvent selection, however the concept is still under development [17].
- With respect to supercritical fluid extraction, extraction and separation steps are faster and more efficient compared to solvent extraction [17].
- Utilization of supercritical fluid comes with chemical and processing costs. Extra operations such as purification, recycling, solvent recovery, and compression make the process more complex [17].

26. MATHEMATICAL MODELING OF OIL SHALE PYROLYSIS

The transformation process of solid kerogen to liquid and then to gaseous products is fairly complex [18]. Several physical and chemical reactions take place simultaneously [18]. Main components of the oil shale pyrolysis process are outlined in Fig. 9.16 [18]. Governing reaction mechanisms are shown in Fig. 9.17 [18]. Results acquired from oil shale pyrolysis phenomena using COMSOL multiphysics simulation software are shown in Figs. 9.18–9.20 [18]. As can be seen, increasing the temperature and heating rate reduces optimal time, however if products are exposed to heating for a longer period of time (isothermal) or to higher temperatures (nonisothermal), the final results are coke and gases [18].

- Residence time is much longer at lower temperatures [17].
- Fluid loss due to adsorption can be significant [17].

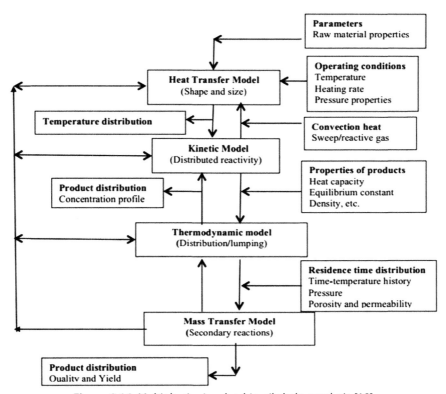

Figure 9.16 Multiphysics involved in oil shale pyrolysis [18].

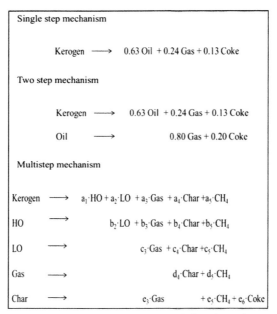

Figure 9.17 Reaction mechanisms in shale oil pyrolysis [18].

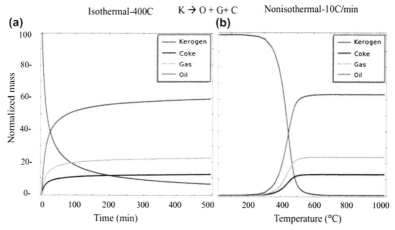

Kerogen decomposition (single particle) and product formation profiles using single step mechanism under (a) isothermal (400°C) and (b) nonisothermal (10°C/min).

Figure 9.18 One-step mechanism in kerogen decomposition [18].

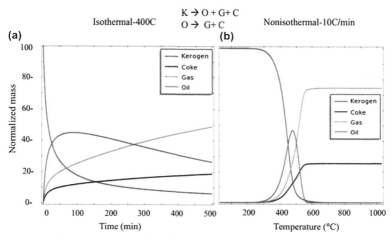

Kerogen decomsposition (single particle) and product formation profiles using two step mechanism under (a) isothermal (400°C) and (b) nonisothermal (10°C/min).

Figure 9.19 Two-step mechanism in kerogen decomposition [18].

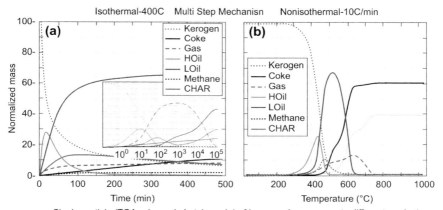

Single particle (TGA scheme in batch mode) of kerogen decomposes to different products using multiple step reactions mechanism under (a) isothermal (400°C) and (b) nonisothermal (10°C/min) pyrolysis. The small window shows the material profiles at long time scale (a log scale).

Figure 9.20 Multistep mechanism in kerogen decomposition [18].

27. PARAMETRIC STUDY OF OIL SHALE PYROLYSIS

Weight loss generated during pyrolysis of various oil shale is shown in Table 9.3 [19]. It dictates the overall weight loss that takes place during heating, meaning the weight loss by the kerogen and the decomposing minerals [19]. This is attributed to contraction of shale; the larger the extent of

Table 9.3 Shale Oil Weight Loss [19]

Overall Weight Loss by Oil Shales Heated to 700°C at 60°C/min^{-1}; Particle Size 355 × 710 μm, Atmosphere 0.1 MPa N_2

Oil shale	Grade (L/mg)	Weight loss (%)
Colorado	252	31.0
	104	15.5
	88	13.0
Kentucky sunbury	53	12.0

Overall Weight Loss by Oil Shales Heated to 800°C at 60°C/min^{-1}, Particulate Size 74 μm, Atmosphere 0.1 MPa N_2

Oil shale	Grade	Weight loss (%)
Joadja	High	50.0
Rundle	I501/mg	42.5
Asturian	Low	5.0

contraction, the larger the weight loss [19]. The contraction is directly related to the decomposition of kerogen to oils, gases, and residue products [19]. The conversion to kerogen leads to the formation of partial voids or cavities in the shale matrix [19]. Gases produced in the process of kerogen decomposition get trapped within the closed cavities of shale matrix, leading to the development of high pressure regions, causing the shale bed to expand [19]. The higher the kerogen content in shale oil, the larger will be the extent of the shale bed [19]. Therefore, it can be concluded that the weight loss data closely correlates with the shale type (i.e., the shale grade) [19]. As well, organic matrix of shale has a direct influence on the structural changes that take place during pyrolysis [19]. It is the decomposition of this organic matrix that results in significant changes in volume [19]. Evidence has indicated that higher grade shale, in some cases, softens and forms hard coke upon pyrolysis [19]. Plasticity mechanism of kerogen is shown in Fig. 9.21 [19].

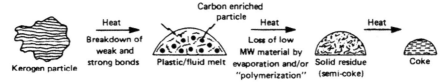

Figure 9.21 Kerogen coke formation [19].

28. ECONOMIC CONSIDERATIONS IN OIL SHALE PROCESSING

Each additional barrel produced domestically will reduce the trade deficit by replacing a barrel of oil imports. A fully developed plant with 100,000 bbl/day capacity has costs including the average minimum economic of $38/bbl, surface mining of $47/bbl, and modified in situ prices of $62/bbl as well [20–22].

It is estimated that by 2030, production of 3 million barrels per day of shale oil could lead to a 3–10% decrease in global oil prices. Over the next 30 years, $325 billion could be avoided being spent on imports because of domestic production. A commercial-scale project would yield 10,000–100,000 bbl/day for a surface retort so that full-scale in situ projects would reach 300,000 bbl/day.

The process technology and the quality of the resource will affect the capital and operating costs so that operating costs are estimated to be in the range of $12–20 per barrel of shale oil produced. On the other hand, the capital costs vary from $40,000–55,000 per stream day barrel of daily capacity. It should be noted that mining retorting and upgrading have been considered in these costs. The cost changes depending upon the mining, retorting, and upgrading. Using the gross revenue from the potential oil shale production, the economic module estimates the gross domestic product. The potential oil shale production is also applied to evaluate the impact on the trade deficit. To calculate the labor cost, the labor component of all major cost elements is isolated. Technology and resource quality play a notable role in minimum economic price for shale oil programs. This minimum economic price is defined as the world crude oil price required to obtain a 15% rate of return on the project.

The shale oil production is predicted to become 500,000 bbl/day until 2020 and it is expected to remain unchanged by 2035 [20–23]. Excess $15 to $48 billions of direct revenues could be added to federal treasury due to the reachable 12.8 billion barrels of cumulative shale oil production.

Employment is a bold consequence of this industry while it is expected that up to 300,000 new high-paying careers can be created owing to oil shale development (see Table 9.4) [20–23].

Table 9.4 Typical Costs for Various Stages of Oil Production/Processing

Item	Unit	Business as Usual	With Targeted Tax Incentive	With RD&D
Production	Billion Bbls	3.2	7.4	12.8
Direct federal revenues	Billion $	15	27	48
Direct local/state revenues	Billion $	10	21	37
Direct public sector revenues	Billion $	25	48	85
Contribution to GDP	Billion $	310	770	1300
Value of imports avoided	Billion $	70	170	325
New jobs	FTE (thousand)	60	190	300

29. THEORETICAL, PRACTICAL, AND ECONOMIC CHALLENGES IN OIL SHALE PROCESSING

A few of the oil shale processing challenges are outlined here.
- Advanced production techniques like hydraulic fracturing are politically controversial, and not well accepted by the public [20]. It is not regarded as a safe extraction technique and has been criticized to pose adverse environmental ramifications [20]. Many critics simply want to stop hydraulic fracturing operations wherever they are taking place, and do not want to start any new ones [20].
- Shale oil has high paraffin content, not typically found in conventional crude oil. These variations in the quality and physical properties of shale oil require refiners to continually adapt, since thorough understanding of unique characteristics and physical properties have to be investigated. As a consequence to this dramatic variability, there is a significant impact on the overall profitability and on-stream time [21].
- Refinery equipment fouling is often experienced, and coke and inorganic solids are the main culprits [21]. This is mitigated with operational and mechanical adjustments, raising maintenance costs [21].
- Depending on the properties and concentrations of amine in the shale ore, shale deposit—related corrosion or salt build-up develops in overhead systems [21]. Traditional neutralizers do not help to prevent this build-up [21]. Hence, total system amine management for equipment reliability is a challenge [21].

- Presence of olefins and diolefins in combination with high nitrogen content makes shale oil difficult to refine [22]. Presence of other elements such as arsenic, iron, and nickel are also known to interfere with refining [22].
- High levels of aromatic compounds, low hydrogen-to-carbon ratio, low sulfur levels, and solid particles are other characteristics of shale oil that makes further processing a must for shale oil [22].
- Nitrogen compounds in shale oil cause poisoning of refining catalysts during downstream processing of shale oil [22].

30. HOT TOPIC RESEARCH STUDIES IN OIL SHALE PROCESSING AND EXTRACTION TECHNOLOGIES

The main research studies in this area are:
- Raw shale, like crude petroleum, is highly paraffinic and contains very high levels of sulfur, nitrogen, oxygen, and olefins [2,22]. Substantial upgrading is required in order to make raw shale oil suitable for refinery feed [2,22]. Sulfur removal must be reduced to less than few parts per million, and this removal protects multimetallic reforming catalysts [2,22]. Likewise, removal of nitrogen is a must as it poisons cracking catalysts from condensed heterocyclic [2,22]. Therefore, more research needs to be done to understand oil shale from a molecular level [2,22].
- Although there are many available analytical methods are utilized in relation to petroleum analytical chemistry, their application on shale still remains questionable [2,22]. In order to better understand shale technology, better analytical techniques will have to be developed, including new shale utilization technology [2,22]. Chemical characterization of organic heteroatoms, their bonding into basic carbon structure, as well as model compounds for sulfur and nitrogen sources are areas that can be investigated [2,22].
- Additionally, electrical and conductive measurements or other conventional methods should be leveraged to characterize physical properties of oil shale [2,22]. Physical properties correlations with shale oil conversion are valuable in design and development of commercial retort, and are also useful in engineering design calculations [2,22].
- Considering the reaction mechanism of oil shale retorting is not generally studied by investigators; therefore the retorting process can be further researched and improved, with chemical reaction mechanisms fully identified [2,22].

- Shale rocks are found to have low porosity as well as permeability, therefore heat and mass transfer processes (combined) of a retort system have to be identified beforehand [2,22]. The heat and mass transfer conditions have a direct impact on the recovery of oil and gas from a retort. Although process technologies do exist that serve as a means of process transport (method of heat and mass transfer), there is room for further research in this area [2,22]. The mass transfer studies are very useful in mathematical modeling [2,22].
- Shale formations consist of many carbonate and silicate minerals such as Trona Beds in Wyoming, and nahcolite in Utah [2,22]. Commercial favorable recovery processes for such minerals do not quite exist, and recovery of these minerals can be researched in the near future [2,22].

REFERENCES

[1] Oil shale and tar sands programmatic environmental impact statement (PEIS) information center. http://ostseis.anl.gov/guide/oilshale.
[2] Lee S. Oil shale technology. CRC Press; December 11, 1990. p. 20−200. Science.
[3] Riva JP. Oil shale. Encyclopædia Britannica, Inc.; 2014. http://www.britannica.com/EBchecked/topic/426232/oil-shale/308298/Pyrolysis.
[4] Patten JW. Capturing oil shale resources EcoShale in-capsule process. RED Leaf Resources Inc. http://www.costar-mines.org/oss/30/presentation/Presentation_08-4-Patten_Jim.pdf.
[5] Shawabkeh AQ, Abdulaziz M. Shale hold time for optimum oil shale retorting inside a batch-loaded fluidized-bed reactor. Oil Shale June 1, 2013;30(2):173−83. http://www.kirj.ee/public/oilshale_pdf/2013/issue_2/Oil-2013-2-173-183.pdf.
[6] Lee S, Speight JG, Loyalka SK. Handbook of alternative fuel technologies. 2nd ed. CRC Press; July 8, 2014. Nature − 712 pages.
[7] Hubbard AB, Robinson WE. A thermal decomposition study of Colorado oil shale. US Dept. of the Interior, Bureau of Mines; 1950.
[8] Lü X, Sun Y, Lu T, Bai F, Viljanen M. An efficient and general analytical approach to modelling pyrolysis kinetics of oil shale. Fuel November 1, 2014;135:182−7.
[9] Yongjiang X, Huaqing X, Hongyan W, Zhiping L, Chaohe F. Kinetics of isothermal and non-isothermal pyrolysis of oil shale. China University of Geosciences (Beijing), Beijing, China, Petrochina Research Institute of Petroleum Exploration & Development − Langfang, Hebei, China Oil Shale 2011; vol. 28(3):415−24. Estonian Academy Publishers.
[10] Malhotra R. Fossil energy: selected entries from the encyclopedia of sustainability science and technology. Springer Science Business Media; December 12, 2012. Electric engineering − 637 pages.
[11] Shih CC. Technological overview reports for eight shale oil recovery processes. Environmental Protection Agency, Office of Research and Development [Office of Energy, Minerals, and Industry], Industrial Environmental Research Laboratory; 1979.
[12] Diligence application for the Getty Oil Company water system water rights. Chevron USA Inc.; 2014. http://westernresourceadvocates.org/land/SummaryReport2014.pdf.
[13] Guo SH. The chemistry of shale oil and its refining, coal, oil shale, natural bitumen, heavy oil and peat. The chemistry of shale oil and its refining, vol. II. Beijing (China): University of Petroleum. http://www.eolss.net/sample-chapters/c08/e3-04-04-04.pdf.

[14] De Miguel Mercader F. Pyrolysis oil upgrading for co-processing in standard refinery units. Enschede, The Netherlands. 2010. http://doc.utwente.nl/74369/1/thesis_F_de_Miguel_Mercade.pdf.
[15] Ancheyta J, Rana MS. Future technology in heavy oil processing, petroleum engineering — downstream. Mexico City (Mexico): Instituto Mexicano del Petroleo. http://www.eolss.net/sample-chapters/c08/e6-185-22.pdf.
[16] Top-notch technology in production of oils and fats. Chempro Technovation Pvt. Ltd. http://www.chempro.in/processes.htm.
[17] Das K. Solvent and supercritical fluid extraction of oil shale. US Department of Energy, Office of Fossil Energy. http://www.netl.doe.gov/kmd/cds/disk22/G-CO2%20&%20Gas%20Injection/METC89_4092.pdf.
[18] Mathematical modeling of oil shale pyrolysis. University of Utah. http://www.inscc.utah.edu/~spinti/Public/DE_FE0001243_Apr_June_2012_attachments.pdf.
[19] Khan R. A parametric study of thermophysical properties of oil shale. Morgantown (WV): US Department of Energy, Morgantown Energy Technology Center; 1986.
[20] Global Energy Network. The Monterey shale & California's economic future. University of Southern California; 2013. http://gen.usc.edu/assets/001/84955.pdf.
[21] Benoit B, Zurlo J. Overcoming the challenges of tight/shale oil refining, GE water & process technologies. Processing Shale Feed 2014. www.eptq.com.
[22] Speight JG. Shale oil production processes. Gulf Professional Publishing; 2012. Technology & Engineering, pgs v. — 30.
[23] Rühl C. Five global implications of shale oil and gas, EnergyPost.eu. http://www.energypost.eu/five-global-implications-shale-revolution/.

FURTHER READING

[1] Rodgers B. Declining costs enhance Duvernay shale economics. Rodgers Oil & Gas Consulting; 2010. http://www.ogj.com/articles/print/volume-112/issue-9/exploration development/declining-costs-enhance-duvernay-shale-economics.html.
[2] Stark M (Clean Energy Lead), Allingham R, Calder J, Lennartz-Walker T, Wai K, Thompson P, Zhao S. Water and shale gas development. Accenture. http://www.accenture.com/sitecollectiondocuments/pdf/accenture-water-and-shalegas-development.pdf.
[3] Dusseault MB, Collins PM. Geomechanics effects in thermal processes for heavy oil exploitation. In: Heavy oils: reservoir characterization and production monitoring, vol. 13; 2010. p. 287. http://csegrecorder.com/articles/view/geomechanics-effects-in-thermal-processes-for-heavy-oil-exploitation.
[4] Nauroy J-F. Geomechanics applied to the petroleum industry. Editions TECHNIP. Business & Economics; 2011. 198 pages.
[5] Johnson H, Crawford P, Bunger J. Strategic significance of America's oil shale resource. Washington, DC: AOC Petroleum Support Services; 2004. http://www.learningace.com/doc/5378939/f9b4836cb3c90360a25e395f1289d5f0/npr_strategic_significancev1.
[6] SNBCHF.com. Shale oil and oil sands: market price compared to production costs. SFC Consulting. http://snbchf.com/global-macro/shale-oil-oil-sands.
[7] nrcan.gc.ca. Responsible shale development enhancing the knowledge base on shale oil and gas in Canada. In: Energy and mines Ministers' Conference; August 2013. Yellowknife (Northwest Territories). https://www.nrcan.gc.ca/sites/www.nrcan.gc.ca/files/www/pdf/publications/emmc/Shale_Resources_e.pdf.
[8] http://www.iop.pitt.edu/shalegas/pdf/research_unconv_b.pdf.

[9] Pfeffer FM. Pollutional problems and research needs for an oil shale industry. Washington: National Environmental Research Center, Office of Research and Development, US Environmental Protection Agency; for sale by the Supt. of Docs., US Govt. Print. Off. 1974. Technology & Engineering — 36 pages.
[10] Unconventional oil and gas research fund proposal. Shale Gas Roundtable. http://www.mitsubishicorp.com/jp/en/mclibrary/business/vol2.
[11] Lyons A. Shale oil: the next energy revolution — the long term impact of shale oil on the global energy sector and the economy. 2013. http://www.pwc.com/en_GX/gx/oil-gas-energy/publications/pdfs/pwc-shale-oil.pdf.
[12] Reso A. Oil shale current developments and prospects. Houston Geological Society of Bulletin. May 1968;10(9):21. http://archives.datapages.com/data/HGS/vol10/no09/21.htm.

CHAPTER TEN

Shale Oil and Gas: Current Status, Future, and Challenges

1. INTRODUCTION

One of the most important foundations of the modern economy is energy. Despite significant investments and innovations in green energies (e.g., solar and wind) in recent years, still fossil fuels like crude oil, natural gas, and coal constitute the majority of the world's energy (80% energy from: oil [36%], coal [23%] and gas [21%] based on BP Statistical Review, 2012) [1,2]. Over the past decades, the nuclear and hydropower sources have increased their contribution to provide energy as well. Oil is mainly utilized for transportation and petrochemicals, coal for hydropower, nuclear for generating electricity, and gas mostly for heating purposes [1–3]. Hence, the energy in the world is still considerably dependent on fossil fuel energies.

Due to the advancements and innovations made throughout various stages of oil and gas energy, the shale hydrocarbons and other nonconventional reservoirs have become economically accessible. The United States has abundant shale gas formations which have played an important role in revolutionizing its energy economy, leading to a significant decrease in the US gas price [2–4].

With an inexpensive source of energy available, many companies from different industries are growing and benefiting the overall economy of the country. The revolution in the shale oil and gas industry is estimated to create more than 600,000 jobs in US in the near future [2–4]. Such a development also will increase the energy security of US, resulting in a decrease of the US's interest in hydrocarbon-rich regions, such as the Middle East [2,4]. As a consequence, there will be noticeable changes in the trade relationships and political foreign affairs with other countries. In addition, increased shale gas production has lowered the importation of crude oil from other countries from 60% to 39% [2,4]. It is believed that the US is on the way to becoming fully energy-independent as the unconventional hydrocarbon industry advances [2,4]. Being able to control the world's energy sources would give any nation considerable economic and political influence on other countries. Currently, more than 50% of the world's energy supply comes from the Middle East. This dependency on oil and gas energy forces countries to keep their militaries along the transportation path of the oil tankers to secure their required resources, which boosts the overhead costs of energy for the end-users [3,4].

The development of oil and gas shale in two major hydrocarbon importers will affect other countries in terms of energy market, and geopolitical relations. It is clear that exporter countries will lose a part of their benefit. Trinidad and Tobago is the only natural gas exporter for which the trade shock is expected. The fracking process in the US has a potential effect of over 1% on GDP, but other countries including Yemen, Egypt, Qatar, Equatorial Guinea, Nigeria, Algeria, and Peru experience greater influences. Generally, developing countries may have lost an estimated US$1.5 billion in annual gas export revenue due to US shale gas [1–4].

The potential availability and accessibility to huge resources of shale oil in the world goes beyond the petroleum industry taking into consideration the possible effect of raised shale oil production due to limited oil price. Its potential in changing the main parameters such as the world economy and energy security as well as bringing independence and affordability in the long term can be significantly highlighted. However, these benefits should be presented with environmental objectives at local and global frameworks together. Any modification in regulatory rules and policies will result in a crucial influence on the oil industry. Lower oil price will aggravate the whole energy value chain; therefore, any investment plan based on normal increases in oil prices should be reviewed again. The likelihood of a huge impact of shale oil in the market is a driving force and predictably changes the global economy at a higher level. It is therefore essential for stockholders and policymakers to count the strategic implications of these potential changes soon [4,5].

There is certainly a need for political will to change the current approach to the shale oil and gas industry and bring transparency to the environmental impacts of extraction and production of oil and natural gas from shale resources in order to mitigate risks and avoid irreversible harms to natural resources, especially drinking water resources [4,5].

The regulations on the shale oil and gas industry differ by state in the US. Some states do not permit drilling in shale regions until further studies are conducted on potential issues concerning the public and environment [5,6]. Other states have no firm rules/regulations and allow drilling in suitable regions containing shale hydrocarbons [4,6].

The main environmental regulations/Acts concerning oil and gas industry are as follows:
- The National Environmental Policy Act (NEPA) that is requires the exploration and production on federal lands to be thoroughly analyzed with regard to environmental impacts [7].
- The Clean Air Act, National Emission Standards for Hazardous Air Pollutants (NESHAP) is used to set the standards for the release of toxic pollutants in the environment [7]. NESHAP rules are also employed to qualify newly refurbished engines and are leveraged to establish their monitoring and reporting requirements [7].
- The Clean Water Act (CWA) regulates disposal of wastewater which requires tracking of toxic chemicals utilized in fracturing fluids [7].
- The National Pollutant Discharge Elimination System (NPDES) permit program [7].
- The Oil Pollution Act (OPA) has been developed to enforce spill prevention requirements as well as reporting operations [7].
- Comprehensive Environmental Response, Compensation and Liability Act (CERLA) regulations authorize federal government to respond to the release of hazardous substances that can threaten human health or/and the environment [7].
- The Hazardous Materials Transportation Act standardizes the transportation of hazardous materials [7]. Additionally, material safety data should be reported in the event of emergency crisis [7]. This is a part of the Emergency Planning and Community Right to Know Act [7].
- Fracking is excluded from the Underground Injection Control (UIC) program with the help of the Safe Drinking Water Act (SDWA) [7].
- Waste management procedures are regulated with the federal Resource Conservation and Recovery Act [7].
- Groundwater is protected under the State Pollutant Discharge Elimination System (SPDES) [7].

The above regulations are generally based on [1,7]:
- The best available data and sound science;
- A transparent and public process that allows input from stakeholders;
- Appropriate roles for local, state, and federal regulators;
- A stable regulatory environment;
- A consistent and effective enforcement policy.

2. POLITICAL IMPLICATIONS

The geostrategic implications of fossil energy continue to make global headlines for international security. The world's dependency on energy resulted in significant developments of new technologies in exploration, extraction, processing, and drilling of previously nonrecoverable sources like shale oil and gas. The world's energy necessity makes these developments very important. "Shale oil is the most significant development in the energy industry since coal was replaced by oil as the principal fuel for transport in the 1920s" [8].

The political implication of the new development may directly impact US foreign policies as the US is currently importing a large portion of its daily consumption from the Middle East. However, the US has the world's largest recoverable shale oil reserves, which could make it independent from the rest of the world for centuries. The same story is applied to the European Union, as well. New shale oil developments in the Eastern Europe will make the Union independent from Russian's energy resources. Such energy advances will certainly have vital geostrategic impacts all around the world [2,5].

Shale gas exploitation presents a unique and challenging hurdle for policy makers so that it breaches many areas of politics. The central goals of the policies applicable to shale gas production are environmental conservation, economic improvement, diversifying the governing bodies (ensure representatives from industrial, political, and environmental groups are consulted and/or involved) and landowner protection [2,3].

At present, there is no significant legislation that applies directly to shale gas. Most bills are still pending (they have not yet been passed), or have not yet been proposed. Even those proposed and pending bills are all at the state/provincial level, and currently there is no federal legislation governing shale gas exploitation, production and, more specifically, fracking operations. Although there are lobbying groups working hard (on behalf of the oil and gas industry) to prevent new environmental policies that restrict production strategies/possibilities, some proper regulations are required to ensure responsible and sustainable exploitation of this resource [3,9].

Policy differs significantly from region to region, not only between Canada and the US, but from state to state as well. Policies for each region are highly influenced by several factors, such as economic dependency, historical gas production, degree of political competition, and the existence of a major environmental constituency. Some shale gas reserves are placed in areas that have not been historically linked to oil and gas production. These areas have much stricter policies opposing gas production, and their tolerance to change is limited [3,9]. This fact makes some regions much more attractive than others. For example, shale gas production is well under way in Texas, but policy, lobbyists, and other political roadblocks have impeded exploitation in Colorado [3,9,10]. There are many states which do not have policies that cover all aspects of oil and gas production. In particular, the prenotification of landowners prior to drilling is not regulated in all regions. This presents a significant disadvantage for surface owners as mineral rights owners can be granted "reasonable" access without the surface owner's permission. This results in legal battles between surface owners and the companies, attempting to access the oil and gas beneath the surface. Other areas that lack effective regulations restrict possible impacts on wildlife and biodiversity, as well as the environmental review process for proposed fracking operations. The regulation in these areas must be amended and improved to ensure responsible and sustainable exploitation of shale gas [9,10].

In contrast to the United States, Canada faces some unique political challenges. First and foremost, there are significant challenges that arise with First Nations rights and exploitation of traditional lands. The laws and regulations governing land rights, titles, and mineral rights are historically convoluted, and there is significant lobbying that takes place on behalf of the First Nations people in the face of any industrial development on these lands [3,9–11]. There is a considerable resistance to pipelines being constructed in their territory, implying the need for engaging First Nations leaders at all stages of the process, from proposal to development and production. There is also a need to balance the demands of labor unions, environmentalists, and First Nations people to align everyone with a common goal. Therefore, it seems necessary to remind all parties about various matters such as climate change concerns, land rights, perception of declining dependence on natural resources and increase in provincial gross domestic product (GDP). A decrease in dependence on natural resources is only possible through the implementation of equalization payments made by the provinces that are exploiting their natural resources. With the aid of royalties, extensive assessments, and other political tools, these concerns can be abated to allow shale gas production in Canada [3,9–11].

In terms of energy and environmental policy, natural gas would cause lower greenhouse gas (GHG) emissions, leading to lower pressure from the corresponding agencies. Although, natural gas is not a final solution to

GHG emissions, it could act as a bridge between reducing short-term emissions, and developing new technologies such as carbon capture and storage (CCS). This would involve a move from coal to natural gas power plants until CCS becomes a viable technology, at which time coal power production would become an environmentally responsible energy source. This is especially true if a carbon tax (or cap) and trade systems are imposed on the industry. Some hurdles to this switch involve archaic grandfather of coal plants' right to opt out implementation of emission reduction policies (based on energy demand), and trade-off between natural gas supplies (e.g., cheaper conventional sources and more readily available shale gas sources) as the natural gas price fluctuates [10,12,13].

A part of the political implications associated with oil shale development is outlined below.

- According to BP's Energy Outlook 2030 (see Fig. 10.1), global shale production is expected to increase by 9% in the year 2030 [1–3]. The production will remain positive, however high rates will decline post year 2020 [1,2]. Shale production will mostly come from North America; however countries such as Russia, China, Argentina, and Columbia will also start contributing [1,2]. The production rate coming from these countries may not offset the decline post year 2020 [1,2]; however, it significantly affects political and trade relationship between the countries.
- Rising shale oil (and gas) production will significantly impact the existing oil market and respective prices [1–3]. OPEC manages prices and production of this type of hydrocarbon. It will be interesting to see how OPEC would respond to shale supplies [1,2,32]. They may cut their

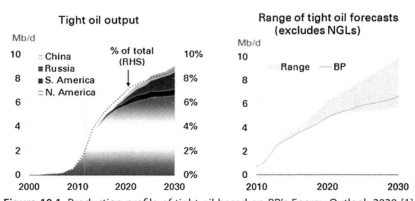

Figure 10.1 Production profile of tight oil based on BP's Energy Outlook 2030 [1].

production and also reduce their prices to stay competitive [1—3]. Hence, such a strategy can negatively affect oil market-based companies associated with OPEC [1—3].
- As the shale resources are geographically distributed, political discussions will evolve in determining where and when shale resources should be accessed [1].
- Over the years, the relationship between the largest oil consumers of the world (e.g., the US) with the world's central oil suppliers (e.g., Middle East) has been well witnessed [1,2]. This rapport can experience geopolitical consequences, and can shift in unpredictable ways [1,2].
- Shale oil and gas production in the US could potentially lead to cheaper energy prices in the US, and this could promote a competitive advantage, and benefit various industries [1]. Therefore, a global macroeconomic impact in a positive way can be expected [1].
- Shale oil and gas production could replace coal-fired electricity generation plants with cheaper shale gas [1]. This will significantly reduce carbon emissions [1]. Hence, countries can consistently comply with climatic policies in place with respect to carbon emitted in the atmosphere [1].

3. FEDERAL AND PROVINCIAL (OR STATE) REGULATIONS

Shale gas production is a part of the oil and gas industry and therefore falls under the same regulatory framework as the rest of the industry. The government regulations set for the shale oil and gas industry are still premature due to the unexpected rapid growth in the industry. There are many factors that should be studied in order to assess the risks of this industry to the public and the environment. Due to the leading role of the US in shale gas production, most of the studies/recommendations concerning environmental regulations have been conducted in the US at the federal and state levels. Current regulations for shale oil and gas production projects which are mostly set by the state government are against the federal government's rules [5,6]. Federal government's regulations normally target the requirements to maintain relatively clean air and water in the US [5,6]. For instance, the Marcellus Shale play is principally located in the state of Pennsylvania and about 90% of the gas production from this particular shale is conducted in this state's territory [5,6]. Under the Oil and Gas Act, companies are required to obtain a permit from the state before drilling a well [5,6].

Detailed plans to mitigate potential environmental pollution throughout various stages (e.g., well excavation and gas processing) must be presented to the Pennsylvania Environmental Protection Department [5,6]. The state then has the right to grant or refuse production in that specific shale region.

In this section, reviewing the regulatory and environmental challenges of shale gas production in the US could help create a pathway at the global level.

In the US, the US Environmental Protection Agency (EPA) is responsible for enforcing and developing environmental regulations that apply to the shale gas industry. Clearly, the most concerning risk factor in shale gas production is water usage and wastewater management due to the hydraulic fracturing operation [5–7].

The federal government set the Clean Air and Water Acts. The Clean Air Act is intended to reduce the amount of pollutants released to the atmosphere [5,6]. It has set threshold levels for each type of pollutant that can be emitted directly to the atmosphere such as methane, ozone, carbon dioxide, and so on. For example, the maximum amount of ozone in the discharged air must be about 70 ppb [5,6]. A carbon tax of around $20/metric ton of CO_2 released to air should be paid by the corresponding companies [5,6]. This law was established to decrease the greenhouse effects due to carbon emissions.

The Clean Water Act (CWA) specifies standard levels for each type of potential pollutant that should be met before produced water or wastewater is discharged to the environment [5,6]. The Resource and Conservation Act set by the federal government determines regulations for discarding waste rock and sludge wastewater from the oil and gas industry [5,6]. At present, the federal government does not strictly enforce these legislations on the oil and gas industry. Enforcement would cause that more than 70% of the oil and gas wells across the US to cease their operations [5,6]. This would lead to catastrophic impacts on the country's economy.

American Water is committed to providing the quality of water and wastewater services, protecting the environment, and using precious natural resources wisely. There is a trustworthy history of complying with, and in many cases surpassing environmental laws and regulations. This is the foundation on which our environmental performance was built [5–7]. Various industrial sectors strive not to just meet, but to exceed environmental regulations and to establish new benchmarks by which industries should be measured [5–7]. Recognizing the economic and environmental potentials of shale oil and gas, American Water supports safe and responsible development of shale gas production pursuant to state and federal regulations that ensure protection of the environment, especially water sources. It is recognized that some states have been regulating natural gas development by the oil and gas industry for decades. In general, they are uniquely positioned to develop and enforce shale gas regulations that protect the environment and reflect local factors such as geology and structure. Many states have already embarked on regulatory processes to update their guidelines/instructions to address shale gas (and oil) development and to protect water quality. In addition, the federal government has the responsibility to ensure that state regulations on water quality meet important criteria and should set national standards for shale oil and gas development on the basis of the federal standards and local considerations [14].

The safety of drinking water is regulated at the federal level under the Safe Drinking Water Act (SDWA), 1974; however the oil and gas industry (particularly the shale hydrocarbons sector) is currently exempted by the EPA. The law was amended twice, once in 1986 and the second time in 1996 to broadly provide protection for all public water supplies from harmful contaminants. This exemption is the result of establishing the 2005 law under the pressure of the oil and gas lobbies that excluded their industry from the SDWA regulation, implying the oil and gas industry is the only industry in USA that is allowed by the EPA to inject known toxic and hazardous materials into underground unchecked drinking water supplies [5—7]. The law is known as the "Halliburton loophole" due to the massive support for lobbying by Halliburton which has great economical interests in hydraulic fracturing technology. The company is also one of the principal producers of chemical fluids used in fracking operation [5—7].

Congress amended SDWA as a part of the Energy Policy Act of 2005 to change the definition of underground injection. As a result, it effectively removed EPA's authority to regulate underground injection of fluids in hydraulic fracturing under SDWA [5—7].

The basic program for regulating public water systems is the state-administrated Public Water Supply Supervision (PWSS) program which except for a few areas has been generally delegated by the EPA to the states. The EPA is also mandated under the SDWA to regulate the underground injection fluids in order to protect underground drinking water sources; however a majority of states have assumed primacy for administrating these regulations, especially for oil and gas drilling operations [5—7].

The gas industry generally refers to a 2004 study by EPA that assessed the potential for contamination of underground sources of drinking water from injection of hydraulic fracturing fluids. The EPA study was accomplished on coal bed methane (CBM) wells and concluded that there were a few or no threats to underground drinking water as the result of injection of hydraulic fracturing fluids into CBM wells. However, some environmentalists, scientists, and EPA officials have called the study scientifically unsound, since some information regarding the risks associated with fracturing to human health had been omitted from the final report. It is interesting to mention that while the 2004 study concluded modest or no risks from fracking fluids injection into water supply reservoirs; very few research works had been performed on environmental impacts of injecting these chemical fluids [5–7]. This has led for renewed calls by concerned parties for inclusion of the shale gas industry and hydraulic fracturing under the SDWA, which will provide a minimum federal floor for drinking water protection in areas that deal with shale gas production [5–7].

Another concern has been the industry refusal for transparency about the chemicals utilized in hydraulic fracturing fluids. Indeed, the chemical formulations of the fracking fluids are considered a highly guarded secret. Also, the mixture of chemicals can be unique for any particular well. Without public access to this information, it is clear that the EPA and other environmental officials cannot trace with an acceptable certainty when, where, or how these chemicals entered the water reservoir [5–7].

There has been renewed pressure on the industry to add chemical tracers to their fracking fluids in order to create measurable accountability for leakage and also the ability to trace back issues in order to fix them as soon as possible.

Since 2005, there have been multiple attempts on regulating the hydraulic fracturing process and shale gas industry to curtail its environmental impacts. Some of the activities include:
- The FRAC Act: The federal government has recently introduced the Fracturing Responsibility and Awareness of Chemicals (FRAC) Act in order to oversee and regulate the types of chemicals utilized in the fracturing fluids [5,6]. This Act/law requires the production company to disclose the different chemicals that make up the fracturing fluids [5,6]. Moreover, prior to instilling this Act, the shale oil and gas industry was not required to abide by the regulations established by the SDWA [5,6];
- Congressional call for EPA study of 2009–2010 (the study was conducted in 2011) [5,6];
- Congressional request for disclosure of FRAC fluid chemicals (2010) [5,6].

Most of these activities have been futile and despite the risks associated with water resources contamination and public interest, there has been little effort by different levels of government to study, control, and regulate the industry.

4. ENVIRONMENTAL ISSUES/ASPECTS

The hydrocarbon production from shale reserves presents its own environmental threats. Some of these are similar to production from conventional reservoirs. The development and production of shale oil and gas require greater volumes of water than conventional reservoirs due to hydraulic fracturing [5,6,15]. Thus, large volumes of wastewater are produced from fracturing fluid withdrawal as well as from the produced water. The wastewater needs to be treated before being discharged to the environment to avoid contamination of natural water resources. Contamination of water supplies would have devastating consequences on both terrestrial and aquatic organisms [5,6,15].

Water supplies might be also polluted during hydraulic fracturing if the fracturing fluid contacts fresh groundwater supplies. The fracturing fluid contains many chemical additives including hydrochloric acid, heavy metals, and radioactive chemicals, which are all extremely toxic to living organisms including humans [5,6,15].

The greenhouse gas emissions from shale hydrocarbon production activities are of major concern to the environment. The principal gas that is emitted to the atmosphere is methane. The large magnitude of methane released to the environment is basically attributed to the step in which the fracturing fluid is withdrawn from the reservoir [15]. Methane is released as the fracturing fluid flows back to the surface. Fig. 10.2 illustrates a comparison between shale gas and conventional gas in terms of methane emissions through a life cycle of the project based on different studies.

Moreover, a considerable magnitude of CO_2 is released from typical production activities and processing plants. These gases contribute to the global warming effect which can have disastrous consequences on all aspects of the environment [6,15].

Hydraulic fracturing can pose the threat of creating seismic waves in the earth's crust, leading to vibrations or earthquakes. When the fracturing fluid is injected into the reservoir to create new fractures or open existing cracks, the stress forces in the formation are altered, which may cause tectonic plates to move, resulting in an earthquake [6,15].

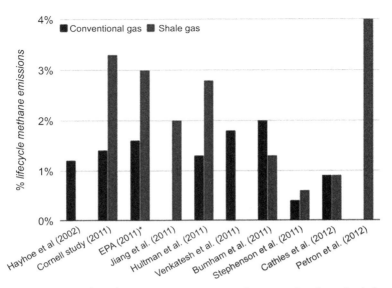

Figure 10.2 Life cycle of methane emissions of conventional and shale gas production [16].

Shale oil and gas projects create a disturbance where production is taking place because processing facilities, access roads, pipelines, and many wells need to be constructed. This results in high amounts of surface intrusions and disruption of the local ecosystems [5,6,15].

In some shale regions, the vegetation needs to be cleared off to allow production. This causes many organisms to lose their habitats so that it will change the traditional ways in which the land was used by its indigenous inhabitants [5,6,15].

All the potential ways in which the environment can be polluted from shale hydrocarbon production are summarized in Fig. 10.3.

Based on current extraction methods, which require a large extent of water, the production of shale gas could significantly affect both ground and surface water resources. The hydraulic fracturing process uses a considerable quantity of water (3—6 million gallons of water per well). After completion of fracturing, the water that will return to the surface contains residual fracturing chemicals, naturally occurring salts, metals, radioactive elements (the most dangerous of which is mercury), and organic chemicals [18,19].

The EPA, as shown in Fig. 10.4, illustrates five main steps in the process of hydraulic fracturing that might cause environmental damage Adapted

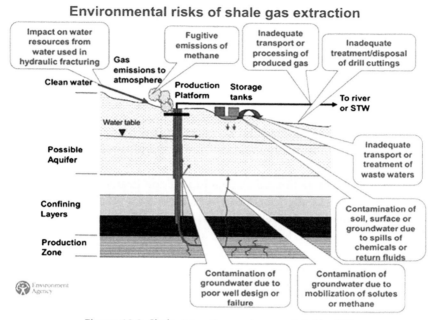

Figure 10.3 Shale extraction environmental risks [17].

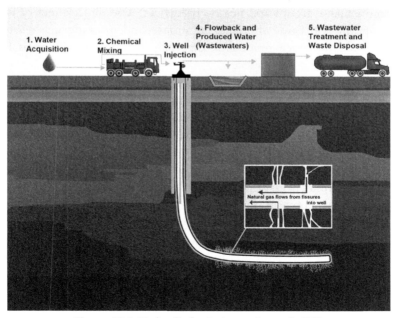

Figure 10.4 Hydraulic fracturing water cycle [22].

from the US Enviromental Protection Agency [18—21], the stages are briefly illustrated below:
- Water acquisition: This process involves extracting and transporting million of liters of water from different sources to the fracturing site.
- Chemical mixing: This includes adding chemical additives and proppant agent to the water. If this process leaks, it can contaminate surface and groundwater.
- Well injection: High-pressure injecting process creates cracks/fractures in the geological formation and can also contaminate surface and groundwater.
- Flowback period: Before oil and gas production begins, flowback occurs a couple of weeks after fracturing, which contains residuals from the fracturing chemicals as well as dissolved materials in the water from the formation (e.g., salts, hydrocarbons, and radioactive substances).
- Wastewater treatment: If the process water is inadequately treated or spills during transportation, water resources could be contaminated.

Due to the cost of shale oil and gas extraction, these resources have only become economically feasible to extract over recent years. No sufficient studies have discussed the long-term impacts of fracturing processes on the environment and natural resources. Based on the nature of the fracturing technologies and available practical experiences, significant demand for fresh water per well during the fracturing process has raised a lot of attention from environmentalist groups and nearby residences. The direct effect of water consumption will cause shortages in surface and groundwater resources which could influence residences, especially farmers, on the areas from where water is being extracted.

Currently, only 20—50% of the injected water can be recycled after hydraulic fracturing. The remaining water stays on the ground and some of it will be flowback in the 10—14 days of fracturing and some of it during the life time of the well. However, some of the injected process water, which contains additive and chemical components, may contaminate the underground water resources and/or land. This can raise serious health concerns to the nearby residences, wildlife, and nature [20—22].

Treating the flowback process water is also important as this contains chemicals and dangerous radioactive matters like mercury. Any spill or leak during the transportation and/or treating process, or from the retention pools, can create serious health and environmental concerns. Also, loss of reputation and social impacts of any leak or spill could be quite considerable for the corresponding companies [17,18].

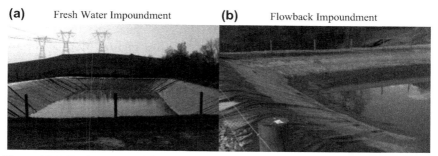

Figure 10.5 Fresh water and flowback retention pounds [18]: (a) fresh water impoundment; (b) flowback impoundment.

Besides water pollution, air pollution is another important concern related to shale oil and gas extraction processes. As a majority of oil and gas fields are located at remote areas, several power generators are required to provide enough power to support drilling and hydraulic operations. The generators are mainly diesel types, which participate in air pollution. Due to the requirement for a tremendous amount of water for the fracturing process, thousands of trucks are required to transport water to the drilling site. These trucks will produce significant exhaust gases. After hydraulic fracturing, the flowback water that contains chemical and radioactive materials will be stored at the retention pounds before treatment, as shown in Fig. 10.5. Evaporating water from these pounds and/or any leak will cause serious health and environmental concerns [19–22].

Furthermore, scientists believe that conducting fracturing processes will cause movement in shale layers under the ground. Performing numerous hydraulic fracturing operations in one oil (or gas) field may cause significant shift on the underground shale layers, leading to the occurrence of earthquakes. To confirm this findings with higher confidence, further research studies are suggested.

5. GEOMECHANICS CHALLENGES

The geomechanical properties of shales are representative of the strength and stiffness of shale, which need to be well understood to diagnose whether they are brittle or ductile. If shales are brittle enough, they originate fractures and then make the fractures open, on the other hand, ductile shales will permit fracture closure and selfsealing [23,24]. Wellbore stability is a rare problem in these rocks because severely reactive shales, such as those replete with smectite, are less likely to happen. This low possibility is related to the thermal maturity of most gas shales (i.e., any original smectite will have changed to illite). In the case of fracturing, stress regime (normal, strike-slip, reverse), magnitude, and orientation of the maximum principal stress direction, especially with regard to fabric elements in the shale due to the anisotropy of shale properties are required to be diagnosed to quantify the in situ stress field. Poisson's ratio, Young's modulus, unconfined compressive strength, cohesive strength, and friction coefficient are the main parameters for the evaluation of shale behavior. The critical issue with tests dealing with geomechanical shale is protection of the core from the moment of retrieval [23,24]. Low-porosity shale dominated by clay will be strengthen if there is loose water and it causes remarkable increases in strength and stiffness (both static and dynamic) parameters [24]. In the case of drying the shale, high capillary pressures (e.g., many MPa in magnitude) would be promoted, which in turn can ruin softer specimens [24,25]. If partially saturated shales such as gas shales are our subject, the sample should be cling-filmed, wrapped in thin foil, and then waxed. Because wax is permeable to air/water on a longer time period, the test should be done in a short time. Geomechanical characteristics of shales also used to determine the likelihood of fractures originating and spreading, propagating in shaly formation [24,25]. Sometimes, brittleness cut-offs are introduced as thresholds to exclude certain stiffness when choosing the best material for hydraulic fracturing operations [23—25]. Using seismic data, we can get the dynamic Young's modulus-associated gas shale sample. It is worth noting that this is different from the static Young's modulus obtained by ultrasonic experiments. The static Young's modulus for shales would be normally obtained using a triaxle test where there are increments with rising effective confining pressure [23—25].

Geomechanical behavior of the shale formation has to be modeled and monitored during production due to the following challenges:
- Rock stiffness inside the shale formation reduces [23]. This happens due to a drastic loss of stiffness as the rock experiences yielding and dilation [23].
- Density changes within the formation take place due to an increase in gas saturations and also dilation [23].
- Pressure, temperature, and saturation variations as a result of exploitation alter the distribution of stresses in the shale formation [26].
- Stress variations as a result of exploitation modify the porous structure of the rock, and consequently its permeability [26].

One of the biggest obstacles in the shale oil and gas extraction is the stability of the shale formation during angled drilling (to eventually produce horizontal wells) and also production stage. Due to the nature of the shale structure, the formations display heterogeneous strength and have stresses which are oriented horizontally. The mechanical stability of the formation is therefore difficult to be predicted due to the variations in rock strengths [26,27].

When drilling at an angle to the shale formation, the natural stresses in the formation will be disturbed and due to the variation in strength, the formation can potentially collapse, plugging the wellbore [26,27]. In addition, the fractures which are originally found in the shale also present a problem in terms of stability during drilling operations. These fractures can collapse/seal hydrocarbon pathways to the wellbore and may result in the drilling pipe being stuck in the formation [26,27]. This reduces the production efficiency and increases the cost of the operation.

The differing rock strength in the shale region in combination with angled drilling causes deviations in the predetermined drilling path [26,28]. Fig. 10.6 illustrates this phenomenon. These deviations increase the drilling expenses as the drilling path becomes longer and the drill pipe might be damaged [26,28].

Once drilling is completed and the well is constructed, the wellbore stability must be continuously monitored on a regular basis to ensure no formation collapse occurs and production is optimal [26,28].

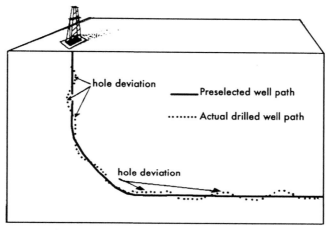

Figure 10.6 Deviations in preselected well path [28].

6. COMPARISON OF CONVENTIONAL AND SHALE OIL AND GAS RESERVES

The rapid exploration and discovery of unconventional resources, such as shale oil and gas, in North America initiated arguments between two groups; those with enthusiasm toward these resources and proclaiming the exceptionally low-risk, manufacturing-style opportunities, and the proponents who question the predicted revenues [2,29]. The fall in regional natural gas has emphasized the marginal nature of these calculations, with their major exposure to commodity-price risk. This induced a debate about how to allocate the scattered capital between unconventional and conventional plans, and consequently, force the industry to review the way they characterize unconventional cases [2,29]. Although the oil and gas industry has adopted the techniques to determine and evaluate risks and opportunities for conventional resources, there is no clear framework to characterize unconventional resources, and deficiency of a general tool for unconventional investment analysis is tangible [2,29]. While assessing the performance, it is only enough to do a full-cycle economics, which can generate returns of 15–25%.

With the advent of shale, the common way of analysis was relied on point-forward, type-well break events, while rates of return were reported in the range of 30–70% (and higher!). This detachment originated from those analyses paying attention to commodity-price and investment dynamics for current plays and separately predicted plans on possible profitability of shale drilling projects [2,29]. Point-forward and average well economics are useful when the development opportunity of the current asset is considered; also, distribution of well performance is well established. However, this practice of using type curves and point-forward analysis contributed to the perception of shales (particularly homogeneous formations) with repeatable results offers close to zero risk [2,29]. The predicted low risk and remarkable development economics motivated buyers to increase the entry price for verified shale plays, therefore inducing companies to enter this play before lifting the access costs. The "no-risk, high-return" mentality jeopardized the projects because it was an unclear way, although high natural gas prices somehow hide this fact [2,29]. To account for this concern, a life cycle of unconventional potential sites should be evaluated and a risk analysis model is necessary. Despite some incomplete practices, translating risk models and conventional life cycle to unconventional reserves are clues to successfully describe these opportunities [2,29]. Due to some experiences, it was understood that unconventional resources and shales have the same magnitude of risk and wide potential exploration as conventional resources. In addition, the characteristics of their risk shows a less definitive manner over their analyzed life cycle, and for residual risks (percent developable) and uncertainties (marginal economics) severity is even greater during the development stage [2,29].

There is a role for unconventional reserves in the corporate portfolio on condition that companies utilize approaches/frameworks for risk assessment which permit them to make a comparison between conventional and unconventional opportunities on a near like-for-like basis. The existence of several parameters and unknown variables in unconventional modeling strategy is vital to find which ones deserve the most upfront scrutiny. Generally, the asset's current life cycle stage can help to achieve this objective [2,29].

There is a temptation to focus on the more well-understood engineering and operational aspects of unconventional reservoirs; however, it is highly believed that strong technical and commercial evaluations of risks and uncertainties are far more important [2,29].

A comparison of conventional and unconventional oil reserves in terms of cumulative production is shown in Fig. 10.7 [30]. Shale production has the potential to offset the oil imports which are expected to increase with growing industrialization [30]. However, the cost of shale production compared to conventional oil is quite high [30,31]. Production cost per barrel is shown in Fig. 10.8 [30,31]. One of the major cost contributors is the high expense of transportation to refineries [30,31].

Conventional oil and gas reserves are much easier to produce because the hydrocarbons are trapped in large pores. The porosity and permeability of the conventional formations are usually high and acceptable [32]. This allows the hydrocarbons to readily flow toward and into the wellbore with the aid of a pressure gradient. On the other hand, unconventional reservoirs have much lower permeability and shale formations (as a part of these reserves) have the lowest values [2,32]. Hence, additional treatment is necessary to increase permeability to a level where hydrocarbons entrapped in the pores are able to travel to the wellbore. As a consequence, developing production from unconventional reserves is much more costly than conventional reserves [32].

In addition to vertical wells, horizontal wells need to be drilled for shale oil and gas reserves. The horizontal wells are difficult to drill, requiring intensive energy techniques and risks of formation collapse as the drilling

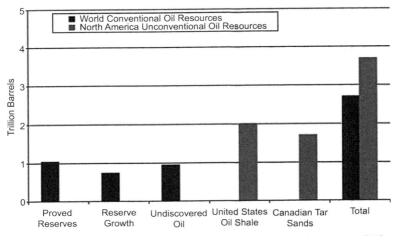

Figure 10.7 Unconventional resources versus conventional resources [30].

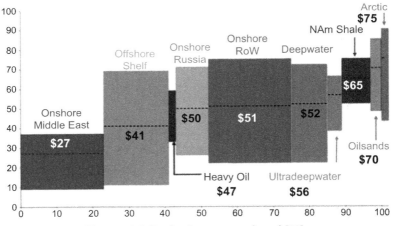

Figure 10.8 Production cost per barrel [30].

rig damage is very high [2,32]. Horizontal well drilling costs about twice as much as that of vertical well. Based on a study on the Marcellus shale play, it was found that the average cost for a vertical well is $663,000, while the cost for a horizontal well is $1,200,000 [2,32,33].

The materials and equipment required for hydraulic fracturing also add a considerable cost to the shale oil and gas operations compared to conventional operations. The fluid must be pumped into the reservoir at high pressures, which is energy-intensive [2,32]. Following fracturing, the fluid should be withdrawn back to the surface. The returned fluid generally contains heavy metals and radioactive chemicals leached out from the formation [32,33].

7. MANAGEMENT RULES IN DEVELOPMENT OF OIL AND GAS SHALE

With respect to shale resource development, federal and provincial (or state) oil and gas regulatory regimes include a wide range of technical prospects [34]. These regulatory regimes are implemented to minimize environmental, health and safety risks, and are applicable to areas of infrastructure development, drilling and production approvals, land management, decommissioning and reclamation, surface casing, cementing, and groundwater protection [34].

The International Energy Agency (IEA) has established a series of golden rules which are outlined as follows:
- All stakeholders should be measured, disclosed, and engaged prior to development [34].
- Drilling area should have minimal impact on the environment [34].
- Wells should be isolated, and leaks should be sealed [34].
- Water should be treated responsibly [34].
- Flaring and other emissions should be eliminated [34].
- Cumulative impacts should be taken into account when coordinating infrastructure [34].
- Continuous improvement initiative should be promoted toward regulations and operating practices [34].

Management rules should be provided for a project to be successful without severely influencing the surrounding environment and public.

There are five main areas in which operators should aim to develop their capabilities to succeed in the current operating landscape. The areas are described in brief as follows.

7.1 Data Management and Compliance

An issue that operators are facing is upscaling their data management through the cycle of shale gas processes because of a high requirement of material flow. Obtaining, saving, and reporting these data need another class of data management to properly employ the data, leading to cumulative environmental impact assessments. To guarantee the efficient handovers of data and compliance supporting, operators should manage the interaction with suppliers and contractors. Larger operators can benefit from a clear economy of scale in the development of such data management systems, particularly if there is consistency within and across basins [34,35].

7.2 Wastewater Disposal

A common technique for managing produced water from oil and gas processes is to inject the disposal into producing reservoir to maintain pressure (or/and enhance oil recovery) or underground injection into EPA-approved Class II Salt Water Disposal (SWD) wells [34,35]. Water conservation measures and lack of disposal capacity have stressed more on research projects about recycling and reuse of produced water, especially on-site treatment and reuse. Although reusing flow back water without much pretreatment for fracking is a short-term solution, more long-term options of producing a large amount of highly saturated brine should be taken into account [34,35]. A variety of treatment techniques have been introduced for treatment of wastewater from shale gas production so that it can be useful for both recycling and reuse. The techniques are available to operators in different arrangements (e.g., with or without pretreatment, from simple filtration to high-end crystallization) at a higher cost than traditional disposal alternatives. Also, some possible partnerships exist with treatment suppliers to properly apply these technologies. Besides the operations to manage wastewater along the whole chain, these partnerships would help operators to find solutions which suitably fit their needs (e.g., volumes and quality of water needed for the specific fracturing fluid and the particular play in which they function).

Finally, these collaborations pave the way for future improvements to recycle more water with the lowest cost along elevated efficiencies (e.g., waste flow, energy inputs demanded) [34,35].

7.3 Water and Emission Intensity Reduction

With increased pressure of public concern on global freshwater sources and addressing of volume of water used in shale gas, attention has move toward reducing the water use. Optimization of the well configurations and the number of wells per site, and maximizing potentials for end-to-end reuse of wastewater are strategies to reduce water consumption [34,35]. Finally, the current research on developing proppants with lower water requirements and alternatives to hydraulic fracturing would shed a light on minimizing water usage of shale gas operations [34,35]. The superiority of GHG emission reductions of shale gas compared to coal and oil motivated operator targets, the public, and governments to emission reductions of shale gas process as a license to operate [34,35]. For instance, the standard method in all US shale gas development is that the companies make an effort to conduct green completion operations so that venting or flaring of methane over well is minimized.

A greater database would make it possible to track emissions from energy use in fracturing and water transportation, and possible emission reduction improvements [34,35].

7.4 Logistics and Operating Models
Considering the scale and intensity of the water movement requirements for shale gas, the following factors are important to shale operators where the existing or new opportunities of shale gas development are evaluated.

7.4.1 Make Logistics a Key Part of the Development Strategy
The role of logistics in shale development is of importance so that it encompasses water supply to supporting fracking operations to wastewater transfer reporting to supporting compliance requirements. Considering this strategic role, a shale-specific logistics framework should be developed. If this strategy was adapted early, it would be verified that any change in logistics implications and partnership opportunities are diagnosed and followed in a schedule [34,35].

7.4.2 Adopt Leading Logistics Practices and Operating Models
Traditional logistics practices are designed for conventional onshore development, whereas the requirement of road transportation for shale development makes it necessary to improve these logistics practices. If operators split the water supply chain from drilling services, they will have more control, as well as an optimized water footprint throughout the life cycle [34,35]. Employing pioneer tools, systems, and logistics practiced by other industries can help control EHS exposure, boost operational performance, and reach cost-effectiveness. Several renowned operators in North America have already adopted these strategies with global perspectives.

7.4.3 Collaboration Opportunities in New Locations
When shale plays are being discovered and developed, there could be a lack of sufficient infrastructure and logistics resources to fulfill the requirement of

large-scale operation. Also, there are always many operators working closely and collaboratively under the same state regulatory environment. Along with the competition for resources and the cost of developing the supply chain infrastructure, a positive collaboration in operators governs. Operators should actively seek potential synergies, such as share logistics management platform, share excess capacity, coordinate local supplier development, and cross-basin infrastructure development, as discussed before. This manner is especially interesting in countries where the shale development foundation is the least developed [34,35].

7.5 Collaboration

Collaboration with other operators and regulators to lessen the intensity of the basin (e.g., sharing infrastructure, sharing excess capacity, and shared logistics) and to treat wastewater (e.g., shared regional facility) is a constructive approach to tackle challenges. Reducing the individual environmental footprint and applying modern practices from the industry and regulatory bodies to reach final goal sustainability and regulatory compliance are the consequences of sharing exposure in these critical regulatory areas [34,35].

8. TECHNICAL AND ECONOMIC CONSTRAINTS

Technological and policy constrains can potentially impact shale energy performance, reliability, efficiency, product quality, and associated economics [30]. A majority of the mining and upgrading shale technologies are yet to be proven at commercial-scale level, and requires further development [30]. Some of these technological barriers are shown in Fig. 10.9 [30].

In addition to technological barriers, economic barriers are the next major barriers to the shale industry [30]. Massive capital investments are required for shale commercialization [30]. Also, uncertain operating costs including environmental costs continue to be a risk for project investors [30]. Fig. 10.9 also gives more details about the economic challenges.

Hydraulic fracturing for shale gas production faces many technical challenges which should be resolved to develop the resources in a responsible manner.

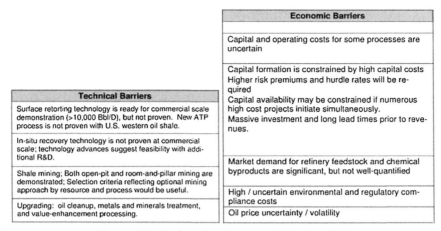

Figure 10.9 Technical and economic barriers [30].

8.1 Water Management

Shale gas production using hydraulic fracturing uses very large quantities of water. A study conducted by researchers in Texas shows that the Barnett Shale (the largest in the state) has used 145 Mm^3 of water to date—equivalent to 1% of Texas's water usage over the past decade [36]. They expect shale gas plays to consume another 4350 Mm^3 of water over the next 50 years which could put a strain on local water resources. A well consumes about 13,500 m^3 of water (Table 10.1) if both drilling and fracturing are considered. It should be noted that water usage can vary highly between wells [36,37].

Scarcity and regulations pose significant challenges to shale gas production [36–38]. In addition, fresh water is generally used for hydraulic fracturing because fracturing additives degrade in saline water [36,37].

8.1.1 Produced Water

After hydraulic fracturing is performed, the well is depressurized and the fracture fluid returns to the surface. This fluid is known as "flowback." The flowback period lasts around 2 weeks and allows for the recovery of 10–40% of the fracturing fluid [38,39]. The recovery of low volumes (2–8 m^3/day) of injected fluids continues throughout the period of gas production. This wastewater is known as "produced water" and contains light hydrocarbons and very high TDS (total dissolved solids) concentrations

Table 10.1 Total Water Consumption for Different Formations [36,37]

Formation	Volume of Water Per Well (m³)
Barnett	10,968
Fayetteville	12,430
Haynesville	15,030
Marcellus	15,761

[38,39]. Minerals and organic constituents originally present in the reservoir dissolve into the retreating fracturing fluid (see Table 10.2). The reported data are from typical wells in the Marcellus shale for flowback (early) and produced water (late) [38,39]. In Table 10.2, TSS represents the total suspended solids. It should be also noted that the hardness and alkalinity occur as $CaCO_3$.

The best technique to manage the challenge of produced water will depend on regulation and the suitability of treatment techniques for particular water mixtures.

8.1.2 Underground Injection

Currently, 40% of produced water is disposed by injection into deep underground wells [40]. The disposal of fluid into wells is regulated in the United States by the EPA. These "Class II" wells are designed not to allow injected fluids to migrate into underground sources of drinking water (Fig. 10.10). This figure shows the fluid injection in a real case (e.g., Ellenburger Formation).

Texas has 11,000 approved Class II disposal wells; however Pennsylvania has only seven, which could limit production in the Marcellus Shale [38,39]. The majority of approved disposal wells are in Texas, California, and Kansas far away from many shale reserves such as Marcellus and Bakken [40,41].

Table 10.2 Impurities in Flowback Water [38,39]

Constituent	Flowback (mg/L)	Produced Water (mg/L)
TDS	66,000	261,000
TSS	27	3200
Hardness	9100	55,000
Alkalinity	200	11,000
Chloride	32,000	148,000
Sodium	18,000	44,000

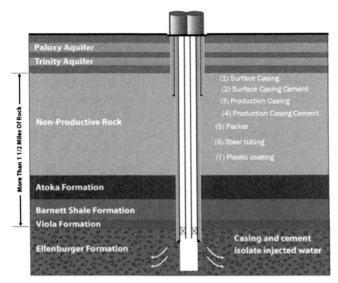

Figure 10.10 Deep well disposal design in the Barnett Shale [40].

Injecting water into wells costs $4.70/m^3$ of water which is still inexpensive compared to the transportation costs. Transporting 13,000 m^3 of produced water would require 650 trucks driving long distances across the country [42]. Shale gas production will continue to occur in areas that do not have sufficient produced water disposal capacity and other solutions for produced water management are required.

8.1.3 Reverse Osmosis

Reverse osmosis (RO) is widely used in industry to purify water and is a well-understood process. Produced water under high pressure is flowed through a semipermeable membrane and treated water of high purity is produced along with concentrated waste of up to 20% of the original volume [43,44].

However, RO is an energy-intensive process and is not believed to be economically feasible for waters containing higher than 40,000 mg/L TDS, which would exclude most of the flowback wastewater (see Table 10.2) [43,44].

For high TDS waters, a newer technology called VSEP (vibratory shear-enhanced processing) has shown some results. The technique uses flat membranes arranged as parallel disks and shear is created by a "leaf" element tangent to this surface. The shear reduces fouling on the membrane allowing for operation at higher TDS levels.

8.1.4 Distillation and Crystallization

Distillation can remove up to 99.5% of dissolved impurities from a produced water stream. Distillation requires the wastewater to be evaporated and thus uses a large amount of energy. The payback is in the form of reduced treatment and disposal costs for the wastewater—up to 75% in some shale gas plays [43,44]. Distillation is economic for mixtures of up to 125,000 TDS but suffers from low flow rates—typically around 300 m^3/day which is one-tenth of the produced water flow rate in some wells in the Marcellus Shale [38,39]. As a result, distillation will involve building large storage tanks to temporarily hold the produced water.

Crystallizers which rely on mechanical vapor compression have gained popularity in produced water treatment because they recycle heat from vapor streams to reduce energy costs by 95% (Fig. 10.11).

Crystallization can create zero liquid discharge and produce solid salts that can be used as industrial feedstock. Crystallizers can take in waters of up to 300,000 TDS but have high capital costs. In Fig. 10.12, the produced water is from Devon's operations in the Barnett. Devon Energy, a Canadian

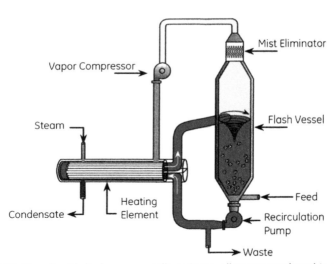

Figure 10.11 Zero liquid discharge crystallizer (note: all waste produced is solid) [43].

Figure 10.12 Produced water (left) and treated water from an evaporator [42].

company operating in the Barnett Shale uses a large crystallizer (18 × 18 m) and produces 300 m^3/day of distilled water [42,43].

8.1.5 On-Site Reuse

The most economically feasible solution to the challenge of produced water is to reuse as much as possible on-site. Flowback water can be impounded at the surface and diluted with make-up water until it is suitable for reuse. This method is especially useful if disposal capacity is limited as it reduces the volume of water required for hydraulic fracturing.

A sequential precipitation process can be used to remove ions from the produced water. The first precipitation will target iron, barium (a toxic heavy metal), and suspended solids. Recovered barium sludge may have uses in drilling mud. The second and third stages would remove other scale-forming ions such as calcium, magnesium, manganese, and strontium to form a nontoxic solid sludge.

The effectiveness of additives may be reduced in high TDS waters. In addition, divalent cations may precipitate in the wellbore forming stable carbonates and sulfates (especially of barium and strontium) which reduce the production rate of gas by clogging fractures [38]. The produced water have to be treated to remove such ions before it can be reinjected (see Fig. 10.13).

Existing commercial technologies could treat 1000 m^3/day of high TDS wastewater (4300 mg/L Ba^{2+}, 31,300 mg/L Ca^{2+}, 250,000 mg/L TDS)

Figure 10.13 Barium precipitate from Marcellus Shale produced water [38].

and make it suitable for reinjection. The proposed site will have a footprint of 2000 m^2 and equipment cost of ~ $3.5 million [45,46].

Regulations and economics are driving the industry to recycle an increasing proportion of wastewater. Chesapeake Energy claims to be saving up to $12 million/year by recycling 40,000 m^3/day of flowback water [45,46].

8.1.6 Brine-Resistant Additives

Produced water recycling is necessitated in shale gas production because of the immense volumes of water used and new regulations on water disposal. Flowback water is currently recycled to remove impurities such as iron [47]. However, removing hardness and dissolved salts remains a challenge because of the difficulty of building water treatment infrastructure on-site.

One factor limiting the usage of recycled water is the commercially available friction-reducing additives which cannot function in highly saline water [47]. Friction reducers are copolymers of polyacrylamide and typically have a charge density of ~ 30% making them vulnerable to salts (especially to salts containing multivalent ions such as $CaCl_2$). Hardness can lead to irreversible polymer conformation changes and will require more friction-reducing agent to be used. Manufacturing the friction-reducing agents as anionic

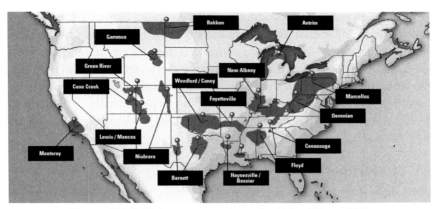

Figure 10.14 Shale gas reservoirs in the United States [48].

emulsion polymers has been shown to make them more resistant to salts and will allow for increased recycle of produced water [47].

8.2 Geology

Shale gas production is a relatively new activity and shale reservoir characterization presents an important challenge. Every shale is characterized by low permeability but there is no industry standard definition [48]. Furthermore, shale rock comprises more than half the earth's sedimentary rock. Key reservoir characteristics such as geology, geochemistry, and geomechanics differ from shale to shale and even within the same shale. Therefore, knowledge of local conditions will determine the best production method.

Furthermore, vital reservoir properties for shale gas production are total organic carbon, reservoir maturity, natural fractures, and mineralogy and not permeability and porosity (as is the case for conventional reservoirs) [49].

The unique nature of producing from shales requires knowledge of every basin, play, and well. The Barnett, Woodford, Haynesville, Fayetteville, and Marcellus shales are all different and require different approaches from operators (see Fig. 10.14).

9. ECONOMIC CHALLENGES

The emergence of new extraction technologies has allowed shale reservoirs to be viable sources of natural gas, so that it will make the US a global leader in oil and gas production. In fact, some estimate that the Western Hemisphere will be energy-independent by 2030 and shale gas will account for 70% of all natural gas production [50]. As noted before, shale reservoir development and extraction require a larger capital investment to complete drilling and provide the necessary infrastructure. There are also increased operating expenses associated with pumping the water slurry into the reservoir at high pressures. Although, shale gas has drastically altered the energy map in North America, its economic viability is precariously dependent on several factors, namely quality of the reservoir, production rates, and the market price of natural gas [50].

9.1 Production Rates

Since shale gas extraction is a relatively new technology, there are limited historical data (usually less than 5 years) available for forecasting models. Consequently, current estimations on the operating life and production rate of known shale deposits vary significantly. It is agreed, however, that shale gas wells deplete significantly faster than conventional wells. These concerns are compounded considering that nearly 70% of available shale reserves have been exploited in the first year of production [51]. A new production forecast model was completed on the Barnett Shale Formation in Texas by the University of Texas. The projected future output is exhibited in Fig. 10.15.

A total of 15,000 wells have already been drilled in this field and an anticipated 13,000 more will be drilled by 2030. According to this figure, the

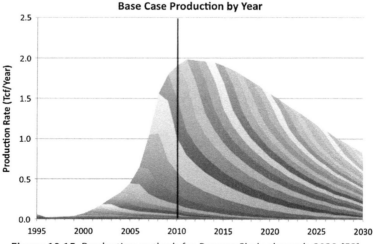

Figure 10.15 Production outlook for Barnett Shale through 2030 [52].

shale field has reached its production peak and will see a decline in 2030. At that time, the production would be less than 50% of current rates [51,52]. This poses a challenge to sustaining current or higher levels of gas production from shale. In order to do so, exploration and drilling will have to occur continuously and rapidly [51,52].

Fig. 10.16 presents a forecast of production rate over the lifetime of a single well. There is a precipitous drop in output in the first several years of operation. Production is heavily dependent on the shale formation and the rock properties so this particular trend cannot be extrapolated to other wells even in this formation. However, these forecasts provide a sobering dose of reality check on whether shale production could be sustainable [51,52].

Currently, operational fracking wells are all drawing from shale formations with relatively favorable rock properties including thickness and porosity. Within several years, drilling in poorer-quality rock will be required and this will translate into higher overall costs [51,52]. Unlike conventional gas wells, where the reservoir can be characterized based on core samples and the viability of the well can be determined, fracking operations require complete drilling (both vertical and horizontal) before this can be known. Combining these two factors, the initial investments in exploration will continue to increase as quality reservoirs are exhausted.

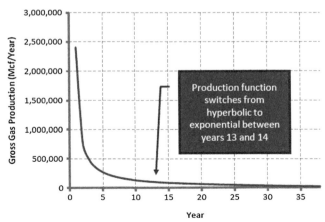

Figure 10.16 Modeled production rate from Haynesville #1 well over a 40-year period [51].

9.2 Natural Gas Price

The price of natural gas is the determining factor for investment in shale operations. The influx of natural gas supply from the new shale source has pushed the natural gas price to their lowest levels in decades. The current price of approximately $4.00/mcf is just above the breakeven point for fracking operations. In order to guarantee economically viable wells, the price of natural gas has to be $8.00/mcf. This imposes a limitation on the fraction of wells that could be explored under current market conditions [50,51].

Since this is a new and emerging area of development, it is likely that the cost of operation will decrease over time as advancements take place. Also, the viability of shale gas cannot only be determined based on domestic North American markets. The price of gas in other jurisdictions overseas is dictated by regulatory agencies and can be as high as $8.00/mcf and $17.00/mcf in Europe and Asia, respectively [50,51]. If shale gas is exported to these markets, the current domestic glut of supply could be curbed. Of course, natural gas exports would require an entirely different cost–benefit analysis, especially if liquefied natural gas (LNG) is considered. There are many economic uncertainties currently clouding shale development. These challenges have to be mitigated in order to make gas shales a commercially sustainable source of energy.

10. RESEARCH NEEDS IN OIL AND GAS SHALE

North American governments were ill-prepared for the sudden growth of shale oil and gas production due to lack of research on the industry. There are many areas concerning the shale oil and gas where research is further needed.

Indeed, more research and practical investigations should be accomplished on the regional distribution of shale hydrocarbons and a detailed map should be produced, showing the exact perimeter of the containing-shale region [2,53]. Research efforts also need to be directed toward studying the properties of the shale hydrocarbons, and physical and chemical characteristics of the shale formations. This will allow the prediction of how shale formation and existing hydrocarbons will react under various physical and chemical conditions [2,53].

Not much is known about how the shale oil and gas production impacts the water systems within the premises. Hence, more research is needed on the surface water bodies as well as underground water supplies to determine the flow patterns and vulnerability to production activities at all main shale production regions [2,53].

Climate changes in and around the shale plays should be studied to determine future impacts at proposed sites of production [2,53].

Research also needs to be conducted on optimizing production and hydrocarbon processing procedures and minimizing the use of clean water for hydraulic fracturing processes. Currently, many shale oil and gas companies are examining deep aquifers containing water for possible applications [2,53].

Conducting more in-depth research studies to analyze the environmental influences of exploration fracturing technologies is curtailed. There are still numerous unknown technical and practical aspects that require further investigation and analysis prior to commercialization of the shale oil technologies [2,53].

Joint research works of government, academia, and industry should be conducted to help the development of new technologies, study the environmental impacts of extraction of shale oil and gas resources, and develop new production techniques. Due to the scarcity and importance of fresh water sources, investing in new extraction technologies to diminish the dependency on the fresh water resources is crucial [2,53].

In order to assure the public about the health and environmental concerns, research results should be communicated with the public and policymakers to increase their awareness about the methodologies and development. Policymakers can assist significantly in regulating further developments [2,53].

Following the IEA, seven golden rules for developing shale oil and gas in a sustainable way would be strongly recommended. These include: (1) measure, disclose, and engage; (2) watch where you drill; (3) isolate well and prevent leaks; (4) treat water responsibly; (5) eliminate venting, and minimize flaring and other emissions; (6) be ready to think big; and (7) ensure a consistently high level of environmental performance [2,53,54].

In conclusion, it seems important to conduct studies concerning shale development in which some of them are listed below:
- Ongoing research activities related to shale are not being conducted fully with respect to the knowledge needs of policymakers as well as public health and environmental impacts [20,55].
- Studies to identify continuous monitoring and stability guidelines for shale sites ensuring no portion of dump reaches out to water saturation [20,55].

- Performing complete assessment of total environmental influences throughout the course of oil shale development [20,55].
- Development of pollution control guidelines for shale mining sites as well as disposal of spent shale [20,55].
- Research on control and treatability to prepare effluent limitations for in situ generated retort water [20,55].
- Grouting requirements and groundwater seepage standards for abandoned mining sites [20,55].

11. PAST AND CURRENT STATUS OF OIL AND GAS SHALE

Technical barriers prior to 2005 had made extraction very difficult. As a result of technological advancements, extraction has became considerably more economical, justifying the sharp incline in shale gas production since 2009, as presented in Fig. 10.17 for the US shale gas production trend [56]. Extrapolating the available data, the future supply of shale gas has been determined, and an upward trend can be seen over time (refer to Fig. 10.17) [55,56]. A similar future trend is expected for shale oil as shown in Fig. 10.18.

Shale reserves were formed millions of years ago and humans were knowledgeable of their existence for many decades but there was no technology available in the past in order to economically extract the hydrocarbons from the shale reserves. With hydraulic fracturing and the ability to drill horizontal wells, shale hydrocarbon reserves have become more economically accessible recently. An increase in shale gas production has caused

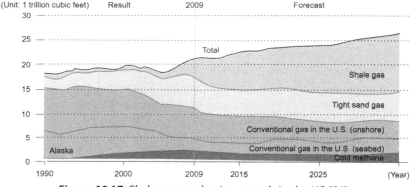

Figure 10.17 Shale gas production growth in the US [56].

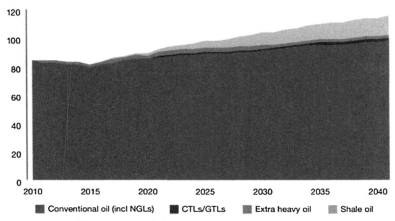

Figure 10.18 Shale oil production in the US [56].

the prices of natural gas on the US market to drop significantly and reduced the dependence on crude oil importation [56,57]. The natural gas prices dropped from $8—9/MBtu in 2008 to $4/MBtu in 2010 [56,57]. Natural gas importation dropped from 103.5 bcm (billion cubic meters) to 106 bcm from 2007 to 2010, respectively [56,57].

The US currently produces about 0.6 trillion cubic meters (tcm) of gas from shale each year [56,57]. It is estimated that the amount of natural gas trapped within shale formations in the US alone ranges between 24.4 and 26 tcm [56,57]. This magnitude of gas ensures the future energy security of the US for another 41 years [56,57].

Europe is starting to develop production of natural gas from shale formations but at a much slower pace than the US. Europe has abundant shale gas reserves and the shale gas industry is expected to boom in the near future. For example, due to increased production of shale gas in Europe, the price of natural gas declined from 11.5 to 8/MBtu in the time period of 2008—2010 [56,57]. At the current state, the technology for shale oil and gas extraction is much more developed in the US. Fig. 10.19 shows the top shale gas reserve holders in the world [56,57].

China, the US, Argentina, and Mexico are the top reserve holders in the world. Considering that China possesses the most, it is predicted that 31 tcm of natural shale gas exists in China's shale formations [57,58]. The shale gas industry is still in its premature stages in China. In 2013, the annual natural gas production from shale made only about 0.2% of China's cumulative gas production [57,58]. China is aiming to increase its level of production to 30 bcm annually [57,58].

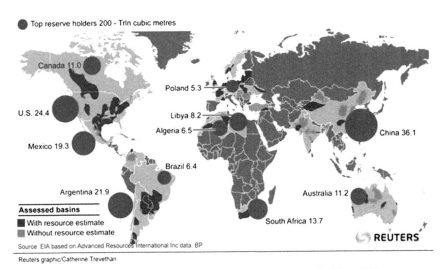

Figure 10.19 Global shale gas basins, top reserve holders [58].

Large changes in energy markets have been observed. For instance, US oil imports from Africa have dropped significantly and US gas imports have collapsed over the last 5–10 years, as tight oil and shale gas production in the US have increased. It is anticipated that US imports of oil and gas may have been 50% lower as a result of fracking in 2012. Hence, it is expected that Chinese imports of gas might be 30–40% lower in 2020 as a consequence of fracking [58,59].

In the case of a reduction in US gas imports, the available analysis suggests that Trinidad and Tobago would suffer export revenue loss which is equivalent to more than 3% of GDP, and other countries affected include Yemen, Egypt, Qatar, Equatorial Guinea, Nigeria, Algeria, and Peru. In total, developing countries are estimated to lose US$1.5 billion in annual gas export revenues because of the rise in fracking [58,59].

A larger number of countries are exposed to a potential trade shock emerging from a change in US oil imports, including Angola, Congo, and Nigeria. An increase in fracking in China with the same size in the trade shock would double the effect. The total estimated effects from a reduction in US oil imports from African countries would be around US$32 billion. The net impacts on exporters will depend on their ability to find other markets, and the conditions under which they do so [58,59].

The fracking revolution is also likely to have major geopolitical impacts. The US and China stand to benefit from the prospect of greater energy independence. Upon developing various alternatives for fracking, Europe has an extensive plan to reduce its dependence on energy imports from Russia and elsewhere. Russia, the Middle East, and OPEC are expected to lose in terms of political aspects. For nonoil-exporting countries, the economic impacts seem broadly positive, through growth effects and reductions in the cost of importing energy [58,59].

12. FUTURE PROSPECTS OF OIL AND GAS SHALE

Shale has become a viable industry in recent times. It is common knowledge that every major petroleum company has put together a shale department staffed by geologists and engineers [58,60]. Additionally, substantial funds and resources have been allocated for both research and land positions including investment on other company's interests and reservoir evaluation [58,60]. A number of shale reserves are under private ownership, and commercial development using mining and retorting methods have been already practiced [58,60]. Considering extensive ongoing activities and optimism associated with this natural resource, there is no doubt that shale has the potential to become an important future source of fuel energy.

The exploitation of shale provides a top reserve holder with long-term security for their energy requirements. Moreover, as gas production from shale increases, the natural gas price will continue to decrease making it accessible to many industries and people. The influences on the environment will increase with increasing production, as well. It is projected that gas from shale formations will make about half of the total natural gas production in the US. This is depicted in Fig. 10.20 [57–59].

It is estimated that US shale gas production may have led to a decline in US imports of gas by around 50% since 2007. The decrease in oil imports is also expected to be around 50% (about 4 MMbbl/day); however, the value for oil is much larger than that for gas. In addition, the gas import might be reduced by 50% in China in the future [57–59].

In the case of suppliers of US oil, a much larger number of countries seem exposed to a potential trade shock induced by fracking. This includes Angola, Congo, and Nigeria. Should an increase in fracking in China generate a similar trade shock, this doubles the effect. The total effects in African countries amount to US$32 billion (of which US$14 billion is in Nigeria, US$6 billion in Angola and US$5 billion in Algeria). A decline in African oil exports to the US of US$23 billion between 2011 and 2012

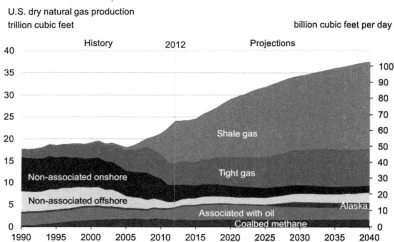

Figure 10.20 Growth projection in shale gas production to 2040 [57].

(or US$27 billion 2007−12) has been already noticed. It is obvious that the net economic loss to those countries would depend on their ability to sell the oil [58,59].

The fracking revolution is also likely to have major geopolitical impacts. The US and China stand to benefit from the prospect of greater energy independence. Europe will be faced with an imperative to reduce its dependence on energy imports from Russia and elsewhere, and will face various options for doing this. Russia, the Middle East, and OPEC are expected to lose in political terms. For nonoil-exporting developing countries the economic impacts can be expected to be broadly positive. This may contribute to a continuance of the phenomenon of "convergence" seen since the early 2000s—with many developing countries growing much faster than the OECD average [58,59].

We conclude that fracking has been an important technological shock with potentially large consequences for developing countries, certainly in terms of trade. Several energy exporters will lose out from lost export revenues, but other developing countries might gain from lower oil prices and faster world growth. Some of this may have already happened, but some may still happen in the future. It is important that developing countries account for this in their future economic projections [57−59].

13. CURRENT PROJECTS FOR OIL AND GAS SHALE

Major shale oil projects since the year 1980 are listed in Table 10.3 [61]. The table includes the qualitative and quantitative information such as estimated cost, actual cost, project size, nature of project, and company involved in each project [61]. The Union Oil Company of California (Unocal) has operated a project with the largest capacity to date, and produced nearly 10,000 barrels/day before the Unocal plant was closed in 1991 [61]. As the price of oil dropped in early 1981, many of the oil shale projects were abandoned by the corresponding companies [61]. These companies feared further drops in the oil price, and shifted their investment toward better opportunities [61].

This section briefly describes some of the recent large projects related to oil and gas shale.

13.1 Project Name: Ecoshale Utah

Companies involved: Red Leaf Resources Inc. and Total S.A.

Description: The project is conducted in the Uintah Basin region of Utah, United States. For oil production, the company utilizes both an economical and environmentally friendly technology known as the Ecoshale In-Capsule Process. The casing surrounding the capsule which prevents leaks to the environment is composed of a natural clay material. Basically, heat is applied to metal pipes extending into the capsule where the shale containing oil surrounds the pipes. Heating allows the oil to separate from the shale. With joint effort of the French-based Total S.A. Company, Red Leaf currently produces about 9800 bbl/day [61,62]. The economical and environmental benefits of using the Ecoshale-In Process Capsule are shown in Fig. 10.21.

13.2 Project Name: Fuling Project

Companies involved: Sinopic.

Description: The project is located in the shale gas fields of Fuling, which is in the southwest of China. The Fuling project is the first shale gas development project to begin in China. Sinopic has invested $322 million on this project. At present, the project is producing about 0.6 billion cubic meters (bcm) annually. The company is aiming to produce 1.8 and 5 bcm by 2014 and 2015, respectively [64].

Table 10.3 Major Shale Projects Since 1980 [61]

Name of Project	Technology	Name of Companies	Estimated Cost	Actual Cost	Capacity (Barrels Per Day)
C-a (Rio Blanco)	Modified in situ plus surface lurgi	Rio Blanco Corp. (Amoco, Gulf)	N/A	$132 million	90,000
C-b (Cathedral Bluffs)	Modified in situ plus surface retort	Occidental, Tenneco	N/A	$156 million	100,000
Clear Creek	STB	Chevron/Conoco	N/A	$130 million	100,000
Colony	Tosco-II	Tosco-Exxon	$5–6 billion	>$1 billion	47,000
Horse Draw	Multi mineral in situ extraction	Multi Mineral Corporation	N/A	N/A	50,000
Logan Wash	Modified in situ	Occidental	N/A	$180 million	N/A
U-ab (White River)	Paraho/Union-B	Sun, Sohio, Philips	$1.6 billion	>$10 million	100,000
Sand Wash	Tosco-II	Tosco	$1 billion	N/A	50,000
Seep Ridge, Utah	Explosive uplift in situ	Geokinetics	N/A	$20 million	70,000
Pacific	Rotating grate	Superior, Sohio, Cleveland Cliffs	N/A	N/A	50,000
Parachute Creek	Union-B	Unocal	$5.4 billion	$1.2 billion	90,000
Parachute Creek	Union-B	Mobile	$8 billion	N/A	100,000
Paraho-Ute	Paraho	Consortium	$1.8 billion	$35 million	40,000

Figure 10.21 Economic and environmental benefits of Ecoshale In-Capsule Process Technology [63].

13.3 Project Name: Al Lajjun

Companies involved: Jordan Energy and Mining Limited.

Description: In Jordan, just southwest of the capital Amman are the shale basins where the Jordan Energy and Mining Limited are conducting the Al Lajjin project. In this project, the company is utilizing the Alberta Tacuik Process Technology and producing oil from the shale formation at a rate of 15,000 barrels/day. Production at the particular shale region is expected to last for 29 years [65].

In addition to the above companies, other companies which perform research and engineering on shale oil and gas include Intertek, AMEC, Royal Dutch Shell Plc., Cheveron Corporation, and Vonoco Inc. [62].

Over the past decades, a majority of developments has occurred in the US; however, shale gas deposits are going to found in many other countries. Canada, Poland, France, South Africa, Argentina, and China are all known to be the countries that are exploiting shale gas technology [19].

13.3.1 Canada
The most geographically distant gas field producing in North America is the commercial-scale gas field of Horn River Basin. The main obstacle to growth in this region is the pipeline infrastructure. The Pacific Trail Pipeline (EOG Resources, EnCana, and Apache) and South Peace Natural Gas Pipeline (Spectra Energy) will collaborate to overcome restrictions in the region. Well-known companies like Talisman, Devon Energy, ExxonMobil, EnCana, Apache, and Petro China are active [19].

13.3.2 Argentina
The thickness of shale gas formation is two or three times broader than North American plays. Government policy has been on keeping the gas prices low and they have initiated the Gas Plus Program to motivate unconventional development. This policy determines the development rate. Until now, Argentina imports liquid natural gas. There are active companies such as Repsol YPF, ExxonMobil, Total, and Apache [19].

13.3.3 South Africa
The Karoo Basin is dealing with the challenges of water resource and infrastructure as the main issues. However, natural gas can be utilized as a substrate to their gas-to-liquids and coal-to-liquids plants due to the potential of shale gas resources. There are some active companies in the region, namely, Royal Dutch Shell, Sasol Ltd., Statoil ASA, and Chesapeake Energy Corp [19].

13.3.4 Poland
A corridor starting from Poland through Germany to the UK is the range of exploration. Because of the dependence on Russian gas, the demand is high. Experience delays and regulatory delays are two bottlenecks on the way to unconventional gas development. These problems stem from logistics, as the majority of the equipment and expertise is only employed in North America, and the environmental concerns of European regulators regarding fracturing. PGNiG, ExxonMobil, Chevron, Marathon, and ConocoPhillips can be addressed as operative there [19].

13.3.5 China
Complex geology and national polices cause some technological and cost issues. However, a Shale Gas Initiative was signed by US President Obama and Chinese President Hu in 2009 to boost environmentally sustainable development of shale gas reserves, and to carry out joint technical investigations/studies to stimulate shale gas resources development in China [19].

REFERENCES

[1] Rühl C. Five global implications of shale oil and gas. EnergyPost.eu., http://www.energypost.eu/five-global-implications-shale-revolution/.
[2] EIA. US Energy information administration — EIA — independent statistics and analysis. EIA; 2013. http://www.eia.gov/forecasts/aeo/er/early_production.cfm.
[3] Khan A. Political implications of shale energy. March 2013.
[4] Belli J. The shale gas 'revolution' in the United States: global implications. April 2013. http://www.europarl.europa.eu/RegData/etudes/briefing_note/join/2013/491498/EXPO-AFET_SP(2013)491498_EN.pdf.
[5] PWC. Shale oil: the next energy revolution. London: PWC UK; 2013.
[6] Sumi L. The regulation of shale gas development: state of play. June 28, 2013. http://www.canadians.org/sites/default/files/publications/OEB%20Sumi.pdf.
[7] Stark M. (Clean Energy Lead), Allingham R, Calder J, Lennartz-Walker T, Wai K, Thompson P, Zhao S. Water and shale gas development. Accenture., http://www.accenture.com/sitecollectiondocuments/pdf/accenture-water-and-shale-gas-development.pdf.
[8] Riley PA. The geostrategic implications of the shale gas revolution. London: The Institude for Statecraft; 2012.
[9] Davis C. The politics of fraccing: regulating natural gas drilling practices in Colorado and Texas. Review of Policy Research 2012;29(2).
[10] Henry DJ, O'Sullivan F. The influence of shale gas on US energy and environmental policy. Cambridge (MA): MIT; 2011.
[11] Lyons A. Shale oil: the next energy revolution e the long term impact of shale oil on the global energy sector and the economy. 2013. http://www.pwc.com/en_GX/gx/oil-gas-energy/publications/pdfs/pwc-shale-oil.pdf.
[12] Medlock KB. Impact of shale gas development on global gas markets. Wiley Periodicals, Inc. — Natural Gas & Electricity; 2011. http://dx.doi.org/10.1002/gas.
[13] Medlock K. Shale gas and US National security. Energy Forum — James A. Barker III Institute for Public Policy. Rice University; 2011.
[14] American Water. American water. December 23, 2014. Retrieved from: Principles for shale gas regulation: www.amwater.com.
[15] Council of Canadian Academics. Environmental impacts of shale gas extraction in Canada. 2014. http://www.scienceadvice.ca/uploads/eng/assessments%20and%20publications%20and%20news%20releases/Shale%20gas/ShaleGas_fullreportEN.pdf.
[16] Peters C. Fugitive emissions from shale gas. The Carbon Brief Blog; May 29, 2012. http://www.carbonbrief.org/blog/2012/05/qa-on-fugitive-emissions-from-fracking/.
[17] Environment Agency. An environmental risk assessment for shale gas exploratory operations in England. 2013. https://www.gov.uk/government/uploads/system/uploads/attachment_data/file/296949/LIT_8474_fbb1d4.pdf.
[18] Scott Institute & Carnegie Mellon University. Shale gas and the Enviroment. Pittsburgh (PA): Wilson E. Scott Institute for Energy Innovation; March 2013.
[19] Linley D. Fracking under pressure: the environmental and social impacts and risks of shale gas development. Toronto: Sustainalytics; August 2011.
[20] Pfeffer FM. Pollutional problems and research needs for an oil shale industry. Washington: National Environmental Research Center, Office of Research and Development, US Environmental Protection Agency; for Sale by the Supt. of Docs., US Govt. Print. Off.; 1974. Technology & Engineering — 36 pages.
[21] Shih CC. Technological overview reports for eight shale oil recovery processes, vol. 1. Environmental Protection Agency, Office of Research and Development, [Office of Energy, Minerals, and Industry], Industrial Environmental Research Laboratory; 1979. Nature — 107 pages.

[22] Adapted from US Enviromental Protection Agency. Retrieved from: The Hydraulic Fracturing Water: EPA.gov; August 18, 2014. www.epa.gov/hfstudy/hfwatercycle.html.
[23] Dusseault MB, Collins PM. Geomechanics effects in thermal processes for heavy oil exploitation. In: Heavy oils: reservoir characterization and production monitoring; 2010;13:287. http://csegrecorder.com/articles/view/geomechanics-effects-in-thermal-processes-for-heavy-oil-exploitation.
[24] Ghorbani A, Zamora M, Cosenza P. Effects of desiccation on the elastic wave velocities of clay-rocks. International Journal of Rock Mechanics and Mining Sciences 2009;46:1267—72.
[25] Horsrud P, Sønstebø EF, Bøe R. Mechanical and petrophysical properties of North Sea shales. International Journal of Rock Mechanics and Mining Sciences 1998;35:1009—20.
[26] Nauroy J-F. Geomechanics applied to the petroleum industry. Editions Technip; 2011. Business & Economics — 198 pages.
[27] Khan S, Yadav A. Integrating geomechanics improves drilling performance. Exploration & Production; January 2014. http://www.epmag.com/item/Integrating-geomechanics-improves-drilling-performance_127118.
[28] PetroWiki. PEH: drilling problems and solutions. 2012. http://petrowiki.org/PEH%253ADrilling_Problems_and_Solutions.
[29] Shale Versus Big Exploration. December 24, 2014. Retrieved from: Unconventional Oil & Gas Center. http://www.ugcenter.com/shale-versus-big-exploration-612776.
[30] Johnson H, Crawford P, Bunger J. Strategic significance of America's oil shale resource. Washington (DC): AOC Petroleum Support Services; 2004. http://www.learningace.com/doc/5378939/f9b4836cb3c90360a25e395f1289d5f0/npr_strategic_significancev1.
[31] SNBCHF.com. Shale oil and oil sands: market price compared to production costs. SFC Consulting. http://snbchf.com/global-macro/shale-oil-oil-sands.
[32] CAPP. Conventional & unconventional. Canadian Association of Petroleum Producers. Canada's Oil and Natural Gas Producers; 2014.
[33] Hefley W. How much does it cost to drill a single Marcellus well? $7.6M. Marcellus Drilling News; 2011. http://marcellusdrilling.com/2011/09/how-much-does-it-cost-to-drill-a-single-marcellus-well-7-6m/.
[34] nrcangcca. Responsible shale development enhancing the knowledge base on shale oil and gas in Canada. In: Energy and mines ministers' conference, Yellowknife, Northwest territories; August 2013. https://www.nrcan.gc.ca/sites/www.nrcan.gc.ca/files/www/pdf/publications/emmc/Shale_Resources_e.pdf.
[35] Melissa Stark RA-W. Water and shale gas development-leveraging the US experience in new shale developments. Accenture; 2012.
[36] Nicot J-P, Scanlon BR. Water use for shale-gas production in Texas, US. Environmental Science & Technology 2012;46:3580—6.
[37] Ground Water Protection Council. Modern shale gas development in the United States: a primer. Oklahoma City: National Energy Technology Laboratory; 2009.
[38] Gregory KB, Vidic RD, Dzombak DA. Water management challenges associated with the production of shale gas by hydraulic fracturing. Elements 2011;7:181—6.
[39] Arthur JD, Bohm B, Layne M. Hydraulic fracturing considerations for natural gas wells of the Marcellus shale. In: 2008 annual forum, Cincinnati; 2008.
[40] Clark CE, Veil JA. Produced water volumes and management in the United States. Oak Ridge: Argonne National Laboratory; 2009.
[41] McCurdy R. Underground injection wells for produced water disposal. Oklahoma City: Chesapeake Energy; 2010.

[42] Kenter P. Waste not. Gas, Oil & Mining Contractor; February 2012. p. 1—3.
[43] All Consulting. Handbook on coal bed methane produced water: management and beneficial use alternatives. Tulsa: US Department of Energy; 2003.
[44] Cline JT, Kimball BJ, Klinko KA, Nolen CH. Advances in water treatment technology and potential affect on application of USDW. In: Underground injection control conference, San Antonio; 2009.
[45] ProChemTech. Marcellus gas well hydrofracture wastewater disposal by recycle treatment process. Brockway: ProChemTech International, Inc.; 2009.
[46] Verbeten S. Recycling flowback can reap rewards. January 2, 2013 [Online]. Available: http://www.gomcmag.com/online_exclusives/2013/01/recycling_flowback_can_reap_rewards.
[47] Paktinat J, O'Neil B, Aften C, Jurd M. Critcal evaluation of high brine tolerant additives used in shale slick water fracs. In: SPE production and operations symposium, Oklahoma City; 2011.
[48] Halliburton. US shale gas: an unconventional resource. Unconventional challenges. Houston: Halliburton; 2008.
[49] Bolle L. Shale gas overview: challenging petrophysics and geology in a broader development adn production context. Houston: Baker Hughes; 2009.
[50] Engdahl W. The fracked-up USA shale gas bubble. Global Research March 2013;13 [Online]. Available: http://www.globalresearch.ca/the-fracked-up-usa-shale-gas-bubble/5326504.
[51] Martin J, Douglas Ramsey J, Titman S, Lake LW. A primer on the economics of shale gas production. Baylor University; 2012.
[52] University of Texas at Austin. New, rigorous assessment of shale gas reserves forecasts reliable supply from Barnett shale through 2030. University of Texas at Austin; February 28, 2013 [Online]. Available: http://www.utexas.edu/news/2013/02/28/new-rigorous-assessment-of-shale-gas-reserves-forecasts-reliable-supply-from-barnett-shale-through-2030/.
[53] National Research Council. Research and information needs for management of oil shale development. Google Books; 1983. http://books.google.ca/books?id=EEcrAAAAYAAJ&pg=PR7&lpg=PR7&dq=research+needs+in+shale+oil+and+gas&source=bl&ots=TMYpOc7KWQ&sig=-B5BM3CxfTGfKgBs1rvzXjrEplA&hl=en&sa=X&ei=Qw7tU97dN8j2yQTQuIKABg&ved=0CE4Q6AEwCQ#v=onepage&q=research%20needs%20in%20shale%20oil%20and%20gas&f=false.
[54] International Energy Administration. Are we entering a golden age of gas?. July 12, 2014. www.worldenergyoutlook.org/goldenageofgas. Retrieved from: World Energy Outlook 2012.
[55] http://www.iop.pitt.edu/shalegas/pdf/research_unconv_b.pdf.
[56] Unconventional oil and gas research fund proposal. Shale Gas Roundtable. http://www.mitsubishicorp.com/jp/en/mclibrary/business/vol2.
[57] Farghaly A. What is the effect of shale gas in the gas prices in the world ?. Linkedin; 2014. https%3A%2F%2Fwww.linkedin.com%2Ftoday%2Fpost%2Farticle%2F20140606232945-39296622-what-is-the-effect-of-shale-gas-in-the-gas-prices-in-the-world.
[58] Faynzilbert I. Shale gas race: political risk in China, Argentina and Mexico. Journal of Political Risk 2014. http://www.jpolrisk.com/shale-gas-race-political-risk-in-china-argentina-and-mexico/.
[59] Zhenbo Hou DG. The development implications of the fracking revolution. (London, UK): Overseas Development institute; April 2014. ODI.
[60] Reso A. Oil shale current developments and prospects. Houston Geological Society Bulletin May 1968;10(9):21. http://archives.datapages.com/data/HGS/vol10/no09/21.htm.

[61] Mackley AL, Boe DL, Burnham AK, Day RL, Vawter RG. Oil shale history revisited, 2012. American Shale Oil LLC, National Oil Shale Association, http://oilshaleassoc.org/wp-content/uploads/2013/06/OIL-SHALE-HISTORY-REVISITED-Rev1.pdf.
[62] Natural-gas processing. Wikipedia. Wikimedia Foundation; July 19, 2016. In: http://en.wikipedia.org/wiki/Natural-gas_processing.
[63] EcoShale™ in-capsule process, http://www.tomcoenergy.uk.com/our-business/ecoshale-in-capsule-process.
[64] Burgess J. China eyes massive production from first shale project. 2014. http://oilprice.com/Energy/Natural-Gas/China-Eyes-Massive-Production-from-First-Shale-Project.html.
[65] Al Lajjun oil shale project. Hatch News; 2014. http://www.hatch.ca/oil_gas/projects/jeml.htm.

INDEX

'*Note*: Page numbers followed by "f" indicate figures and "t" indicate tables.'

A

Aboveground retorting, 296—297
Adsorbed gas, 11, 14—15
Adsorption processes, gas sweetening
   condensate stabilization, 175
   high nitrogen content, 174—175
   mercaptan removal, 176—177, 176f
   NGL extraction and fractionation, 171—174, 173f
   presence of mercury, 171
   sour water stripper unit, 180—181
   sulfur recovery process, 177—180, 179f
Alternating current (AC), 40—41
American Petroleum Institute (API) gravity, 213
Anisotropy, 54
Artificial well stimulation, 131—134, 132f—133f

B

Biogas, 7
Blow-out preventer (BOP), 89
Borehole instability, 96

C

Carbon disulphide (CS_2), 178
Carbonyl sulfide (COS), 178
Casing head gas, 9
Characteristics, shale oil, 254—270, 255t
   adsorption/desorption method, 276—279, 277f—278f, 278t, 279f
   boiling range, 260, 261f
   composition, 256—259, 258t—259t, 260f
   composition determination pyrolysis method, 259—260
   diffusivity parameter, 264—267, 266f—267f
   electrical properties, 268—270
   fracturability, 271—275, 273t
   permeability, 271—275, 273t
   petrology and geochemistry, 279—280
   pore size distribution, 275—279, 276f, 276t
   pore structure, 275—279, 276f, 276t
   porosity, 270—271
   self-ignition temperature, 260—264, 262f—263f
   surface area, 275—279, 276f, 276t
   time-temperature index (TTI), 267—268, 267f
Characterization, shale gas
   background, 29—30, 30f—31f, 32t
   challenges, 70—72, 73f
   composition, 43—44, 44t
   density, 47
   depth, 46, 46f
   engineering and research companies, 76, 76t—77t
   geological description, 58—59, 58t
   hydraulic fracturing technology, 74—76
   methods, 31—36, 33t
   overview, 27—29
   permeability, 60—64, 61f, 63f, 64t
   porosity, 28
   petrology and geochemistry, 55—57, 56f—57f
   shale lithology, 56—57
   petrophysical characteristics, 36—43
      coring rock, 37—39, 37f—38f
      flow type and permeability, 43
      formation pressure data fluid, 43
      gamma ray logging rock, 39
      nuclear magnetic resonance (NMR), 43
      porosity logging volume estimation, 39—40
      resistivity logging fluid, 40—42
   petrophysical data analysis, 27—28
   place volume, gas estimation in, 57—58
   pore size distribution, 68—70, 69f—70f
   porosity, 60—64, 62f—63f, 64t
   porosity/permeability measurements practical methodologies, 64—68, 65f

Characterization, shale gas (*Continued*)
  pressure, 44—46
  PVT behavior, 54—55, 55f
  research, 73—76
  shale reservoir, 74
  temperature, 46
  thermal properties, 47—54, 49f, 51t
    anisotropy, 54
    composition, 52—53
    heat capacity, 50—52
    porosity, 53
    pressure, 53—54
    temperature, 53
    thermal conductivity, 50
  viscosity, 47
  well logging methods, 33—36
    hybrid workflow, 35—36
    seismic data, 35
    seismic waveform, 36
    well log data, 33—35, 34f
Chevron U.S.A. Inc., 225
Coal bed methane, 8, 107
CO_2 flooding, 290
Cold separator, 173—174
Composition, 52—53
Conventional gas, 6

D

Deep natural gas, 7
Density, 47
Density log, 40
Depth, 46, 46f, 213
Dielectric constant, 213
Directional drilling, 90—97, 91f—92f, 134—135
Dissolved gas, 11
Distal areas, 59
Distillation, 245—246
Drilling fluids, 92—94, 93f—94f
Drilling mud, 42
Drilling technologies
  advantages, 91—92
  borehole instability, 96
  challenges, 95
  costs, 97
  directional drilling, 90—97, 91f—92f
  disadvantages, 91—92
  drilling fluids, 92—94, 93f—94f
  equipment, 88—90
  fluid loss, 95—96, 95f
  horizontal drilling, 90—97, 91f—92f
  procedures, 88—90
  risks, 95
  stuck pipe, 96—97
  types, 88—89, 89f—90f
  vertical drilling, 90—97, 91f—92f

E

Electrical properties, shale oil, 268—270, 269f
  dielectric constant, 269—270, 269f
Energy dispersive X-ray (EDX) analysis, 29—30
Exploration techniques, 81—82, 82f
  advantages, 82—85, 83f
  challenges, 85
  costs, 85—87
  disadvantages, 82—85, 83f
  risks, 85
  stages, 83—85
    geophones, 85
    onshore drilling rig, 85, 86f
    seismic vibrator, 84, 84f
  surface mining, 87, 87f
  underground mining, 87, 88f
Exploration wells
  abandonment, 117
  casing and perforating wells, 112—113
  completion equipment, 113—114
  construction, 111—112, 112f
  costs, 114
  design, 111—112, 112f
  equations, 115
  normalized rate of penetration (NROP), 115
  rate of penetration (ROP), 116
  reclamation, 117
  research and development, 117—119, 118f
    technological evolution, 118
    transverse fractures, 118—119
  shale formations borehole instability, 114—115
  shale reservoirs, 114

Ex situ retorting techniques, 332—337
  advantages, 337—338
  chevron retort system, 337, 338f
  conduction through wall, 334
  disadvantages, 337—338
  externally generated hot gas, 334
  hot recycled solids, 333—334, 334f
  internal combustion, 333
  mines gas combustion retort, US Bureau, 335—336, 337f
  plasma gasification, 335
  reactive fluids, 335
  union oil retorting process, 335, 336f
External corrosion, 186
ExxonMobil, 222

F

Field emission-scanning electron microscopy (FE-SEM), 29—30
Fourier transform infrared spectroscopy (FTIR), 29
Free gas, 11, 16—17
FTIR. *See* Fourier transform infrared spectroscopy (FTIR)

G

Gas adsorption, 65
Gas chromatography analysis, 246—248, 247f
Gas chromatography-mass spectrometry (GC-MS), 248—249, 249f—250f
Gas dehydration process, 160—166
  compression, 163
  cooling, 163
  cooling below initial dew-point, 163
  deliquescent systems, 164
  liquid desiccant, absorption of water, 163
  physical absorption process, 164—166, 165f—166f
  solid desiccant, absorption of water, 163
  TEG dehydration system, 160—162, 161f
  traditional and new technologies, 162—166
Gas injection, 290
Gas liquid separation, 137—139
  design, 137—138, 138f
  material selection, 138—139

Gas reservoir, 3—5, 4f
Gas sweetening
  adsorption processes, 169—181, 170f
  process description, 167
  traditional and new technologies, 167—181, 168f
GC-MS. *See* Gas chromatography-mass spectrometry (GC-MS)
Geophones, 85
Geopressurized zones, 8

H

Heat capacity, 50—52
Helium porosimetry, 65
Horizontal drilling, 90—97, 91f—92f, 292, 292f
H_2S gas, 180
Hydraulic fracturing technology, 74—76, 293—294, 294f
  challenges, 109—110
  characteristics, 105—106, 106f
  completion, 127—129, 127f, 128t
  defined, 97—98, 97f—98f
  equipment, 98, 99f
  fluid rheology, 103—104
  fluids and additives, 101—102, 102f
  fracturing proppant, 102—103, 103f
  modeling and simulators, 104—105
  processes, 107—109, 108f
  risks, 109—110, 110f
  rock properties, 106—107
  theory, 99—101, 100f—101f
  treatment design and optimization, 104
Hydrocarbon liquid, 175

I

In situ method, 296—297
In situ oil shale
  air emissions, 314
  radionuclides, 313
  seismic effects, 314
  water contamination, 313
In situ retorting, 296
Internal corrosion, 186

K

Kerogen
  composition, 194, 201–204, 202f, 208f
  history, 235–238
  humic, 236
  planktonic, 236
  residue, 236–238
  sapropelic, 236
  structure, 235–238
  types, 235–238, 237f, 240f, 242f–243f

L

Lacustrine shale, 12
Lean gas, 9, 173–174
Level of maturity (LOM), 33–35
Linear flow, 142–143, 142f

M

Marine shale, 12
Mathematical formulas, 142–144
  linear flow, 142–143, 142f
  radial flow, 143–144, 143f
MDEA, 169
Mercury injection capillary pressure (MICP), 242
Mercury porosimetry, 65
Merox process, 176–177, 176f
Methane hydrates, 9
MICP. *See* Mercury injection capillary pressure (MICP)
Mineralogy, 16
Molecular sieve dehydrators, 165–166
Multilateral wells, 91
Multiple fracturing, 292, 293f

N

Natural gas
  biogas, 7
  coal bed methane, 8
  deep natural gas, 7
  gas reservoir, 3–5, 4f
  geopressurized zones, 8
  methane hydrates, 9
  shale gas, 7–8
  tight gas, 8
  types of, 5–9, 6f
Neutron log, 40
NGL recovery processes, 171–174
Normalized rate of penetration (NROP), 115
Nuclear Magnetic Resonance (NMR), 43, 244

O

Onshore drilling rig, 85, 86f
OPEC, 280–281, 281f
Optimization, 135–136
Organic maturity, 15

P

Parameters, shale gas
  adsorbed gas, 14–15
  depth, 14
  fluid in place, 16
  free gas quantification, 16–17
  mineralogy, 16
  organic maturity, 15
  permeability, 15
  porosity, 15
  productibility, 17
  reservoir thickness, 15
  thermal maturity, 16
  total organic content (TOC), 15
  type, 14
  viscosity, 16
Pay zone, 124
Permeability, 15, 60–64, 61f, 63f, 64t, 213–214, 271–275, 273t
  measurements, practical methodologies, 64–68, 65f
Petrophysical data analysis, 27–28
Petrophysical shale gas
  characteristics, 36–43
    coring rock, 37–39, 37f–38f
    flow type and permeability, 43
    formation pressure data fluid, 43
    gamma ray logging rock, 39
    nuclear magnetic resonance (NMR), 43
    porosity logging volume estimation, 39–40
    resistivity logging fluid, 40–42
Polymer flooding, 290

Porosity, 15, 28, 53, 130, 213, 270–271
  logging volume estimation
    density log, 40
    neutron log, 40
    sonic log, 40
  measurements, practical methodologies, 64–68, 65f
Potential Gas Committee (PGC), 5
Pressure, 44–46, 53–54
Primary recovery, 289
Process, shale gas
  background, 154–156, 155f
  corrosion, 185–188, 187f
  costs, 188–190, 189f, 190t
  description, 156–158
  design, 181–183, 182f
  equipment modeling/optimization, 183–185, 184t
  flow scheme, 158, 159f
  gas dehydration process, 160–166
    compression, 163
    cooling, 163
    cooling below initial dew-point, 163
    deliquescent systems, 164
    liquid desiccant, absorption of water, 163
    physical absorption process, 164–166, 165f–166f
    solid desiccant, absorption of water, 163
    TEG dehydration system, 160–162, 161f
    traditional and new technologies, 162–166
  gas sweetening
    adsorption processes, 169–181, 170f
    process description, 167
    traditional and new technologies, 167–181, 168f
  hydrate formation and inhibition, 158–160
  overview, 153–154, 154f
  steps, 156, 157f
  transport, 185
  utilities/storage and off-site facilities, 181–183

Process, shale oil, 330–332
  advantages, 341–342, 346–347
  carbonate decomposition, 324–325
  CO_2 extraction, 344–345, 345f
  continuous supercritical extraction, 346
  description, 322
  disadvantages, 341–342, 346–347
  economic consideration, 352t, 351
  ex situ retorting techniques, 332–337
    advantages, 337–338
    chevron retort system, 337, 338f
    conduction through wall, 334
    disadvantages, 337–338
    externally generated hot gas, 334
    hot recycled solids, 333–334, 334f
    internal combustion, 333
    mines gas combustion retort, US Bureau, 335–336, 337f
    plasma gasification, 335
    reactive fluids, 335
    union oil retorting process, 335, 336f
  externally generated hot gas, 339, 340f
  Exxonmobil electrofrac, 340
  hot topic research studies, 353–354
  isothermal analysis, 332, 333f
  kerogen decomposition, 324, 325f
  nonisothermal analysis, 331–332, 332f
  optimal retorting conditions, 326–327, 327f–328f
  pyrolysis, mathematical modeling, 347, 347f–349f
  pyrolysis, parametric study of, 349–350, 350f, 350t
  pyrolysis/retorting, 325–326
    apparatus, 325–326, 326f
    experiments, 325–326, 326f
    kinetics, 328–330
    methodology, 325–326, 326f
  refining and upgrading processes, 342–343
    advantages, 343–344
    disadvantages, 343–344
    hydrocracking process, 343
    moving bed hydroprocessing reactor, 342–343
    thermal cracking process, 342

Process, shale oil (*Continued*)
  retorting
    chemistry, 322, 323t—324t
    defined, 322, 323f
    in situ retorting techniques, 338
    supercritical extraction, 344
    supercritical methanol, 345—346
    theoretical/practical and economic challenges, 352—353
    volumetric heating, 340—341
      Dow Chemical Company, 341
      equity oil company process, 341
      Talley Energy System, 341
    wall conduction, 339, 339f
    water extraction, 345—346
Productions, shale gas
  abandonment, 129
  casing and perforating, 125—127
  corrosion issues, 139—140, 139f
  defined, 123—124
  drilling, 125, 125f
    methodology, 135, 136f
  economic considerations
    breakeven gas price, 149—150, 149t
    pricing, 148—149, 148f
  future prospects, 146—148, 147f
  gas liquid separation, 137—139
    design, 137—138, 138f
    material selection, 138—139
  hydraulic fracturing and completion, 127—129, 127f, 128t
  limitations, 136—137
  mathematical formulas, 142—144
    linear flow, 142—143, 142f
    radial flow, 143—144, 143f
  methods, 131—135, 132f
    artificial well stimulation, 131—134, 132f—133f
    directional drilling, 134—135
  optimization, 135—136
  overview, 123—124
  reclamation, 129
  research and development, 145—146
    current status, 145—146, 145f—146f
  road and well pad construction, 125
  rock properties, 129—130
    production effect, 130, 131f
  shale gas production, 144, 144f
  storage, 141
  transportation, 140—141
Productions, shale oil
  advantages, 307—308
  costs, 310, 311t
  economic aspect, 308—309, 310t
  environmental/public support issues, 311—314
    in situ oil shale, 313—314
    surface mining, 311—312
  equations, 301—303
  future prospects, 315—316
  history, 287—288, 288f
  implication, 301
  limitations, 300—301, 307—308
    oil price, 300
    public opposition, 300—301
  modeling and optimization, 303—306, 305f
  oil reservoirs
    CO_2 flooding, 290
    gas injection, 290
    polymer flooding, 290
    primary recovery, 289
    secondary recovery, 289—290
    steam flooding, 290
    tertiary recovery, 290
    water injection, 289
  opposing views, 314—315
  overview, 285—286, 285f—286f
  research and technology development, 316—317
  rock properties, shale oil, 297—298
  screen criteria, 306—307
  technical aspect, 308
  techniques, 291—297, 291f
    aboveground retorting, 296—297
    horizontal drilling, 292, 292f
    hydraulic fracturing, 293—294, 294f
    multiple fracturing, 292, 293f
    in situ method, 296—297
    in situ retorting, 296
    surface mining, 295, 295f
    surface retorting, 294—296
    underground mining, 295—296

wellhead and gathering, 298—300, 298f—299f
Properties, shale oil
  characteristics, 254—270, 255t
    adsorption/desorption method, 276—279, 277f—278f, 278t, 279f
    boiling range, 260, 261f
    composition, 256—259, 258t—259t, 260f
    composition determination pyrolysis method, 259—260
    diffusivity parameter, 264—267, 266f—267f
    electrical properties, 268—270
    fracturability, 271—275, 273t
    permeability, 271—275, 273t
    petrology and geochemistry, 279—280
    pore size distribution, 275—279, 276f, 276t
    pore structure, 275—279, 276f, 276t
    porosity, 270—271
    self-ignition temperature, 260—264, 262f—263f
    surface area, 275—279, 276f, 276t
    time-temperature index (TTI), 267—268, 267f
  extraction processes
    ultrasonic method, 249—253
  formations, 233—234
    central Queensland region, 233—234, 233f
    Green River Formation, 234, 235f
  kerogen
    history, 235—238
    humic, 236
    planktonic, 236
    residue, 236—238
    sapropelic, 236
    structure, 235—238
    types, 235—238, 237f, 240f, 242f—243f
  methods, 238—249, 244t
    distillation, 245—246
    gas chromatography analysis, 246—248, 247f
    gas chromatography-mass spectrometry (GC-MS), 248—249, 249f—250f
    mercury injection capillary pressure (MICP), 242
    NMR, 244
    Raman spectroscopic analysis, 244—245, 245f—246f
    scanning electron microscope (SEM), 242
    ultrasonic, 244
    X-Ray CT, 238—240
  OPEC, 280—281, 281f
  overview, 231—232, 232f
  utilization, 232—233
Proximal areas, 59
PVT behavior, 54—55, 55f

R

Radial flow, 143—144, 143f
Raman spectroscopic analysis, 244—245, 245f—246f
Rate of penetration (ROP), 116
Reclamation, 129
Recoverable gas, 5
Recoverable oil, 5
Reservoir rock, 4, 196
Reservoirs, shale oil, 214—218, 215f, 216t
  Africa, 216—217
  Asia, 215—216
  Europe, 217
  Middle East, 217
  North America, 218
  Oceania, 217
  production history, 218—220, 219t
  South America, 217—218
Residue gas, 9
ROCK-EVAL, 30
Rock properties, 129—130

S

Scanning electron microscope (SEM), 29—30, 242
Schlumberger Inc., 225
Seal, 4
Secondary recovery, 289—290
Seismic vibrator, 84, 84f
Self-ignition temperature, 213

Shale gas, 107
　applications, 25
　challenges, 24
　conventional exploration, 374—377, 376f
　current projects, 397—400
　　Al Lajjun, 399—400
　　Argentina, 400
　　Canada, 400
　　China, 400
　　Ecoshale Utah, 397
　　Fuling Project, 397
　　Poland, 400
　　South Africa, 400
　defined, 10—11, 11f
　discovery and exploitation, 2—3
　environmental issues/aspects, 367—371, 368f—369f
　federal and provincial regulations, 363—367
　future prospects, 395—396, 396f
　geology, 387, 387f
　　natural gas price, 390
　　production rates, 388—389, 388f—389f
　geomechanics challenges, 371—373, 373f
　greenhouse gas (GHG) emissions, 3
　history, 12—13
　importance, 24—25
　management rules, 377—380
　　collaboration, 380
　　compliance, 377—378
　　data management, 377—378
　　logistics models, 379—380
　　operating models, 379—380
　　wastewater disposal, 378
　　water and emission intensity reduction, 378—379
　natural gas
　　biogas, 7
　　coal bed methane, 8
　　deep natural gas, 7
　　defined, 7—8
　　gas reservoir, 3—5, 4f
　　geopressurized zones, 8
　　methane hydrates, 9
　　tight gas, 8
　　types of, 5—9, 6f
　occurrence, 12—13, 14f
　origin, 11—12
　overview, 1—3
　parameters
　　adsorbed gas, 14—15
　　depth, 14
　　fluid in place, 16
　　free gas quantification, 16—17
　　mineralogy, 16
　　organic maturity, 15
　　permeability, 15
　　porosity, 15
　　productibility, 17
　　reservoir thickness, 15
　　thermal maturity, 16
　　total organic content (TOC), 15
　　type, 14
　　viscosity, 16
　past/current status, 392—395, 392f—394f
　political implications, 360—363, 362f
　production trend, 23, 23f
　research, 390—392
　reserves, 17—22, 18t
　　Canada, 21—22, 22f, 22t—23t
　　USA, 19—21, 20f—21f
　　world/global, 19, 19f
　shale, 9—10, 10f
　technical and economic constraints, 380—387, 381f
　　brine-resistant additives, 386—387
　　distillation and crystallization, 384—385, 384f—385f
　　on-site reuse, 385—386, 386f
　　produced water, 381—382, 382t
　　reverse osmosis (RO), 383—384
　　underground injection, 382—383, 383f
　　water management, 381—387, 382t
　types, 11—12
Shale lithology
　composition and color, 57
　texture, 56
Shale oil
　companies, 221—225, 222t—224t
　　Chevron U.S.A. Inc., 225
　　ExxonMobil, 222
　　Schlumberger Inc., 225
　　Shell Oil Company, 225
　composition, 198—200, 198t—201t

conventional exploration, 374—377, 376f
current projects, 397—400
    Al Lajjun, 399—400
    Argentina, 400
    Canada, 400
    China, 400
    Ecoshale Utah, 397
    Fuling Project, 397
    Poland, 400
    South Africa, 400
defined, 196—197, 198f
energy implication, 225—228, 227f—228f
environmental issues/aspects, 367—371, 368f—369f
factors and parameters, 213—214
    American Petroleum Institute (API) gravity, 213
    depth, 213
    dielectric constant, 213
    permeability, 213—214
    porosity, 213
    self-ignition temperature, 213
federal and provincial regulations, 363—367
future prospects, 395—396, 396f
geology, 387, 387f
    natural gas price, 390
    production rates, 388—389, 388f—389f
geomechanics challenges, 371—373, 373f
history, 206—212, 209f—210f
    Central Queensland region, 211, 211f
    Green River Formation, 212, 212f
importance, 221
kerogen
    composition, 201—204, 202f, 208f
management rules, 377—380
    collaboration, 380
    compliance, 377—378
    data management, 377—378
    logistics models, 379—380
    operating models, 379—380
    wastewater disposal, 378
    water and emission intensity reduction, 378—379
occurrence, 206—212
overview, 193—195
past/current status, 392—395, 392f—394f
political implications, 360—363, 362f
recoverable estimates, 220—221, 220f
research, 390—392
reservoirs, 214—218, 215f, 216t
    Africa, 216—217
    Asia, 215—216
    Europe, 217
    Middle East, 217
    North America, 218
    Oceania, 217
    production history, 218—220, 219t
    South America, 217—218
technical and economic constraints, 380—387, 381f
    brine-resistant additives, 386—387
    distillation and crystallization, 384—385, 384f—385f
    on-site reuse, 385—386, 386f
    produced water, 381—382, 382t
    reverse osmosis (RO), 383—384
    underground injection, 382—383, 383f
    water management, 381—387, 382t
types, 195—196
    heavy/sticky oils, 196
    light/volatile oils, 195
    nonfluid oils, 196
    nonsticky oils, 195
types and source, 204—206
Solid desiccant, 163
Sonic log, 40
Source rock, 4
Sour gas, 9
Steam flooding, 290
Stratigraphic traps, 5
Structural.traps, 5
Stuck pipe, 96—97
Sulfur recovery, 177—180
Surface mining, 87, 87f, 295, 295f
    area usage, 311—312
    gas emissions, 312
    waste materials, 312
    water usage, 312
Surface retorting, 294—296
Surfactants, 133
Sweet gas, 9

T

Tail gas treating (TGT), 178—179
TEG dehydration system, 160—162, 161f
Temperature, 46, 53
Terrestrial shale, 12
Tertiary recovery, 290
Thermal conductivity, 50
Thermal maturity, 16
Thermal properties, shale gas, 47—54, 49f, 51t
  anisotropy, 54
  composition, 52—53
  heat capacity, 50—52
  porosity, 53
  pressure, 53—54
  temperature, 53
  thermal conductivity, 50
Thermo-analytical methodologies, 48
Tight gas, 8, 107
Tight oil, 107
Time-temperature index (TTI), 267—268, 267f
Total organic carbon content (TOC), 15, 28
Trap, 4

U

Ultrasonic method
  column chromatographic separation, 252—253, 253f
  GC/MS qualitative analysis, 253, 254f
  sample preparation, 249—251
  sulfur removal, 251, 252f
  supercritical extraction, 253—254, 255f, 255t
Unconventional gas, 6
Unconventional reservoirs, 6, 130
Underground mining, 87, 88f, 295—296
United States Geological Survey (USGS), 9

V

Vertical drilling, 90—97, 91f—92f
Viscosity, 16, 47

W

Wall conduction, 339, 339f
Water extraction, 345—346
Water injection, 289
Well logging methods, 33—36
  hybrid workflow, 35—36
  seismic data, 35
  seismic waveform, 36
  well log data, 33—35, 34f
Wells, exploration
  abandonment, 117
  casing and perforating wells, 112—113
  completion equipment, 113—114
  construction, 111—112, 112f
  costs, 114
  design, 111—112, 112f
  equations, 115
  Normalized rate of penetration (NROP), 115
  rate of penetration (ROP), 116
  reclamation, 117
  research and development, 117—119, 118f
    technological evolution, 118
    transverse fractures, 118—119
  shale formations borehole instability, 114—115
  shale reservoirs, 114
Wet gas, 9

X

X-Ray CT, 238—240
X-ray powder diffraction (XRD), 29
XRD. See X-ray powder diffraction (XRD)